METHODS IN MOLECULAR BIOLOGY

For further volumes:
http://www.springer.com/series/7651

For over 35 years, biological scientists have come to rely on the research protocols and methodologies in the critically acclaimed *Methods in Molecular Biology* series. The series was the first to introduce the step-by-step protocols approach that has become the standard in all biomedical protocol publishing. Each protocol is provided in readily-reproducible step-by-step fashion, opening with an introductory overview, a list of the materials and reagents needed to complete the experiment, and followed by a detailed procedure that is supported with a helpful notes section offering tips and tricks of the trade as well as troubleshooting advice. These hallmark features were introduced by series editor Dr. John Walker and constitute the key ingredient in each and every volume of the *Methods in Molecular Biology* series. Tested and trusted, comprehensive and reliable, all protocols from the series are indexed in PubMed.

The Surfaceome

Methods and Protocols

Edited by

Csaba Matta

Anatomy, Histology and Embryology, University of Debrecen, Debrecen, Hungary

Ali Mobasheri

Research Unit of Health Sciences and Technology, Faculty of Medicine, University of Oulu, Oulu, Finland

Roland Takács

Anatomy, Histology and Embryology, University of Debrecen, Debrecen, Hungary

Humana Press

Editors
Csaba Matta
Anatomy, Histology and Embryology
University of Debrecen
Debrecen, Hungary

Ali Mobasheri
Research Unit of Health Sciences
and Technology, Faculty of Medicine
University of Oulu
Oulu, Finland

Roland Takács
Anatomy, Histology and Embryology
University of Debrecen
Debrecen, Hungary

ISSN 1064-3745 ISSN 1940-6029 (electronic)
Methods in Molecular Biology
ISBN 978-1-0716-4433-1 ISBN 978-1-0716-4434-8 (eBook)
https://doi.org/10.1007/978-1-0716-4434-8

Cover Caption: A photomicrograph immunostaining of normal human epidermal melanocytes (NHEM).

This Humana imprint is published by the registered company Springer Science+Business Media, LLC, part of Springer Nature.
The registered company address is: 1 New York Plaza, New York, NY 10004, U.S.A.

Preface

Biological membranes are critical for maintaining the integrity of cells, regulating their internal environment, facilitating communication and interactions with other cells, and enabling cellular processes within the cell. The plasma membrane is a crucial component of cellular architecture. Its composition and organization are critical for maintaining cellular homeostasis.

The surfaceome, which represents a dynamic interface between cells and their environment, plays crucial roles in various processes, including signalling, adhesion, and transport. To study this dynamic entity, innovative methods such as biotinylation and proximity labelling have been developed and fine-tuned, which enable the selective capture and analysis of surface proteins. Förster resonance energy transfer (FRET) is a popular tool for studying molecular interactions between individual cell surface proteins.

Biophysical profiling of the protein corona on extracellular vesicles offers a unique window into the interactions between these vesicles, and their surroundings. Furthermore, the application of Fourier transform infrared (FT-IR) and attenuated total reflection (ATR) spectroscopy, as well as acoustic sensing, provides a powerful toolkit for characterizing the molecular make-up of extracellular vesicles.

The analysis of membrane proteomes is challenging due to the hydrophobic nature of membrane proteins. A metal organic layers-based sample preparation method has been developed to overcome this challenge, enabling the efficient analysis of membrane proteomes. The Langmuir monolayer approach has also been employed to study floating films of biological importance at the liquid-gas interface, shedding light on the behaviour of biological molecules in these environments. The integration of CITE-seq (Cellular Indexing of Transcriptomes and Epitopes by Sequencing) has revolutionized the field of single-cell analysis, enabling the simultaneous profiling of transcriptomes and surface proteins, providing a comprehensive view of cellular identity and function.

The electrophysiological properties of cells are also largely determined by components of the surfaceome such as ion channels and other transporters. The functional characterization of single ion channels in planar lipid bilayers provides a platform for understanding the intricate mechanisms of ion transport across biological membranes. The patch clamp technique can be utilized for the direct assessment of the temperature sensitivity of ion channels on isolated, inside-out membrane patches. Dielectrophoresis (DEP) is a powerful tool for studying the electrophysiological properties of cells. It involves the manipulation of cells using non-uniform electric fields, which can provide valuable information about the 'electrome' of the cells.

This methods book is designed to bring together a collection of methodologies to explore the complexities of biological membranes and the associated surfaceome. With detailed, step-by-step instructions and expert insights, this book aims to facilitate the adoption of these innovative methods, ultimately advancing our understanding of the biology of membranes and their role in health and disease.

Debrecen, Hungary ***Csaba Matta***
Oulu, Finland ***Ali Mobasheri***
Debrecen, Hungary ***Roland Takács***

Contents

Contributors

KOTB ABDELMOHSEN • *Laboratory of Genetics and Genomics, Baltimore, MD, USA*

AUDREY ASTORI • *Princess Margaret Cancer Centre, University Health Network, Toronto, ON, Canada; Department of Medical Biophysics, University of Toronto, Toronto, Canada*

ADAM BARTOK • *Department of Biochemistry, Semmelweis University, Budapest, Hungary; HCEMM-SE Molecular Channelopathies Research Group, Budapest, Hungary; HUN-REN-SE Ion Channel Research Group, Budapest, Hungary*

TÍMEA BEBESI • *Hevesy György PhD School of Chemistry, Eötvös Loránd University, Budapest, Hungary; HUN-REN Research Centre for Natural Sciences, Institute of Materials and Environmental Chemistry, Biological Nanochemistry Research Group, Budapest, Hungary*

TAMÁS BEKE-SOMFAI • *Biomolecular Self-assembly Research Group, Institute of Materials and Environmental Chemistry, HUN-REN Research Centre for Natural Sciences, Budapest, Hungary*

JIANG BO • *State Key Laboratory of Medical Proteomics, National Chromatographic R. & A. Center, CAS Key Laboratory of Separation Science for Analytical Chemistry, Dalian Institute of Chemical Physics, Chinese Academy of Sciences, Dalian, China*

DAVID J. BOOCOCK • *John van Geest Cancer Research Centre, Nottingham Trent University, Nottingham, UK*

ANNA CHACHAJ-BREKIESZ • *Faculty of Chemistry, Jagiellonian University, Kraków, Poland*

KRISTA S. P. CLARKE • *Centre for Biomedical Engineering, School of Mechanical Engineering Sciences, University of Surrey, Guildford, Surrey, UK*

PAULINE M. L. COULON • *Australian Institute for Microbiology and Infection, University of Technology Sydney, Ultimo, NSW, Australia*

CLARE COVENEY • *John van Geest Cancer Research Centre, Nottingham Trent University, Nottingham, UK*

GIOVANNI D'ATRI • *IRCCS Istituto Ortopedico Rizzoli, Laboratorio di Immunoreumatologia e rigenerazione tissutale, Bologna, Italy*

STEVEN P. DJORDJEVIC • *Australian Institute for Microbiology and Infection, University of Technology Sydney, Ultimo, NSW, Australia*

PATRYCJA DYNAROWICZ-LATKA • *Faculty of Chemistry, Jagiellonian University, Kraków, Poland*

ANIKÓ GAÁL • *Biological Nanochemistry Research Group, Institute of Materials and Environmental Chemistry, HUN-REN Research Centre for Natural Sciences, Budapest, Hungary*

ELENA GABUSI • *IRCCS Istituto Ortopedico Rizzoli, Laboratorio di Immunoreumatologia e rigenerazione tissutale, Bologna, Italy*

MYRIAM GOROSPE • *Laboratory of Genetics and Genomics, Baltimore, MD, USA*

GABRIELA GRAZIANI • *Department of Chemistry, Materials and Chemical Engineering "Giulio Natta", Politecnico di Milano, Milan, Italy*

OREOLUWA V. GRIFFITHS • *Centre for Biomedical Engineering, School of Mechanical Engineering Sciences, University of Surrey, Guildford, Surrey, UK*

GAGAN D. GUPTA • *Department of Chemistry and Biology, Toronto Metropolitan University, Toronto, ON, Canada*

TIBOR HAJDÚ • *Department of Anatomy, Histology and Embryology, Faculty of Medicine, University of Debrecen, Debrecen, Hungary*
KAI HOETTGES • *Department of Electrical Engineering and Electronics, University of Liverpool, Liverpool, UK*
MIA L. HUANG • *Department of Chemistry, Scripps Research, La Jolla, CA, USA*
MELISSA IAZZI • *Department of Chemistry and Biology, Toronto Metropolitan University, Toronto, ON, Canada; Princess Margaret Cancer Centre, University Health Network, Toronto, ON, Canada; Department of Medical Biophysics, University of Toronto, Toronto, Canada*
VERONICA M. JAROCKI • *Australian Institute for Microbiology and Infection, University of Technology Sydney, Ultimo, NSW, Australia*
ZSÓFIA KERESZTES • *HUN-REN Research Centre for Natural Sciences, Institute of Materials and Environmental Chemistry, Functional Interfaces Research Group, Budapest, Hungary*
PATRIK KOVÁCS • *Department of Anatomy, Histology and Embryology, Faculty of Medicine, University of Debrecen, Debrecen, Hungary*
KATALIN KUFFA • *Protein Bioinformatics Research Group, Institute of Molecular Life Sciences, Research Centre for Natural Sciences, Hungarian Research Network, Budapest, Hungary; Doctoral School of Biology and Institute of Biology, ELTE Eötvös Loránd University, Budapest, Hungary*
FATIMA H. LABEED • *Department of Biology, United Arab Emirates University, Al Ain, UAE*
TAMÁS LANGÓ • *Protein Bioinformatics Research Group, Institute of Molecular Life Sciences, Research Centre for Natural Sciences, Hungarian Research Network, Budapest, Hungary*
REBECCA LEWIS • *Department of Comparative Biomedical Sciences, School of Veterinary Medicine, University of Surrey, Guildford, Surrey, UK*
ZHANG LIHUA • *State Key Laboratory of Medical Proteomics, National Chromatographic R. & A. Center, CAS Key Laboratory of Separation Science for Analytical Chemistry, Dalian Institute of Chemical Physics, Chinese Academy of Sciences, Dalian, China*
GINA LISIGNOLI • *IRCCS Istituto Ortopedico Rizzoli, Laboratorio di Immunoreumatologia e rigenerazione tissutale, Bologna, Italy*
CAMERON B. LLOYD • *Department of Biophysics and Cell Biology, Faculty of Medicine, Debrecen, Hungary*
CSABA MATTA • *Department of Anatomy, Histology and Embryology, Faculty of Medicine, University of Debrecen, Debrecen, Hungary*
KRYSTYNA MAZAN-MAMCZARZ • *Laboratory of Genetics and Genomics, Baltimore, MD, USA*
GÁBOR MÉSZÁROS • *HUN-REN Research Centre for Natural Sciences, Institute of Materials and Environmental Chemistry, Functional Interfaces Research Group, Budapest, Hungary*
JUDITH MIHÁLY • *HUN-REN Research Centre for Natural Sciences, Institute of Materials and Environmental Chemistry, Biological Nanochemistry Research Group, Budapest, Hungary; Department of Chemistry, Eszterházy Károly Catholic University, Eger, Hungary*
ALI MOBASHERI • *Research Unit of Health Sciences and Technology, Faculty of Medicine, University of Oulu, Oulu, Finland; Department of Regenerative Medicine, State Research Institute Centre for Innovative Medicine, Vilnius, Lithuania; Department of Joint Surgery, First Affiliated Hospital of Sun Yat-Sen University, Guangzhou, China; World Health Organization Collaborating Centre for Public Health Aspects of Musculoskeletal Health and Aging, Université de Liège, Liège, Belgium*
RACHEL B. MUNK • *Laboratory of Genetics and Genomics, Baltimore, MD, USA*

Minh Ngoc Nguyen • *Department of Anatomy, Histology and Embryology, Faculty of Medicine, University of Debrecen, Debrecen, Hungary*

Matthew P. Padula • *School of Life Sciences, Faculty of Science, University of Technology Sydney, Ultimo, NSW, Australia*

Jia Meng Pang • *Skaggs Graduate School of Chemical and Biological Sciences, Scripps Research, La Jolla, CA, USA; Department of Chemistry, Scripps Research, La Jolla, CA, USA*

Jiang Qianqian • *State Key Laboratory of Medical Proteomics, National Chromatographic R. & A. Center, CAS Key Laboratory of Separation Science for Analytical Chemistry, Dalian Institute of Chemical Physics, Chinese Academy of Sciences, Dalian, China; University of Chinese Academy of Sciences, Beijing, China*

Brian Raught • *Princess Margaret Cancer Centre, University Health Network, Toronto, ON, Canada; Department of Medical Biophysics, University of Toronto, Toronto, Canada*

István Rebenku • *Department of Biophysics and Cell Biology, Faculty of Medicine, Debrecen, Hungary*

Maria C. Rezcallah • *Australian Institute for Microbiology and Infection, University of Technology Sydney, Ultimo, NSW, Australia*

Loránd Románszki • *HUN-REN Research Centre for Natural Sciences, Institute of Materials and Environmental Chemistry, Functional Interfaces Research Group, Budapest, Hungary*

Martina Rossi • *Translational Gerontology Branch, National Institute on Aging (NIA) Intramural Research Program (IRP), National Institutes of Health (NIH), Baltimore, MD, USA*

Sara Sadeghi • *Department of Chemistry and Biology, Toronto Metropolitan University, Toronto, ON, Canada*

Priyanka Singh • *HUN-REN Research Centre for Natural Sciences, Institute of Materials and Environmental Chemistry, Biomolecular Self-assembly Research Group, Budapest, Hungary; Hevesy György PhD School of Chemistry, Eötvös Loránd University, Budapest, Hungary*

Tasvilla Sonallya • *Biomolecular Self-assembly Research Group, Institute of Materials and Environmental Chemistry, HUN-REN Research Centre for Natural Sciences, Budapest, Hungary; Hevesy György PhD School of Chemistry, ELTE Eötvös Loránd University, Budapest, Hungary*

Jonathan St-Germain • *Princess Margaret Cancer Centre, University Health Network, Toronto, ON, Canada; Department of Medical Biophysics, University of Toronto, Toronto, Canada*

Imola Cs. Szigyártó • *Biomolecular Self-assembly Research Group, Institute of Materials and Environmental Chemistry, HUN-REN Research Centre for Natural Sciences, Budapest, Hungary*

Andras Szollosi • *Semmelweis University, Department of Biochemistry, Budapest, Hungary; HUN-REN-SE Ion Channel Research Group, Budapest, Hungary; HCEMM-SE Molecular Channelopathies Research Group, Semmelweis University, Budapest, Hungary*

Roland Takács • *Department of Anatomy, Histology and Embryology, Faculty of Medicine, University of Debrecen, Debrecen, Hungary*

Michael Thompson • *Department of Chemistry, University of Toronto, Toronto, ON, Canada*

ADAM V. TOTH • *Department of Biochemistry, Semmelweis University, Budapest, Hungary; HCEMM-SE Molecular Channelopathies Research Group, Budapest, Hungary; HUN-REN-SE Ion Channel Research Group, Budapest, Hungary*

DIMITRIOS TSITSIPATIS • *Laboratory of Genetics and Genomics, Baltimore, MD, USA*

GÁBOR E. TUSNÁDY • *Protein Bioinformatics Research Group, Institute of Molecular Life Sciences, Research Centre for Natural Sciences, Hungarian Research Network, Budapest, Hungary*

ZOLTÁN VARGA • *Biological Nanochemistry Research Group, Institute of Materials and Environmental Chemistry, HUN-REN Research Centre for Natural Sciences, Budapest, Hungary; Department of Physical Chemistry and Materials Science, Budapest University of Technology and Economics, Budapest, Hungary*

GYÖRGY VEREB • *Department of Biophysics and Cell Biology, Faculty of Medicine, Debrecen, Hungary; HUN-REN-UD Cell Biology and Signaling Research Group, Debrecen, Hungary; Faculty of Pharmacy, University of Debrecen, Debrecen, Hungary*

ZAK VILEN • *Skaggs Graduate School of Chemical and Biological Sciences, Scripps Research, La Jolla, CA, USA; Department of Chemistry, Scripps Research, La Jolla, CA, USA*

ZOLTÁN WIENER • *Department of Genetics, Cell and Immunobiology, Semmelweis University, Budapest, Hungary*

ANITA WNĘTRZAK • *Faculty of Chemistry, Jagiellonian University, Kraków, Poland*

ZHANG YUKUI • *State Key Laboratory of Medical Proteomics, National Chromatographic R. & A. Center, CAS Key Laboratory of Separation Science for Analytical Chemistry, Dalian Institute of Chemical Physics, Chinese Academy of Sciences, Dalian, China*

EMESE ZSIGRAI • *Department of Pediatrics, Faculty of Medicine, University of Debrecen, Debrecen, Hungary*

Chapter 1

Introduction: Methods for Studying the Plasma Membrane and the Surfaceome

Roland Takács, Ali Mobasheri, and Csaba Matta

Abstract

The cell surface, or surfaceome, plays a critical role in cell communication, adhesion, and signaling. This volume provides a comprehensive collection of protocols designed to explore the dynamic nature of the surfaceome. Encompassing both established and emerging techniques, this collection equips researchers with methods for isolating and characterizing surface proteins, analyzing protein–protein interactions, and investigating posttranslational modifications. It also includes protocols for advanced approaches, such as single-channel recordings and mass spectrometry–based analyses. Detailed instructions and troubleshooting tips empower researchers of all levels of experience to investigate the intricacies of the surfaceome. This collection serves as a resource for researchers in cell biology, biochemistry, and medicine aiming to understand the complexities of biological membranes and surfaceomes in health and disease.

Key words Surfaceome, Cell surface, Membrane biology, Membrane proteins, Plasma membrane, Laboratory protocols

1 Introduction

Biological membranes, particularly the plasma membrane, are essential for maintaining cellular integrity and function. Biological membranes act as a barrier that compartmentalize living tissue, allowing for the regulation of material transport into and out of the cell, and play a critical role in processes such as endocytosis, exocytosis, exosome formation, cell adhesion, and signal transduction [1]. The lipid bilayer structure of the plasma membrane is fundamental to its function as a barrier, while its proteins are crucial for various cellular activities, including material transport, enzyme localization, and attachment [2].

The cell surface subproteome, also referred to as the *surfaceome*, is a collection of proteins expressed on the cell surface that mediate interactions between the cell and its environment [3, 4]. The surfaceome includes integrated membrane proteins,

Csaba Matta et al. (eds.), *The Surfaceome: Methods and Protocols*, Methods in Molecular Biology, vol. 2908,
https://doi.org/10.1007/978-1-0716-4434-8_1,

which are embedded within the lipid bilayer, as well as surface-associated proteins, which are peripherally associated with the membrane [5]. The surfaceome also includes transmembrane proteins, which span the lipid bilayer and have domains exposed on both sides of the membrane [6]. Cell surface proteins include glycoproteins, which are proteins that have covalently attached glycans, and play important roles in cell–cell interactions and signaling [3]. Extracellular vesicle (EV) surface proteins are found on the surface of EVs and play roles in intercellular communication and transport of biomolecules. EV surfaceomes include both membrane-anchored and surface-associated proteins [7].

The surfaceome is highly dynamic and complex as it serves as a gateway for cellular communication and interaction with the extracellular environment [5]. This subproteome mediates processes such as cellular communication and signaling, translating extracellular signals into intracellular responses, cell adhesion, and managing vital cellular functions by transporting metabolites, ions, and other solutes across the plasma membrane. The surfaceome also includes markers that can be used to identify cells and define their phenotype in health and disease [4].

The surfaceome, with its diverse array of proteins, is an emerging target for immune therapy and drug development in various human disorders, including cancer [3]. The cell surface subproteome presents a crucial resource of biomarkers and potential drug targets for various disease states. This is well illustrated by the fact that more than 60% of currently approved pharmaceuticals target cell surface proteins. For this reason, profiling of the surfaceome has become increasingly important and has led to the discovery of new biomarkers and drug targets, for example, in oncology and immunology [8, 9].

However, studying the cell surface has its own challenges. Most cell surface proteins are trans- or intramembrane proteins that have poor solubility in aqueous buffers, they are often highly modified (e.g., glycosylated) and expressed at low levels [10]. Advanced methods such as mass spectrometry–based proteomics have enabled detailed characterization of the surfaceome, despite challenges posed by the complexity of cellular lysates [11].

To address the challenges of studying the cell surface, cutting-edge techniques have been developed. In this collection, investigators working in the field of cell surface biology and pathology research around the globe have shared the step-by-step methodology they are using and tailored to their research projects in their laboratory to answer questions related to biological membranes and the surfaceome in health and disease. By making these cutting-edge methods accessible, we aim to empower researchers worldwide to tackle and answer critical questions in their own investigations. This collection is organized into three sections, each addressing a distinct aspect of surfaceome research through a series of cutting-edge methodologies.

2 Mass Spectrometry–Based Methodologies to Study the Qualitative and Quantitative Composition of the Surfaceome

Part I describes methods that employ mass spectrometry–based methodology to identify the cell surface subproteome in various models. Mass spectrometry–based approaches for studying the surfaceome currently face several challenges. One key challenge is that the limited quantity of proteins in the cell surface would necessitate techniques that are specific for surface-bound proteins. Biotin-based methods can provide high specificity for surface proteins by exploiting biotin–avidin interactions to capture and enrich surface-bound proteins. This specificity is particularly useful when studying surface proteins that are difficult to distinguish from intracellular proteins using other methods [12]. Opening this section, Langó, Kuffa, and Tusnády describe an optimized biotinylated peptide detection method for characterizing the cell surface proteome (*see* Chap. 2). The authors present a high-throughput method that utilizes biotinylation of accessible primary amino groups within the extracellular regions of cell surface proteins by sulfo-NHS-SS-biotin. The method can be used for side-selective protein biotinylation on biological barrier-forming epithelial cells, allowing the qualitative and quantitative characterization of proteins in apical and basolateral membranes.

However, conventional biotinylation methods can still suffer from background contamination due to the potential for biotin to cross the cell membrane and label intracellular proteins. This can be minimized by using additional steps to remove intracellular contaminants or by employing improved biotinylation reagents and optimized protocols [13, 14]. To help mitigate these challenges, two different proximity labeling-based techniques are being introduced next. Vilen, Pang, and Huang provide a proximity labeling-based technique to covalently label proteins at the cell surface with a biotin handle, enabling downstream streptavidin-based enrichment and manipulation in a variety of modalities, including fluorescence imaging, western blotting, and mass spectrometry–based proteomics (*see* Chap. 3). Using cell surface engineering, the authors of this chapter incorporate modified cholesterol derivatives into the cell membrane of live cells, before conjugating an engineered ascorbate peroxidase (APEX2) to the lipid via an engineered sortase enzyme which catalyzes the conjugation of APEX2 onto the cell surface. APEX2 biotinylates electron-rich residues within a ~ 20 nm radius, labeling cell surface proteins with biotin regardless of posttranslational modifications. To map the cystic fibrosis transmembrane conductance regulator (CFTR) interactome in a human airway model (bronchial epithelial cells grown at the air–liquid interface), Iazzi and colleagues describe the methodology of proximity-dependent biotin identification (BioID), a method that can be used to identify physiologically relevant protein–protein interactions in a living cell (*see* Chap. 4). The technique employs the fusion of an abortive *Escherichia coli* biotin ligase mutant (BirA*) to a protein of interest, allowing for the covalent biotinylation of endogenous proximal (within ~10 nm) polypeptides in mammalian cells.

The next two chapters in this section by Kovács et al. (*see* Chap. 5) and Hajdú et al. (*see* Chap. 6) outline a comprehensive approach for characterizing the cell surface subproteome of chondroprogenitor cells and primary human epidermal melanocytes, respectively, based on aminooxy biotinylation followed by analysis with mass spectrometry. In addition to providing a detailed step-by-step guide to establish the experimental models, they also reveal their optimized technique consisting of selective labeling of cell surface proteins with aminooxy biotin on living cells, ensuring the specific tagging of glycoproteins on the outer membrane. Subsequently, glycocapture technique is employed to enrich the glycosylated fraction of the cell surface proteins. Following multiple wash steps to reduce contamination with detergents and nonsurface proteins, shotgun mass spectrometry is applied for the quantitative and qualitative analysis of the enriched subproteome, allowing for the identification and characterization of surface proteins.

Abdelmohsen and colleagues share their integrated protocol to exploit the presence of cell surface proteins on subsets of senescent cells to study their heterogeneity at the single-cell level (*see* Chap. 7). Their approach starts with identifying senescence-associated surface proteins by mass spectrometry and then performing cellular indexing of transcriptomes and epitopes sequencing (CITE-seq) single-cell analysis. Using this approach, the authors of this chapter have identified unique transcriptomic programs associated with surface protein markers expressed in some senescent cells. This methodology can also be applied to other biological scenarios where cells with unique transcriptomic profiles can be studied individually due to the presence of specific cell surface proteins that distinguish them from other cells within the same population.

Given that mass spectrometry-based identification of membrane proteins poses several challenges due to the intrinsic properties of membrane proteins such as high hydrophobicity and low abundance, Qianqian, Bo, Yukui, and Lihua share their cutting-edge protocol on a metal organic layers (MOL)-based sample preparation method for membrane proteome analysis in living cells (*see* Chap. 8). This protocol overcomes the previously mentioned limitations by a cell surface engineering process through the exquisite interaction between MOL and phospholipid bilayers on the membrane and the following biomembrane fusion process, providing a simple and robust tool for the molecular dissection of membrane proteome with high specificity. This chapter details the preparation of such metal organic layers and enrichment workflow of HeLa membrane proteins.

Finally, the protocol by Jarocki et al. in Chap. 9 enables the accurate representation of a bacterial surfaceome. Their workflow includes enzymatic surface shaving, cell surface biotinylation, solid phase extraction, liquid chromatography coupled with tandem mass spectrometry (LC-MS/MS), and data analysis. The resulting data provides a robust framework for understanding host–pathogen interactions and developing novel vaccines and therapeutic agents. Their protocol can be applied to several bacterial species with Gram-negative and Gram-positive cell wall compositions, including *Mycoplasma*, *Staphylococcus*, *Klebsiella*, *Pseudomonas*, and *Proteus*.

3 Methods and Protocols Related to Bioelectricity

Part II is devoted to methods that rely on the membrane potential of live cells. The ion channels and transporters operating on the cell surface are crucial for maintaining cellular homeostasis and function by defining the electrophysiological properties of cells. Ion channels regulate the flow of ions across the plasma membrane, influencing various cellular processes such as signal transduction, nutrient absorption, and cell volume regulation [15, 16]. Techniques such as single-channel recording and patch clamp are used to study the kinetics and functional properties of these channels [17]. The planar lipid bilayer technique, shared here by Andras Szollosi (*see* Chap. 10), is a powerful tool to address this issue. In this protocol, a purified ion channel protein is incorporated into a membrane bilayer separating two compartments in an experimental chamber. The ionic currents through the channel of interest are recorded between the two compartments. The setup described here allows the characterization of gating and permeation properties of ion channels without the influence of auxiliary components present in native membranes. Next, Toth and Bartok share their protocol on an electrophysiological configuration that allows the direct assessment of the temperature sensitivity of ion channels that act as warm or cold sensors such as the transient receptor potential M2 (TRPM2) channel on isolated, inside-out membrane patches (*see* Chap. 11). The underlying principle of the method is that it detects the temperature dependence of microscopic and macroscopic currents, apparent ligand binding affinities, and the gating parameters of the channel under the precise control of temperature and activating ligand concentrations. The technique described here can be readily adapted to investigate the temperature-dependent properties of diverse ion channels.

The actual interaction between cell surface proteins is often more important than their expression levels. As demonstrated by Rebenku et al. in Chap. 12, Förster (or fluorescence) resonance energy transfer (FRET) is a popular tool for studying molecular interactions, since it is highly sensitive to proximity in the range of 2–10 nm. Spectral spillover-corrected quantitative (3-filterset) FRET is a cost-effective and versatile approach, which can be applied in flow cytometry and fluorescence microscopy, where it provides pixel-by-pixel quantitative FRET efficiency values as well as FRET-corrected fluorescent intensities corresponding to the expression level of the two labeled interacting molecular species.

Dielectrophoresis (DEP) is an emerging approach for studying global electrophysiological properties of live cells. DEP allows for the rapid, label-free determination of dielectric parameters, which can reveal changes in membrane conductance and capacitance in response to ion channel activity [18]. DEP is the motion of a polarizable particle in a nonuniform electric field. The nonuniform field induces a dipole, the orientation of which is determined by the relative permittivity of the particle to the medium and dictates the direction of the force from the field on the particle. DEP has been used extensively to investigate the electrophysiological differences between healthy and diseased cells, providing insights into the role of ion channels in conditions such as osteoarthritis and multidrug-resistant leukemia [19]. Here, Clarke et al. present a detailed protocol for using the 3DEP platform, which rapidly characterizes three key electrophysiological parameters of up to 20,000 cells (*see* Chap. 13). They demonstrate that the 3DEP platform plots the relative DEP force experienced by cells across a range of AC frequencies, which generates a spectrum through which membrane conductance, membrane capacitance, and cytoplasmic conductivity can be derived, allowing researchers to produce an electrical fingerprint of the cell populations studied that can be used to differentiate between neoplastic and healthy cell types or cells in different developmental stages.

4 Techniques for Investigating the Biophysical Characteristics of Biological Membranes

Part III focuses on modern techniques for investigating the biophysical characteristics of biological membranes in both intact cells and extracellular vesicles (EVs). Sonallya et al. provide a protocol on the biophysical profiling of the protein corona on red blood cell-derived extracellular vesicles (REVs) (*see* Chap. 14). The composition of the protein corona in EVs strongly depends on their origin and the biological environment. In this protocol, the authors share their step-by-step guide for characterizing the changes in the protein corona content on REVs by using linear dichroism (LD) spectroscopy with microfluidic resistive pulse sensing (MRPS). This combined methodology can be employed to offer a reliable and reproducible approach for the preliminary profiling of the surface proteins on EVs. In Chap. 15 , Szigyártó and colleagues present their protocol on the use of two nonconventional techniques, i.e., ATR-infrared spectroscopy (ATR-IR) and electromagnetic piezoelectric acoustic sensing (EMPAS) in the characterization of blood and pancreatic ductal adenocarcinoma derived EVs. IR spectroscopy, a fast and label-free technique, gives biochemical insight into the sample composition, while EMPAS serves information on vesicle quantity and membrane integrity changes during storage. The authors of this chapter also provide guidelines on sample collection, isolation, and separation of EVs.

Wnętrzak and colleagues give detailed insight into floating films of biological importance at the liquid–gas interface by using a Langmuir monolayer approach (*see* Chap. 16). Langmuir films are self-assembled monolayers of water-insoluble molecules formed on aqueous subphases. This technique enables the investigation of the orientation of various bioactive compounds at the interface and interactions with lipids present in membranes. Here they present their methodology on an example of a chain-oxidized cholesterol derivative using Langmuir monolayers. The protocol also describes the application of related complementary in situ techniques, such as Brewster angle microscopy (BAM), electric surface potential change, DV measurements, and polarization modulation-infrared reflection-absorption spectroscopy (PM-IRRAS). Finally, Graziani and colleagues share their method on protein analysis using Fourier transform infrared spectroscopy (FT-IR), a powerful tool for the study of inorganic and organic materials (*see* Chap. 17). Their protocol offers high sensitivity, requires small sample volumes, has fast acquisition time, and can analyze a broad range of compounds. FT-IR provides a fingerprint of the analyzed material in terms of quantity and organization, furnishing important information for studies on tissue pathophysiology, disease progression, and therapeutic evaluation.

5 Summary

We believe the cutting-edge protocols detailed in this book will inspire and accelerate research on biological membranes, leading to new discoveries and innovative approaches. Biological membranes are dynamic interfaces that have a deep impact on our understanding of health and disease. By leveraging the latest advances in membrane biophysics, biochemistry, and imaging techniques, researchers are now poised to uncover the intricate details of membrane organization and function in both normal and pathological contexts. It is essential that we continue to integrate these diverse approaches to gain a deeper understanding of the complex interplay between membrane components and their role in shaping cellular behavior. These protocols are essential to develop innovative novel therapeutic strategies and improve our ability to diagnose and treat a wide range of diseases.

Acknowledgments

We would like to thank the contributors who shared their protocols in this collection. We are especially grateful to Dr. John Walker, series editor, for the guidance in helping us to assemble this edition of the Surfaceome: Methods and Protocols.

Funding Information

CM was supported by the Young Researcher Excellence Programme (grant number: FK-134304) of the National Research, Development and Innovation Office, Hungary. CM was also supported by the Bolyai János Research Fellowship of the Hungarian Academy of Sciences. This research was supported by the UNKP-23-5 and UNKP-23-5 New National Excellence Program of the Ministry for Culture and Innovation from the source of the National Research, Development and Innovation Fund, Hungary, awarded to CM (UNKP-23-5-DE-487) and RT (UNKP-23-4-II-DE-378). AM acknowledges financial support from the European Structural and Social Funds through the Research Council of Lithuania (Lietuvos Mokslo Taryba), according to the Program Attracting Foreign Researchers for Research Implementation (Grant No. 01.2.2-LMT-K-718-02-0022). AM also acknowledges funding from Horizon Europe (PROTO-Advanced PeRsOnalized Therapies for Osteoarthritis-TACKLING INFLAMMATION TO IMPROVE PATIENT OUTCOMES, Grant agreement ID: 101095635, https://cordis.europa.eu/project/id/101095635 and ENgineered CArtilage from Nose for the Treatment of Osteoarthritis [ENCANTO], Grant agreement ID: 101137315, https://cordis.europa.eu/project/id/101137315) and the Academy of Finland through the Profi6 336,449 grant awarded to the University of Oulu. CM, RT, and AM also acknowledge financial support from the European Cooperation in Science and Technology (COST) Association Action CA21110—Building an open European Network on OsteoArthritis research (NetwOArk); https://www.cost.eu/actions/CA21110/).

References

1. Lucy JA (1972) Functional and structural aspects of biological membranes: a suggested structural role for vitamin E in the control of membrane permeability and stability. Ann N Y Acad Sci 203:4–11. https://doi.org/10.1111/j.1749-6632.1972.tb27849.x
2. Mouritsen OG (1986) Phase transitions in biological membranes. Ann N Y Acad Sci 491:166–169. https://doi.org/10.1111/j.1749-6632.1987.tb30051.x
3. Leung KK et al (2020) Broad and thematic remodeling of the surfaceome and glycoproteome on isogenic cells transformed with driving proliferative oncogenes. Proc Natl Acad Sci USA 117(14):7764–7775. https://doi.org/10.1073/pnas.1917947117
4. Pauwels J, Fijalkowska D, Eyckerman S, Gevaert K (2022) Mass spectrometry and the cellular surfaceome. Mass Spectrom Rev 41(5):804–841. https://doi.org/10.1002/mas.21690
5. Bausch-Fluck D, Hofmann A, Wollscheid B (2012) Cell surface capturing technologies for the surfaceome discovery of hepatocytes. Methods Mol Biol 909:1–16. https://doi.org/10.1007/978-1-61779-959-4_1
6. da Cunha JP et al (2009) Bioinformatics construction of the human cell surfaceome. Proc Natl Acad Sci USA 106(39):16752–16757. https://doi.org/10.1073/pnas.0907939106
7. Rai A et al (2021) Proteomic dissection of large extracellular vesicle surfaceome unravels interactive surface platform. J Extracell Vesicles 10(13):e12164. https://doi.org/10.1002/jev2.12164
8. Ferguson ID et al (2022) The surfaceome of multiple myeloma cells suggests potential immunotherapeutic strategies and protein markers of drug resistance. Nat Commun 13(1):4121. https://doi.org/10.1038/s41467-022-31810-6
9. Matta C et al (2019) Molecular phenotyping of the surfaceome of migratory chondroprogenitors and mesenchymal stem cells using biotinylation, glycocapture and quantitative LC-MS/MS proteomic analysis. Sci Rep 9(1):9018. https://doi.org/10.1038/s41598-019-44957-y
10. Li Y, Qin H, Ye M (2020) An overview on enrichment methods for cell surface proteome profiling. J Sep Sci 43(1):292–312. https://doi.org/10.1002/jssc.201900700
11. Naik A, Srivastava S, Wiita AP (2023) "Cell surface capture" workflow for label-free quantification of the cell surface proteome. J Vis Exp 193. https://doi.org/10.3791/64952
12. Elia G (2008) Biotinylation reagents for the study of cell surface proteins. Proteomics 8(19):4012–4024. https://doi.org/10.1002/pmic.200800097
13. Jeremiasse B et al (2020) Alterations in the chondrocyte surfaceome in response to pro-inflammatory cytokines. BMC Mol Cell Biol 21(1):47. https://doi.org/10.1186/s12860-020-00288-9
14. Long J et al (2016) Plasma membrane profiling reveals upregulation of ABCA1 by infected macrophages leading to restriction of mycobacterial growth. Front Microbiol 7:1086. https://doi.org/10.3389/fmicb.2016.01086
15. Barrett-Jolley R, Lewis R, Fallman R, Mobasheri A (2010) The emerging chondrocyte channelome. Front Physiol 1:135. https://doi.org/10.3389/fphys.2010.00135
16. Mobasheri A et al (2019) The chondrocyte channelome: a narrative review. Joint Bone Spine 86(1):29–35. https://doi.org/10.1016/j.jbspin.2018.01.012
17. Mortensen M, Smart TG (2007) Single-channel recording of ligand-gated ion channels. Nat Protoc 2(11):2826–2841. https://doi.org/10.1038/nprot.2007.403
18. Abdallat R et al (2021) Dielectrophoresis as a tool to reveal the potential role of ion channels and early electrophysiological changes in osteoarthritis. Micromachines (Basel) 12(8). https://doi.org/10.3390/mi12080949
19. Duncan L et al (2008) Dielectrophoretic analysis of changes in cytoplasmic ion levels due to ion channel blocker action reveals underlying differences between drug-sensitive and multidrug-resistant leukaemic cells. Phys Med Biol 53(2):N1–N7. https://doi.org/10.1088/0031-9155/53/2/N01

Part I

Mass Spectrometry-Based Methodologies to Study the Qualitative and Quantitative Composition of the Surfaceome

Chapter 2

Optimized Biotinylated Peptide Detection Method for Characterizing the Cell Surface Proteome

Tamás Langó, Katalin Kuffa, and Gábor E. Tusnády

Abstract

Cell surface proteins (CSPs) play a pivotal role in cellular processes and are crucial for differentiating various cell types. Despite the significance of the cell surface-localized proteome, it is often characterized using low-throughput techniques that provide limited information from the surface-exposed protein segments of individual cells. If such information is available, it is often scattered across numerous publications, making it time-consuming to gather. To address these challenges, we present a high-throughput method that utilizes biotinylation of accessible primary amino groups within the extracellular regions of CSPs to simultaneously generate several hundred CSP-specific peptides in a single experiment. The steps of the method have been meticulously refined to capture an increasing number of cell type-specific biotinylated peptides localized in cell surface accessible protein regions.

The protocol developed is readily applicable to the analysis of both soluble and adherent cell surface proteins as described in detailed in this paper. Additionally, the method can be used for side-selective protein biotinylation on biological barrier-forming epithelial cells, allowing the qualitative and quantitative characterization of proteins in both their apical and basolateral membranes. The identified amino-modified surface peptides derived from CSPs have potential as targets for various molecular biology research applications, including antibody staining, inhibitor development and monitoring of host–pathogen interactions. Furthermore, label-free quantification of CSPs via mass spectrometry (based on identified peptide list) provides valuable insights into their relative abundance in different cell surface or plasma membrane regions.

Key words Biotinylated peptides, Cell surface accessible protein region, NanoHPLC peptide separation, Tandem mass spectrometry, Membrane proteins, Surfaceome, Peptide sequencing

1 Introduction

In multicellular organisms, cells are isolated from the external environment by lipid bilayers, forming biological membranes that contain several cell surface proteins (CSPs), including transmembrane and other surface-anchored proteins. These CSPs serve as primary sensors of the extracellular stimuli and orchestrate diverse cellular functions, including signal transduction, nutrient uptake, cell–cell adhesion, migration, and molecular transport.

Csaba Matta et al. (eds.), *The Surfaceome: Methods and Protocols*, Methods in Molecular Biology, vol. 2908,
https://doi.org/10.1007/978-1-0716-4434-8_2,

Furthermore, almost two-thirds of approved drug targets are CSPs [1], due to their accessible extracellular segments that can be manipulated to modulate signaling pathways (G protein–coupled receptors (GPCRs), ion transporters), cancer immunotherapy, and vaccine development against infectious diseases. Their significance is particularly evident in biological barriers, such as the blood–brain barrier, renal tubules, and endothelial walls, where single epithelial cell layers form physical barriers between the external environment and neighboring cells. Within these barriers, cells possess two distinct plasma membrane domains, i.e., apical and basolateral compartments, separated by tight junctions, each harboring functionally side-specific proteins [2, 3].

Despite the aforementioned importance of CSPs, their comprehensive characterization in their native environment is a major challenge, primarily due to their unique physicochemical properties (often amphipathic, with both hydrophilic and hydrophobic regions) and their relatively low abundance. Antibody-based low-throughput methods are often used (such as immunostainings detected by flow cytometry or microscopy), but these usually provide information on only one protein and the precise target sequence is unknown (due to manufacturing limitations). Furthermore, these techniques are heavily dependent on the availability of antibody collections, and the resulting data are scattered throughout the literature, making them time-consuming to gather systematically. To address these challenges, more robust methods have been developed in recent years for characterizing CSPs and/or their extracellular regions. The majority of these methods rely on chemical derivatization of accessible segments on the surface of live cells, followed by modification specific affinity purification and sequencing of the peptides of the proteins using tandem mass spectrometry. Detecting chemical modifications on the reactive side chains of the targeted residues provides an opportunity to simultaneously identify surface-exposed peptides from over a hundred CSPs, both qualitatively and quantitatively [4–6].

Our group has been developing more high-throughput methods [7–10] to identify an increasing number of cell surface peptides from CSPs using residue-specific biotinylation agents. The most successful protocol to date is presented in the current paper (Fig. 1), using the most widely used primary amino group-specific chemical reagent, sulfo-NHS-SS-biotin, for cell surface protein labeling. This protocol is highly optimized to identify as many labeled or biotinylated peptides as possible from individual cell surfaces in a single nanoHPLC–MS/MS run. The strategy is readily applicable to both soluble cells and adherent cells (like HL60 and MDCKII, respectively, in the present work). MDCKII cells are often used to model biological barriers by growing them on transwell inserts while forming an intact/confluent monolayer (*see* Fig. 1). Our method can selectively characterize the proteins of

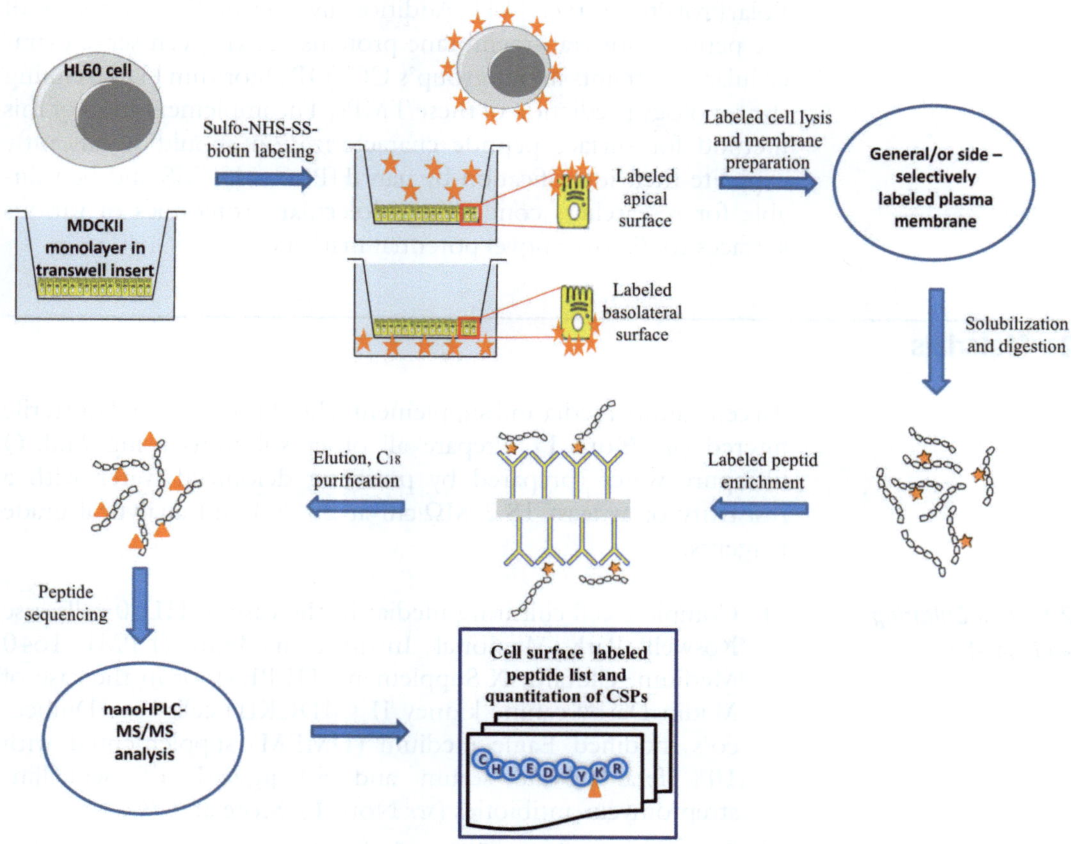

Fig. 1 Flowchart of the developed method to identify the cell surface accessible segments of cell surface proteins (CSPs). Isolated HL60 cells and MDCKII monolayers (where apical and basolateral surface are biotinylated separately) are labeled with a membrane-impermeable, primary amino group-specific labeling agent (sulfo-NHS-SS-biotin). The cells are lysed, and the various labeled plasma membranes are isolated and solubilized. The solubilized membranes are digested with MS grade trypsin to generate peptides. The biotinylated peptides are immobilized on a high-capacity neutravidin agarose resin, then eluted with reducing agent, desalted by a C_{18} spin column and subjected to tandem mass spectrometry (MS/MS) analysis. During the applied nanoHPLC–MS/MS analysis, the resulted covalent modifications on the side chains of lysine residues or on the amino groups of N-termini of the proteins/peptides are identified. Finally, both the surface-accessible segments and abundance of the CSPs are determined in the appropriate membrane compartments. The orange stars represent the sulfo-NHS-SS-biotin labeling, and the orange triangles are the remnant region from the labeling agent after reduction

the apical and/or basolateral membranes by using side-selective biotinylation.

In summary, the data generated by this method is essential for expanding our knowledge of the surfaceome of any individual cell, which can lead to improvements in theoretical surface-detectable peptide lists [11] and surfaceome predictions [12]. When applied to polarized cells (growing on transwell inserts), the intensity of peptides/proteins derived from different cell sides can be incorporated into the PolarProtDb [13], which is linked to the

PolarProtPred server [14]. Additionally, the biotinylated sites of the peptides for transmembrane proteins (TMPs) can serve extracellular constraints in our group's CCTOP algorithm [15], refining the topology prediction of these TMPs. The implementation of this method for surface peptide characterization should significantly expedite their identification by nanoHPLC–MS/MS and be valuable for researchers conducting molecular proteomics of various surfaces to discover novel potential markers.

2 Materials

All cell culture media and supplements should be sterilized or sterile filtered (*see* **Note 1**). Prepare all other solutions using Milli-Q ultrapure water (prepared by purifying deionized water with a resistivity of at least 18.2 MΩ·cm at 25 °C) and analytical grade reagents.

2.1 Cell Culturing and Isolation

1. Complete cell culturing media: In the case of HL60 cells, use Roswell Park Memorial Institute medium (RPMI 1640 Medium, GlutaMAX Supplement, HEPES), or in the case of Madin-Darby canine kidney II (MDCKII) cells, use Dulbecco's modified Eagle medium (DMEM) supplemented with 10% fetal bovine serum and 50 μg/mL of penicillin-streptomycin antibiotics (*see* **Note 1**). Store at 4 °C.
2. Freezing medium: 50% of the appropriate cell culturing medium, 40% sterile-filtered fetal bovine serum (FBS), and 10% dimethyl sulfoxide (DMSO) cryoprotective agent (for ten vials: 5 mL cell-specific medium, 4 mL FBS, 1 mL DMSO). Always use freshly prepared cryogenic medium.
3. Water bath.
4. Phosphate-buffered saline (PBS) solution: 137 mM NaCl, 2.7 mM KCl, 10 mM Na_2HPO_4, and 1.8 M KH_2PO_4; pH 7.4. Weigh 4 g NaCl, 100.64 mg KCl, 709.8 mg Na_2HPO_4, and 122.47 mg KH_2PO_4; dissolve in 400 mL water, adjust pH to 7.4 (if necessary), and make up to 500 mL with water. Store at room temperature or at 4 °C.
5. Trypsin-EDTA solution: 0.5 g/L porcine trypsin and 0.2 g/L EDTA•4Na in Hank's balanced salt solution with phenol red. Store at −20 °C.
6. Ethanol solution: 70% (v/v) ethanol for sterilization. Mix 350 mL absolute ethanol with 150 mL water. Store at room temperature.
7. Cell culture plastics: T25 or T75 flasks; 6- or 24-well plates, polyester or polycarbonate transwell inserts (standing or hanging) with 0.4 μm pore size, compatible with 6-, 12-, and

24-well plates; 1.5–2.5 mL Eppendorf tubes, 15–50 mL centrifuge tubes, and cryovials for long-term storage of cells under liquid nitrogen.

8. Electric pipette with disposable serological pipet tips (2, 5, 10, and 25 mL pipet tips).
9. Bürker chamber (hemocytometer).
10. Eppendorf New Brunswick Galaxy 170R CO_2 humidified incubator.
11. ESCO Class II Biosafety Cabinet (BSC) laminar flow box.
12. Genomic DNA isolation kit.
13. NanoDrop 2000 spectrophotometer.
14. Compounds for mycoplasma test (such as specific primer pairs, Taq polymerase, MyTaq HS Mix (2×), nuclease-free water, etc., *see* details at **Note 2**).
15. Eppendorf centrifuge 5804 R, A-4-44 swing-bucket rotor.
16. Electrical resistance system for measurement of transepithelial electrical resistance (TEER) [16].

2.2 Cell Surface Biotinylation/Labeling

1. PBS solution (pH 8.0): Weigh the components for the PBS solution for 50 mL (*see* Subheading 2.1, **item 4**). Add water to 40 mL. Mix and adjust pH with 5 M NaOH to pH 8.0. Make up to 50 mL with water and store it at room temperature (prepare freshly and equilibrate to room temperature for at least 1 h before the labeling experiment).
2. Sulfo-NHS-SS-biotin solution: Immediately before use, prepare ~2 mM (1.2 mg/mL) sulfo-NHS-SS-biotin stock solution in PBS (pH 8.0) solution (*see* **Note 3**).
3. Tris-buffered saline (TBS): 0.025 M Tris–HCl base, 0.15 M NaCl, pH 7.2. Weigh 1.514 g Trizma base and 4.38 g NaCl and transfer to a 1 L plastic graduated beaker. Add water to 400 mL, mix, and adjust pH with HCl. Make up to 500 mL with water. Store the solution at 4 °C.

2.3 Labeled Cell Isolation, Lysis, and Membrane Preparation

1. Cell scraper for isolation of the adherent cells.
2. Lysis buffer: 0.02 M Tris–HCl, 0.01 M KCl, 0.02 M sucrose, 10 mM iodoacetamide (IA), pH 7.4. Weigh 3.42 g sucrose and transfer into a 500 mL graduated cylinder, and add 58.8 mL of 0.17 M Tris–HCl stock solution (pH 7.4) and 31.25 mL of 0.16 M KCl stock solution (*see* **Note 4**). Add water to a volume of 400 mL, mix, and if necessary, adjust pH with NaOH or HCl. Make up to 500 mL with water. Store at 4 °C. Immediately before the cell lysis, add 18.5 mg IA into 10 mL lysis solution to reach the desired IA concentration.
3. 26-gauge ½ inch needle and 1 mL syringe for mechanical lysis.

4. Micro pestle, also for mechanical lysis.
5. Washing buffer: 10× diluted lysis buffer without iodoacetamide, set the pH to 7.7 with 1.7 M Tris base solution. Store at 4 °C.
6. L7–55 ultracentrifuge (Beckman Coulter) with a 70.1 Ti fixed rotor.
7. 10.4 mL polycarbonate tubes.
8. Potter-Elvehjem PTFE pestle and compatible glass tubes (with 3 mL working volume).

2.4 Solubilization and Fragmentation of Membrane-Associated Proteins

1. Ammonium bicarbonate (AmBic, NH_4HCO_3) buffer: 0.5 M AmBic. Weigh 494.1 mg AmBic and dissolve in 12.5 mL water (*see* **Note 5**).
2. Rapigest stock solution: Dissolve 1 mg Rapigest in 200 μL of 10× diluted 0.5 M AmBic buffer, resulting in 0.5 (w/v) % stock solution. Store at −20 °C.
3. Ultrasonic bath.
4. IA stock solution: Weigh 23.12 mg IA into the 1.5 mL Eppendorf tube and add 250 μL water resulting in 0.5 M IA solution.
5. 2,2′-Thiodiethanol stock solution: Dilute 10 μL of 2,2′--Thiodiethanol to 1 mL with 990 μL water, resulting in 0.1 M Thiodiethanol solution.
6. Dry-block thermostat.
7. Proteomics grade trypsin (20 μg/vial).
8. PNGase F (15,000 units).
9. N_α-Tosyl-L-lysine chloromethyl ketone hydrochloride (TLCK) stock solution: 0.135 M TLCK in 0.2 M sodium-acetate buffer. 100 mg TLCK dissolve in 2 mL of 0.2 M sodium-acetate buffer. Divide into 200–200 μL aliquots and store at −20 °C.

2.5 Biotinylated Peptides Purification on Affinity Column

1. High capacity Neutravidin agarose.
2. 1 mL polypropylene Spin Columns—Snap Cap with polyethylene filter (30 μm pore size).
3. Microspin.
4. AmBic washing buffer: 50 mM AmBic, 10× diluted 0.5 M AmBic buffer (*see* **Note 5**).
5. 60 °C AmBic washing buffer: Preheat 50 mM AmBic washing buffer for 60 °C in dry-block thermostat (*see* **Note 5**).
6. NaCl washing buffer: 1 M NaCl. Weigh 29.2 g NaCl and dissolve in 400 mL water and make up to 500 mL. Store at 4 °C.

7. $NaHCO_3$ washing buffer: 100 mM $NaHCO_3$, pH 10.0. Add 4.2 g $NaHCO_3$ to 400 mL water and adjust pH with 5 N NaOH, thereafter make up to 500 mL with water. Store at 4 °C.
8. Dithiothreitol (DTT) stock solution: 500 mM DTT. Dissolve 7.7 mg in 100 μL water and store at −20 °C.
9. Elution buffer: 50× diluted DTT stock solution in 50 mM AmBic buffer (such as 10 μL of DTT stock solution +490 μL of 50 mM AmBic).
10. Alkylation buffer: 500 mM IA. Weigh 46.24 mg IA and solve in 400 μL water. Mix, then make up to 500 μL with water.
11. Microcron-10, nominal mass cutoff 10 kDa.

2.6 Purification of the Eluted Peptides

1. C_{18} spin column.
2. Activation buffer: 50% (v/v) methanol (MeOH) in HPLC grade water. Add 5 mL MeOH into a 10 mL graduated cylinder and make up to 10 mL with water. Store at 4 °C.
3. Equilibration buffer I: 0.5% (v/v) trifluoroacetic acid (TFA), 5% (v/v) acetonitrile (ACN) in HPLC grade water. Mix 9450 μL water, 50 μL TFA, and 500 μL ACN. Store at 4 °C.
4. Equilibration buffer II: 0.1% heptafluorobutyric acid (HFBA) in HPLC grade water. 10 μL TFA diluted with 9990 μL water. Store at 4 °C.
5. Elution buffer I: 70% ACN, 0.1% TFA in HPLC grade water. Mix 7 mL ACN, 10 μL TFA, and 2990 μL water, then store the solution at 4 °C.
6. Elution buffer II: 70% ACN, 0.1% FA in HPLC grade water. Mix 7 mL ACN, 10 μL FA, and 2990 μL water, then store the solution at 4 °C.
7. Vacuum concentrator (Barnstead Genevac vacuum concentrator).

1. NanoHPLC–MS/MS
2. Resolving buffer of the dried peptide pellet: 2% (v/v) ACN and 0.1% (v/v) FA in HPLC grade water. Mix 20 μL ACN, 1 μL FA, and 979 μL water. Store at 4 °C.
3. Dionex UltiMate 3000 RSLCnano System, Acclaim PepMap100 C_{18} Nano-Trap column (5 μm particle size, 100 Å pore size, 100 μm × 20 mm), and Acquity UPLC M-Class Peptide BEH130 C_{18} column (1.7 μm particle size, 130 Å pore size, 75 μm × 25 cm).
4. CaptiveSpray nanoBooster ionization source.
5. Maxis II QTOF.

6. Solvent A: 0.1% (v/v) FA in HPLC grade water. Mix 0.5 mL FA with 490 mL water and make up to 500 mL with water. Sonicate for 30 min and store at room temperature.
7. Solvent B: 0.1% (v/v) FA in ACN, dissolve 0.5 mL FA in 490 mL ACN and make up to 500 mL with water. Sonicate for 30 min and store at room temperature.

2.7 Quantification of the Protein Content of the Membrane Preparations

1. BSA standard: 1 mg/mL stock solution. Weigh 10 mg BSA and transfer to a 15 mL centrifuge tube, mix with 10 mL water. Divide into 500–500 μL aliquots in 1.5 mL Eppendorf tubes and store at −20 °C.
2. Buffer A: 100 mM NaOH, 188.7 mM Na_2CO_3 in water. Weigh 4 g NaOH and 20 g Na_2CO_3, solve them in 900 μL water, finally adjust volume to 1 L with water. Store at 4 °C.
3. Buffer B: 86.7 mM sodium-tartrate in water. Weigh 2 g Na-tartrate × 2 H_2O and dissolve in 100 mL water. Store at 4 °C.
4. Buffer C: 40 mM copper sulfate in water. Add 1 g $CuSO_4$ × 5 H_2O to 100 mL water. Store at 4 °C.
5. Folin-Ciocalteu's phenol reagent. Store at room temperature in dark. Dilute at a 1:1 ratio in water before use.
6. Microplate reader or spectrophotometer (using 96-well plates or cuvettes).

2.8 Biotinylation Efficiency Monitoring by Immunoblots

1. PVDF membranes (0.2 or 0.45 μm pore size).
2. Methanol.
3. Sample loading dye (4×): 250 mM Tris–HCl, 10% (w/v) SDS, 0.008% (w/v) bromophenol blue, and 40% (v/v) glycerol, pH 6.8. Add 302.85 mg Tris, 1 g SDS into a centrifuge tube, dissolve in 4 mL water, and then add 0.8 mL of 1 mg/mL bromophenol blue stock solution and 4 mL glycerol. Mix and adjust the pH with HCl and make up to 10 mL with water. Make 500–500 μL aliquots and store at −20 °C.
4. SDS-PAGE gels (4% stacking gel and 12% separation gel).
5. Electrophoresis buffer: 25 mM Tris, 192 mM glycine, and 0.1% (w/v) SDS. Add 2.25 g Tris, 10.8 g glycine, and 3.75 mL 20% (w/v) SDS solution (10 g SDS in 50 mL water) to 700 mL water. Mix with continuous stirring until completely dissolved and make up to 750 mL with water.
6. GRS Protein Marker Multicolour (GRiSP).
7. PowerPac Universal power supply with Mini-Protean Tetra vertical electrophoresis cell.
8. Coomassie Brilliant Blue G-250 dye.

9. Transfer buffer: 25 mM Tris, 192 mM glycine, and 15% (v/v) methanol. Add 6.057 g Tris and 28.8 g glycine to 1.5 L water, then add to the mixture 300 mL methanol, finally adjust the volume to 2 L with water. Store at 4 °C.
10. TBS-T buffer: 25 mM Tris–HCl, pH 7.4, 2.7 mM KCl, 137 mM NaCl, 0.1% (v/v) Tween-20. Weigh 3.028 g Tris base, 201.3 mg KCl, and 8.006 g NaCl. Dissolve in 900 mL water and add 1 mL Tween-20, adjust pH with HCl. Make up to 1 L with water and store at 4 °C.
11. Blocking buffer: 5% (w/v) BSA diluted in TBS-T buffer. Weigh 500 mg BSA and dissolve in 9 mL TBS-T buffer. Make up to 10 mL with TBS-T. Store at 4 °C.
12. Bio-Dot microfiltration apparatus.
13. Sponge and Whatman filter papers.
14. Plastic container.
15. HRP conjugated avidin (1 mg/mL stock solution).
16. Enhanced chemiluminescence reagent (mix the two components (luminol and peroxide solution) in 1:1 ratio).
17. ChemiDoc XRS+ Imaging system.

3 Methods

Carry out all procedures at room temperature unless otherwise specified.

3.1 Maintaining Adherent and Nonadherent Cell Cultures

1. Remove cryovials of the appropriate cells from the liquid nitrogen container. The vials should be thawed approximately 1–2 min in a 37 °C water bath and transfer into 9 mL cell-specific media (achieving a tenfold dilution of freezing medium, which can help to avoid the cell damage). Centrifuge the solution at 300 g for 3 min at 4 °C.
2. Discard the supernatant and resuspend the cell pellet using cell-specific medium. Maintain HL60 cells in complete RPMI medium and maintain MDCKII cells in complete DMEM medium inside a humidified incubator with 5% CO_2 at 37 °C.
3. Replace the used medium every 2–3 days in a laminar box (working surface pretreated by 70% ethanol solution) using electric pipette with disposable serological pipet tips and cell culture flasks. In case of HL60 cells, transfer the solutions into a centrifuge tube, spin at 300 g for 3 min at 4 °C and resuspend the cell pellet in 2 mL fresh RPMI medium, divide the solution into flasks according to the minimum recommended 2×10^5 cells/mL concentration (using Bürker chamber to determine the exact cell count). Make up to 5 or 10 mL with RPMI

medium depending on the size of the used flask (T25 or T75) then put the flasks back into the incubator. In the case of MDCKII cells, the whole process is quite similar, but after discarding the used medium, firstly the cells are washed with PBS solution (one or two times) and then incubated with 0.5 mL (T25 flask) or 1.5 mL (T75 flask) Trypsin-EDTA solution for 15–20 min at 37 °C in order to detach the cells from the surface of the flasks. After the incubation, the solution is five times diluted with fresh DMEM medium, then centrifuge and count the cells similarly, finally add appropriate volume of cell suspension to the new culture vessel to reach 2×10^4 cells/cm^2.

4. Every week, harvest 10^6 cells/cell type in 1.5 mL Eppendorf tubes (at least once) in order to examine the eventual possibility of the mycoplasma contamination with polymerase chain reaction after the genomic DNA isolation (*see* **Note 2**). Mycoplasma test is especially important when the cells are frozen in freezing medium for long-term storage (to preserve mycoplasma-free cell populations).
5. MDCKII cells are also cultured on transwell inserts placed in compatible 6- or 24-well plate (basolateral and apical culture media volumes are 2.5 mL and 1.5 mL, and 600 μL and 100 μL, respectively) in order to create a cell monolayer on the semipermeable polyester membrane. This model system is often used to characterize the transcellular transport of the various compounds facilitating the monitoring of diffusion dynamics through a biological barrier. Apply TEER measurements to determine the optimal starting cell number for the appropriate transwell surface and optimal culture time, which is required for construction of the intact cell monolayer (pore-free monolayer, *see* **Note 6**).

3.2 Sulfo-NHS-SS-Biotinylation of the Surface of the Cells

3.2.1 Nonadherent HL60 Cell Isolation for Labeling

1. Pipet the cells from the flasks into a centrifuge tube. Separate the used growing medium of the HL60 cells by centrifugation at 300 g for 3 min at 4 °C. Discard the supernatant.
2. Gently wash the cell pellet with ice-cold PBS buffer (using 3 mL PBS to 10^7 cells), then spin down again and remove the PBS solution. Repeat this washing step two more times (*see* **Note 7**).
3. Wash the cells with PBS (pH 8.0) at room temperature to condition the surface of the cells before the labeling step. Centrifuge again and discard the supernatant (using 3 mL PBS to 10^7 cells, *see* **Note 7**).

3.2.2 Adherent MDCKII Cell Preparation for Labeling

1. Remove the used DMEM medium from the plates containing transwell inserts: first from the basolateral chamber, then from the apical chamber (inside part of the transwell insert; remove media carefully, do not touch the cell monolayer with the end of the pipet tip).
2. Gently pipet PBS (pH 7.4) into the apical and the basolateral chamber (using the appropriate working volumes depending on the size of the transwell). Carefully discard the solution.
3. Wash the cell monolayer with PBS (pH 8.0) similarly as above, and then discard it too.

3.2.3 Biotinylation of Harvested Adherent or Nonadherent Cells

1. Immediately before use, dissolve the sulfo-NHS-SS-biotin powder (*see* **Note 3**) in PBS (pH 8.0) to reach the ~2 mM final concentration of the labeling reagent.
2. Add the biotinylation solution to the above prepared cells: In the case of HL60 cells, use 3 mL sulfo-NHS-SS-biotin solution to 10^7 cells; for the apical or basolateral surface labeling of the MDCKII cells, use 0.5 mL or 1 mL solution (**Note 8**). Incubate the reaction mix with gently shaking at room temperature for 20 min.
3. Quench the labeling process with ice-cold TBS solution (add 3 mL TBS/10^7 cells in the HL60 experiments, and 0.5 mL and 1 mL TBS to the apical and basolateral side in the MDCKII experiments). Discard the solutions and repeat this step three more times (*see* **Note 9**).

3.3 Biotinylated Cell Isolation, Cell Lysis, and Membrane Preparation

1. The whole cell lysis process must be prepared on ice or in a cold room.
2. Discard the last TBS washing step from HL60 cells by centrifugation (at 300 *g* for 3 min at 4 °C). In the case of MDCKII cells, after the last quenching solution is removed, scrape off the cells from the surface of the transwell inserts into the lysis buffer using cell scraper (*see* **Note 10**).
3. Incubate the labeled cells in 5–5 mL ice-cold lysis buffer in 5 mL Eppendorf tubes for 10 min at 4 °C in the presence of 10 mM iodoacetamide (*see* **Note 11**). Centrifuge the solutions at 1700 *g* for 10 min at 4 °C. Place 3–3 mL of the supernatant into new 15 mL centrifuge tubes.
4. Administer mechanical lysis to the cell pellets (in the presence of 2–2 mL of lysis solution) using micro pestle (180° rotation in both directions at least 40 times in the Eppendorf tubes, *see* **Note 12**).
5. After that, facilitate the cell lysis with 26-gauge ½ inch needle using 1 mL syringe, by aspirating the solution up and down 40 times.

6. Transfer the cell solutions into the appropriate 15 mL centrifuge tubes, wash the 5–5 mL Eppendorf tubes with 1–1 mL fresh lysis buffer two times, and pool the contents into the centrifuge tube. Make up to 10–10 mL with lysis buffer.
7. The intact cells, cell debris, and nuclei are pelleted at 1700 g for 7 min at 4 °C (*see* **Note 13**). Subsequently, transfer the supernatant to a 10.4 mL polycarbonate tube. Balance the opposite tubes using a digital analytical balance (the maximal difference must not exceed 0.003 *g*).
8. Centrifuge the supernatant-containing polycarbonate tubes at 40,000 rpm for 1 h at 4 °C using a 70.1 Ti fix rotor in a L7–55 ultracentrifuge. After that, white/yellow membrane pellets are visible at the bottom of the tubes.
9. Afterwards, carefully discard the supernatant (do not touch the pellet to avoid the membrane loss), then wash the pellet with washing buffer (10 mL/polycarbonate tube), and ultracentrifuge again at 40,000 rpm for 1 h at 4 °C, discard the supernatant.
10. Resuspend the resulting membrane pellet in washing buffer and homogenize by 25 strokes of Potter-Elvehjem PTFE pestle in a 2 mL glass tube on ice and quantify final volume. Store at −20 °C.
11. Measure the protein concentration of the membrane preparations by the method of Lowry et al. [17] using a serial dilution of the 1 mg/mL BSA standard to generate a standard curve (*see* **Note 14**). Based on the measurement results, divide into aliquots (with 50–100 μg protein content) and store at −80 °C until later analysis.

3.4 Solubilization and Digestion of the Biotinylated Membrane Preparations

1. Thaw an aliquot of the membrane preps containing 50 μg protein and add 0.5 M AmBic buffer to the solution to reach the 50 mM AmBic final concentration (the volume is often set to 200 μL).
2. Add 50 μL 0.5 (w/v) % Rapigest stock solution and complete with 0.5 μL and 2.5 μL IA and 2,2′-thiodiethanol stock solution, respectively. Mix and sonicate for 1 min in an ultrasonic bath (repeat this step four more times) in a cold room.
3. Incubate the samples on ice for 25 min (vortex every 5 min for 10 s) and denature by heating for 5 min at 95 °C in a dry-block thermostat, then cool back on ice.
4. Add 1–1 μg MS grade trypsin (achieving the 1:50 (w:w) enzyme to protein ratio) and digest the samples for 16 h at 37 °C.
5. Stop the digestion by heat inactivation for 10 min at 95 °C in a dry-block thermostat, then cool back, add 2 μL tenfold diluted

TLCK (trypsin inhibitor) stock solution (achieving the 50 μM final concentration).

6. Before removing the N-linked oligosaccharides, spin down the mixtures at 10,000 *g* for 5 min in a benchtop centrifuge, place the supernatant into a new Eppendorf tube, add 250 unit PNGase F to every sample, and incubate for 2 h at 37 °C.

3.5 Biotinylated Cell Surface Peptide Isolation on Neutravidin Affinity Column

1. The optimal amount of neutravidin agarose resin is determined by dot blot analysis (*see* **Note 15**). Add 50 μL neutravidin agarose resin into a Snap Cap spin column (*see* **Note 16**), place the column into a collection tube, then wash with 250 μL AmBic buffer, and repeat this step four more times.
2. Place the plug on the bottom of the column. Add the digested peptide mixture to the column, close the top cap, and incubate for 1 h at room temperature (rotate around every 5 min).
3. Open the top cap, remove the bottom plug, place the column into a collection tube, and spin down (*see* **Note 17**).
4. Wash the column extensively to efficiently remove nonbiotinylated components. Use five consecutive washing steps: (1) 250 μL AmBic washing buffer (4×); (2) 250 μL NaCl washing buffer (4×); (3) 250 μL AmBic washing buffer (4×); (4) 250 μL $NaHCO_3$ washing buffer (4×); and (5) 250 μL 60 °C AmBic washing buffer (4×).
5. Place the column into a new collection tube. Dissociate the immobilized surface peptides from the column by an elution buffer containing a reducing agent (10 mM dithiothreitol) by two consecutive incubations of 30 min, each at 37 °C. After each incubation, open the top cap and remove the bottom plug, and spin down the column at 1000 g for 1 min at 4 °C. Combine the two eluted fractions and quantify the final volumes. Add the appropriate amount of alkylation solution (to achieve the 25 mM iodoacetamide final concentration), then incubate in dark at 37 °C for 45 min.
6. Transfer the alkylated mixture on an ultrafiltration device (Microcron-10) and centrifuge at 14,000 *g* for 20 min at 4 °C (*see* **Note 18**).
7. Dry the samples in a preheated vacuum concentrator (55 °C) and store at −20 °C until further usage.

3.6 Peptide Purification by Solid-Phase Extraction

1. Activate the C_{18} column with 200 μL 50% MeOH and centrifuge the column at 2000 *g* for 1 min (in each subsequent step, centrifuge the column similarly), repeat it once.
2. Condition the column with 200 μL equilibration buffer I (twice), then with 200 μL equilibration buffer II (twice), and transfer the cartridge to a new collection tube.

3. Dissolve the dried peptide samples in 50 μL equilibration buffer II and place onto the column. Spin down and load the flow-through once more onto the column (peptides remain on the column, other components are in the flow-through fraction).
4. Wash the column with 100 μL equilibration buffer II, repeat it once. Transfer the column into a new collection tube.
5. Elute the peptides with 50 μL elution buffer I (twice) then with 50 μL elution buffer II (once).
6. Pool the eluted fractions and dry the samples in vacuum concentrator. Store the dried samples at −20 °C until further use.

3.7 Peptide Separation and Identification by NanoHPLC–MS/MS

1. Dissolve purified peptides in 10 μL resolving buffer containing 98% H_2O, 2% ACN, and 0.1% FA, then inject 6 μL into the instrument setup.
2. The peptide mix is trapped firstly on trap column and separated using an Acquity UPLC M-Class Peptide BEH130 C_{18} column. Use a gradient separation (mixing the solvent A and solvent B); the solvent B content of the elution buffer is the following: initial 4% B from 0 min to 8 min, 4–20% B from 8 min to 98 min, 20–50% B from 98 min to 128 min, 50–90% B from 128 to 130 min, and 90–4% from 130 min to 133 min (the percentage of solvent B increases or decreases linearly between two time points (*see* **Note 19**).
3. Using a CaptiveSpray nanoBooster ionization source, the peptides are ionized, then analyzed by a Maxis II QTOF where the data-dependent analysis is used as data acquisition strategy (*see* **Note 20**).
4. Recalibrate the raw data with Bruker Compass DataAnalysis software and identify the peptides by Byonic software (*see* **Note 21**).
5. Cell surface biotinylated peptides can be identified either by +87.998 Da (thioacyl modification as nonalkylated form) or + 145.020 Da (CAMthiopropanoyl as alkylated form) modification on the lysine residues or on the peptide termini (*see* **Note 22**).

4 Notes

1. Sterilize PBS solutions, Eppendorf tubes, and pipet tips with dry clave for 4 h at 130 °C. 50 mL filtered FBS should be add in all freshly completed medium using Millex-GP syringe filter unit (0.22 μm pore size polyethersulfone membrane, Merck Millipore Ltd) and 50 mL syringe (Henke-Ject Luer-lock).

2. Lyse isolated cells and prepare genomic DNA according to the manufacturer's recommended instructions (DNA isolation kit, Epicenter). Determine the DNA concentration and purity of extracted DNA by NanoDrop 2000 spectrophotometer. Use 500 ng template DNA per sample, mix with 1–1 μM mycoplasma-specific forward and reverse primer (primer sequences are described in [18]) and 2× MyTaq HS Mix (containing the Taq polymerase and dNTPs), then adjust the volume to 10 μL with nuclease-free water. Use the following PCR conditions: initial denaturation at 95 °C for 2 min, 32 cycles for amplification: 30 s denaturation at 95 °C, 40 s annealing at 65 ° C, 16 s elongation at 72 °C (the last extended 1 s/cycle), final elongation after 32 cycles at 72 °C for 10 min. Load the PCR products into the wells of a 1% (w/v) agarose gel (300 mg agarose in 30 mL TAE buffer mixed with 3 μL GelRed dye). After electrophoresis, the bands are visualized by ChemiDoc XRS+ Imaging system.
3. The sulfo-NHS-SS-biotin powder containing tube must be equilibrated to room temperature before use to minimize the moisture condensation, which may impair the effectiveness of the labeling.
4. 0.17 M Tris–HCl (pH 7.4) stock solution: Weigh 10.29 g Trizma base and dissolve it in 400 mL water and adjust pH with HCl, finally make up to 500 mL with water. 0.16 M KCl stock solution: Weigh 5.96 g KCl and dissolve in 500 mL water. These stock solutions can be stored at least for 1 year at 4 °C.
5. Prepare AmBic buffers freshly every time, the pH value of the solution should be 8.0.
6. Various number of MDCKII cells are cultured in parallel on transwell inserts. Perform TEER measurements daily. The day when the TEER value does not increase further is considered as the beginning of the complete cell monolayer formation. The labeling experiments are made after this monolayer formation.
7. The volume of wash buffer should be set depending on cell number; wash more cells with proportionally more wash buffer.
8. During the apical and basolateral labeling process, it is recommended to use primary amino group containing solution on the opposite side to minimize the undesirable protein labeling.
9. At least three TBS washing steps are essential to remove excess labeling agent and the small amount of damaged cells before the cell lysis.
10. Use at least 0.6 mL lysis solution per 6-well plate-compatible transwell insert. If necessary, repeat it several times.

11. Iodoacetamide prevents any further disulfide bond rearrangement during the cell lysis, so we use it to ensure the protection of the disulfide bridge in the spacer arm of the labeling agent.
12. Wash the micro pestle with 0.1–0.2 mL fresh lysis buffer into the original Eppendorf (where the cell disruption occurred) to avoid cell membrane loss.
13. The resulting pellet sometimes should be grinded manually again by using a plastic micro pestle and then passed through a needle as described in the first round (depending on the cell type, this can even be several rounds to maximize the amount of the isolated membrane mass).
14. Lowry measurement: Mix the buffers A, B, and C in 98:1:1 ratio, immediately before use, and place on ice. Pipet 1–1 mL mixture into 1.5 mL Eppendorf tubes and add the standard solution (use BSA standard from 2.5 μL to 75 μL), and/or 5–20 μL membrane preparations (depending on the concentration) into the appropriate tubes. Pipet 100–100 μL of the diluted Folin-Ciocalteu's phenol reagent into the tubes and incubate for 45 min in dark. Measure the absorbance at 660 nm by spectrophotometer and calculate the protein concentration of the membrane preparation based on the standard curve.
15. We have developed a dot blot analysis to verify that all biotinylated components remain on the affinity column. Different amounts of neutravidin agarose have been tested to find the optimal one. Dot blot method (brief description): Collect samples from the before and after column fractions, dilute them in washing buffer (prepare a serial dilution for every sample; at least three dilutions per sample are recommended). Activate PVDF membrane with 5 mL methanol for 2 min in plastic container. Discard the methanol and add 5 mL TBS-T buffer into the container, incubate for 5 min. Insert the PVDF membrane in the Bio-Dot microfiltration device (use TBS-T wetted filter paper under the PVDF membrane to avoid drying) and pipet the samples into the wells. Directly immobilize the biotinylated samples onto the PVDF membrane using a vacuum pump–driven Bio-Dot microfiltration apparatus. After 1 min, turn off the vacuum pump and put the PVDF membrane back into the plastic container under TBS-T solution (rinse it three times with 5 mL TBS-T). Block the membrane with 5 mL blocking buffer for 45 min at room temperature, then discard this and add HRP-conjugated avidin containing blocking buffer (use in 1:50,000 dilution). Incubate again for 45 min in dark, then wash the PVDF three times with TBS-T for 10 min. Visualize the binding efficiency by the enhanced

chemiluminescence reagent and capture the images by a ChemiDoc XRS+ Imaging system.

16. Immediately before use, the neutravidin agarose must be shaken (to achieve homogenous distribution of the resin) and pipet the solution with a cut-off end pipet tip into the column (because a pipette tip with a narrow opening is not ideal for this task).
17. Save the flow-through fractions and store at −20 °C. The biotin content of these fractions could be monitored by the dot blot method (as described in **Note 15**) to certify that the biotinylated peptides remain on the column.
18. To minimize peptide loss on the filter surface, wash the ultrafiltration device at least once with 100 μL AmBic buffer and centrifuge at 14,000 g for 20 min at 4 °C again. Combine with the original flow-through fraction.
19. We applied five different gradients in our previous work [10], and the gradient described here resulted in a moderately large increase in the number of identified biotinylated/labeled protein(s).
20. Collect spectra using a fixed cycle time of 2.5 s (dynamic exclusion 2 min) and acquired at 3 Hz in the 150–2200 m/z mass range while CID is performed at 4 or 16 Hz depending on the intensity of the precursor. The preferred charge states of precursors are set from +2 to +5.
21. Search engine parameters are listed in our previous work [10].
22. Sometimes, when the number of the identified labeled peptides is relatively low, use SDS-PAGE to test the reproducibility of the membrane preparation and use western blot to visualize the labeling efficiency of the various protein bands. SDS-PAGE process (briefly): Use ~10–20 μg protein containing membrane preparations (based on Lowry method), dilute them in 4× sample buffer and boil for 5 min at 95 °C in a dry-block thermostat, then load the samples into the gel wells (load a protein ladder in a separate well as molecular weight standards, GRiSP). Run the gel for 60 min at 140 V in electrophoresis buffer using a PowerPac Universal power supply with a Mini-Protean Tetra vertical electrophoresis cell. After running, wash the gels twice with water and stain with Coomassie Brilliant Blue G-250 dye for 45 min. Afterwards, wash the gels with water for 10 min, repeat it two more times. Detect the blue protein bands by a ChemiDoc XRS+ Imaging system. Western blot for monitoring the biotinylation efficiency of the protein band: Run another gel; after the protein separation, cut off the stacking gel and wash the remaining resolving with water until the front of the bromophenol blue disappears. Equilibrate the gel in transfer buffer for 10 min. During the

incubation time, wet the sponges and filter papers in transfer buffer and activate the PVDF membrane in methanol. Assemble the transfer sandwich (use the following sequence: sponge–three filter papers–gel–PVDF membrane–three filter papers–sponge) and blot at ~70 V for 80 min in transfer buffer on ice using the PowerPac Universal power supply. After that, treat the PVDF membrane similarly as described in the dot blot method (*see* **Note 15**).

Acknowledgments

This work was supported by the Ministry of Innovation and Technology of Hungary from the National Research, Development and Innovation Fund [K132522 and K146314]. Project no. [C1793861] has been implemented with the support provided by the Ministry of Culture and Innovation of Hungary from the National Research, Development and Innovation Fund, financed under the [KDP-2021] funding scheme (K.K.).

References

1. Santos R, Ursu O, Gaulton A, Bento AP, Donadi RS, Bologa CG, Karlsson A, Al-Lazikani B, Hersey A, Oprea TI, Overington JP (2017) A comprehensive map of molecular drug targets. Nat Rev Drug Discov 16:19–34
2. Mellman I, Nelson WJ (2008) Coordinated protein sorting, targeting and distribution in polarized cells. Nat Rev Mol Cell Biol 9:833–845
3. Zihni C, Mills C, Matter K, Balda MS (2016) Tight junctions: from simple barriers to multifunctional molecular gates. Nat Rev Mol Cell Biol 17:564–580
4. Bausch-Fluck D, Hofmann A, Bock T, Frei AP, Cerciello F, Jacobs A, Moest H, Omasits U, Gundry RL, Yoon C, Schiess R, Schmidt A, Mirkowska P, Härtlová A, Van Eyk JE, Bourquin J-P, Aebersold R, Boheler KR, Zandstra P, Wollscheid B (2015) A mass spectrometric-derived cell surface protein atlas. PLoS One 10:e0121314
5. Bausch-Fluck D, Hofmann A, Wollscheid B (2012) Cell surface capturing technologies for the surfaceome discovery of hepatocytes. Methods Mol Biol 909:1–16
6. Hofmann A, Gerrits B, Schmidt A, Bock T, Bausch-Fluck D, Aebersold R, Wollscheid B (2010) Proteomic cell surface phenotyping of differentiating acute myeloid leukemia cells. Blood 116:e26–e34
7. Langó T, Róna G, Hunyadi-Gulyás É, Turiák L, Varga J, Dobson L, Várady G, Drahos L, Vértessy BG, Medzihradszky KF, Szakács G, Tusnády GE (2017) Identification of extracellular segments by mass spectrometry improves topology prediction of transmembrane proteins. Sci Rep 7:42610
8. Müller A, Langó T, Turiák L, Ács A, Várady G, Kucsma N, Drahos L, Tusnády GE (2019) Covalently modified carboxyl side chains on cell surface leads to a novel method toward topology analysis of transmembrane proteins. Sci Rep 9:15729
9. Langó T, Pataki ZG, Turiák L, Ács A, Varga JK, Várady G, Kucsma N, Drahos L, Tusnády GE (2020) Partial proteolysis improves the identification of the extracellular segments of transmembrane proteins by surface biotinylation. Sci Rep 10:8880
10. Langó T, Kuffa K, Tóth G, Turiák L, Drahos L, Tusnády GE (2022) Comprehensive discovery of the accessible primary amino group-containing segments from cell surface proteins by fine-tuning a high-throughput Biotinylation method. Int. J. Mol, Sci, p 24
11. Waas M, Littrell J, Gundry RL (2020) CIRFESS: an interactive resource for querying the set of theoretically detectable peptides for cell surface and extracellular enrichment proteomic studies. J Am Soc Mass Spectrom 31:1389–1397

12. Bausch-Fluck D, Goldmann U, Müller S, van Oostrum M, Müller M, Schubert OT, Wollscheid B (2018) The in silico human surfaceome. Proc Natl Acad Sci USA 115:E10988–E10997
13. Zeke A, Dobson L, Szekeres LI, Langó T, Tusnády GE (2021) PolarProtDb: a database of transmembrane and secreted proteins showing apical-basal polarity. J Mol Biol 433:166705
14. Dobson L, Zeke A, Tusnády GE (2021) PolarProtPred: predicting apical and basolateral localization of transmembrane proteins using putative short linear motifs and deep learning. Bioinformatics 37:4328–4335
15. Dobson L, Reményi I, Tusnády GE (2015) CCTOP: a consensus constrained TOPology prediction web server. Nucleic Acids Res 43: W408–W412
16. Srinivasan B, Kolli AR, Esch MB, Abaci HE, Shuler ML, Hickman JJ (2015) TEER measurement techniques for in vitro barrier model systems. J Lab Autom 20:107–126
17. Lowry OH, Rosebrough NJ, Farr AL, Randall RJ (1951) Protein measurement with the Folin phenol reagent. J Biol Chem 193:265–275
18. Uphoff CC, Drexler HG (2013) Detection of mycoplasma contaminations. Methods Mol Biol 946:1–13

Chapter 3

Proximity Labeling of Cell Surface Proteins via Cell Surface Remodeling

Zak Vilen, Jia Meng Pang, and Mia L. Huang

Abstract

Within the complex interplay of proteins, lipids and carbohydrates at the cell surface is the surfaceome, a dense layer of proteins and their posttranslationally modified counterparts that serves as a hub for cell signaling and signal transduction. The surfaceome plays crucial roles in mediating interactions between cells and the extracellular environment, which combined with their availability at the cell surface make it an attractive therapeutic target. Despite its importance, the development of technologies to selectively target cell surface proteins for empirical identification is challenged by their structural complexity. Here, we describe a proximity labeling-based technique to covalently label proteins at the cell surface with a biotin handle, enabling downstream streptavidin-based enrichment and manipulation in a variety of modalities, including fluorescence imaging, western blotting, and mass spectrometry–based proteomics.

Key words Cell surface, Lipid remodeling, Proximity labeling, Proteomics, Surfaceome

1 Introduction

Cell surface proteins (CSPs), which collectively make up the surfaceome, serve as vital molecular mediators between the extracellular and intracellular environments. CSPs participate in a variety of biological pathways, including signal transduction, antigen display, and cell adhesion [1–4]. Due to their localization to the cell surface and their crucial biological functions, these proteins are often therapeutically relevant, and > 60% of approved drug candidates within the DrugBank database are reported to be CSPs [5, 6]. Additionally, the expression of CSPs is highly dynamic, fluctuating wildly in response to changes in cell state, making them strong biomarkers for disease [7–9].

Despite their importance, the analysis and investigation of CSPs, especially by mass spectrometry (MS)-based proteomics, can be challenging. CSPs account for approximately 30% of the proteome [10], making them difficult to detect in the presence of

Csaba Matta et al. (eds.), *The Surfaceome: Methods and Protocols*, Methods in Molecular Biology, vol. 2908, https://doi.org/10.1007/978-1-0716-4434-8_3,

much more abundant intracellular proteins. Furthermore, the presence of one or more hydrophobic transmembrane domains can introduce insolubility, which can complicate purification and expression [11]. Hydrophobic transmembrane domains are especially problematic in the context of MS-based proteomics wherein protein preparations are generally digested using trypsin, a protease that cleaves proteins at the C-terminal end of arginine and threonine residues. As transmembrane regions generally do not contain these polar residues, large portions of these proteins may not generate appropriate tryptic peptides, resulting in underrepresentation of CSPs compared to their soluble counterparts [12]. These problems are further compounded by the variable expression and localization of CSPs, which can be altered by cell states, cell type, or the addition of various posttranslational modifications [11, 13, 14].

To overcome these challenges, a variety of strategies have been designed to separate CSPs from intracellular proteins via their unique characteristics or to selectively label CSPs with chemical handles, usually biotin, for subsequent enrichment. Membrane proteins can be separated from their soluble counterparts using ultracentrifugation, which separates biomolecules based on sedimentation coefficients [15]. While useful for preliminary experiments, this technique can result in high levels of intracellular protein contaminants and fail to separate true CSPs from proteins within the membranes of intracellular organelles.

The chemical labeling and enrichment of CSPs often achieves greater specificity and is capable of distinguishing between intracellular membrane proteins and true CSPs [11]. While a variety of methods are available, some even commercially, the most popular strategies involve the labeling of either reactive primary lysine residues with cell impermeant probes [16] such as NHS-Biotin, or the labeling of terminal sialic acids attached to cell surface glycoproteins [17] with amino-oxy biotin. While these strategies display improvements over centrifugal separation, a number of groups have reported significant nonspecific labeling using NHS-biotin or its derivatives [11, 18–20], and sialic acid labeling relies on the presence of specific posttranslational modifications which can vary significantly across cell types and metabolic states.

Here, we describe an alternative surfaceome labeling method for installing a biotin moiety on the electron-rich residues of CSPs (Fig. 1). Using cell surface engineering, we incorporate appropriately modified cholesterol derivatives into the cell membrane of live cells [21], before conjugating an engineered ascorbate peroxidase (APEX2) [22, 23] to the lipid via an engineered sortase enzyme (eSrtA) [24]. eSrtA catalyzes the conjugation of APEX2 onto the cell surface. APEX2 then biotinylates electron-rich residues within a ~ 20 nm radius, labeling CSPs with biotin to facilitate enrichment for downstream applications, including MS-based proteomics. This technique is generalizable across numerous cell types and cell

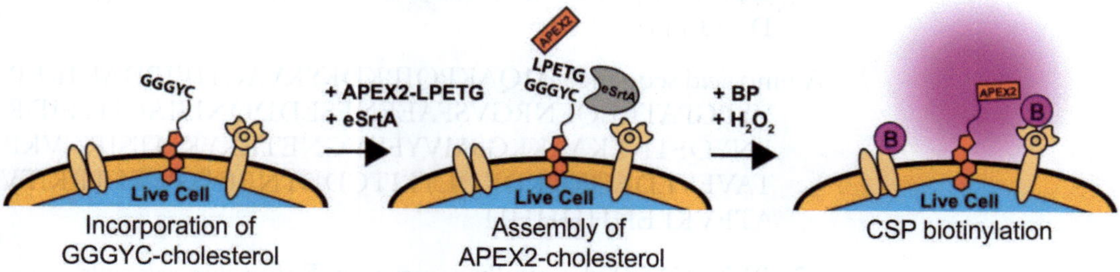

Fig. 1 Cell surface engineering enables tethering of APEX2-LPETG to cell surfaces. Scheme representing the mechanism of the surfaceome labeling procedure. Exogenous GGGYC-cholesterol is introduced to cells, which is then spontaneously incorporated into the cell membrane. The engineered sortase enzyme, eSrtA, catalyzes the covalent linkage of APEX2-LPETG to the incorporated cholesterol via the GGGYC and LPETG recognition sequences. Addition of biotin phenol (BP) and H_2O_2 results in the formation of biotin-phenoxyl radicals which biotinylates electron-rich residues within a ~ 20 nm radius, primarily at the cell surface

culture formats, it labels CSPs regardless of posttranslational modifications and requires reagents that are easily synthesized from commercially available reagents [25].

2 Materials

2.1 General Lab Equipment and Disposables

1. Laminar flow biosafety cabinet.
2. Reversed-phase high-performance liquid chromatography with C-18 column.
3. FPLC (Akta Start, Cytiva, formerly, GE Healthcare Science).
4. Cell culture–coated 6-well and 10 cm plates.
5. Fluorescence capable microscope.
6. Centrifuge.
7. Western blot imager.

2.2 Preparation of APEX2-LPETG and eSrtA

1. APEX2-LPETG (Genscript, cloned in a pETDuet-1 vector using EcoRI and NotI sites)

Amino acid sequence: MHHHHHHGKSYPTVSADYQDAVE KAKKKLRGFIAEKRCAPLMLRLAFHSAGTFDKGTKTGG PFGTIKHPAELAHSANNGLDIAVRLLEPLKAEFPILSYAD FYQLAGVVAVEVTGGPKVPFHPGREDKPEPPPEGRLPD PTKGSDHLRDVFGKAMGLTDQDIVALSGGHTIGAAH KERSGFEGPWTSNPLIFDNSYFTELLSGEKEGLLQLPSD KALLSDPVFRPLVDKYAADEDAFFADYAEAHQKLSELGF ADAGSGGGGSLPETG

2. Sortase A pentamutant (eSrtA, Addgene #75144, a gift from David Liu)

Amino acid sequence: MQAKPQIPKDKSKVAGYIEIPDADIKEP-VYPGPATREQLNRGVSFAEENESLDDQNISIAGHTFIDR PNYQFTNLKAAKKGSMVYFKVGNETRKYKMTSIRNVKP TAVEVLDEQKGKDKQLTLITCDDYNEETGVWETRKIFV ATEVKLEHHHHHH

3. BL21 (DE3) chemically competent *Escherichia coli* cells.
4. LB medium and plates with appropriate antibiotic for vector: Dissolve Luria Broth in ultrapure water per manufacturer's instructions for a working solution. Sterilize by autoclaving and cool to room temperature before the addition of antibiotics. Antibiotics were prepared at 1000× in ultrapure water and stored at −20 °C. Refer to the manufacturer's protocol for specific working concentrations (1×) for the appropriate antibiotics for your constructs.
5. 1000× isopropyl beta-D-thiogalactopyranoside (IPTG): Dissolve IPTG to a final concentration of 1 M in ultrapure water and sterilize using a 0.22 μm filter. Solution may be stored at −20 °C for up to 1 year.
6. DH10B chemically competent *E. coli.*
7. SOC media.
8. Bacterial lysis buffer: Dissolve imidazole in 2× PBS (274 mM NaCl, 5.4 mM KCl, 16 mM Na_2HPO_4, 4 mM KH_2PO_4) to a final concentration of 10 mM. Adjust pH to 7.4 and store at 4 °C.
9. Nickel wash buffer: Dissolve imidazole in 2× PBS (274 mM NaCl, 5.4 mM KCl, 16 mM Na_2HPO_4, 4 mM KH_2PO_4) to a final concentration of 30 mM. Adjust pH to 7.4 and store at 4 °C.
10. Nickel elution buffer: Dissolve imidazole in 2× PBS to a final concentration of 300 mM. Adjust pH to 7.4 and store at 4 °C.
11. PD-10 column.

2.3 Synthesis of GGGYC-PEG2000-Cholesterol

1. GGGYC peptide (synthesized by Genscript).
2. Cholesterol-PEG2000-maleimide.
3. MOPS buffer: Dissolve MOPS in diH_2O to 0.2 M, and supplement with 10 mM Na_2EDTA and 50 mM NaOAc. Adjust pH to 7.0.
4. 1 kDa, MWCO Micro Dispodialyzer.
5. 1× PBS.

2.4 APEX2-LPETG Cell Surface Labeling

1. Appropriate serum-free culture media.
2. Calcium chloride.
3. Biotin-tyramide: Prepare fresh solution by dissolving biotin-tyramide to a stock concentration of 500 mM in DMSO.
4. 30% (w/w) H_2O_2.
5. 1× quencher solution: Mix 1 M sodium ascorbate (1 mg in 5 μL of H_2O), 1 M sodium azide (1 mg in 15.3 μL of H_2O), and 0.5 M Trolox (1 mg in 8 μL of DMSO) with PBS to make 1× quencher solution. (Concentration of each component corresponds to 100× of stock solution.)
6. Streptavidin-Cy5.
7. PBST: Add 1% of Tween-20 to 1× PBS.
8. Hoechst 33342.

2.5 Proteome Profiling Using Quantitative Mass Spectrometry–Based Proteomics

1. Cell surface–labeled cell sample (pelleted).
2. LC/MS grade methanol.
3. Urea (freshly prepared, *see* Subheading 3.5, **step 10** for reagent preparation).
4. Reducing agent: Tris(2-carboxyethyl)phosphine, TCEP, and K_2CO_3 (freshly prepared, *see* Subheading 3.5, **step 11** for reagent preparation).
5. Iodoacetamide (freshly prepared, *see* Subheading 3.5, **step 12** for reagent preparation).
6. Streptavidin agarose resin.
7. 10% and 0.2% sodium dodecyl sulfate, SDS in PBS.
8. 1 M triethylammonium bicarbonate, TEAB, pH 8.5.
9. Sequencing grade porcine trypsin with $CaCl_2$.
10. LC/MS grade acetonitrile, ACN.
11. TMT, TMT10PLEX.
12. 50% hydroxylamine.
13. LC/MS grade formic acid.
14. C-18 spin columns.
15. 1.5 mL low-protein-binding microcentrifuge tubes.
16. Sonicator.
17. Shaking incubator.
18. Tube rotator.
19. Vacuum centrifuge.
20. High-resolution mass spectrometer.

2.6 Software

1. Imagelab (Biorad).
2. Proteome Discoverer (Thermo Fisher).

3 Methods

3.1 Preparation of APEX2-LPETG and eSrtA

The tagged ascorbate peroxidase (APEX2-LPETG) and the engineered sortase (eSrtA) are produced in high yields from commercially available plasmids for production in *E. coli*. Constructs containing a 6 × His tag, or alternative tags for affinity purification, may then be purified via affinity chromatography. Alternatively, proteins may be procured via commercial protein production services using amino acid sequences available within materials.

1. Thaw 20 μL of BL21 competent on ice for 10–20 min.
2. Warm LB agar plates with antibiotic appropriate for selection to 37 °C in bacterial incubator.
3. Add 10 ng of plasmid to 20 μL of cells and mix by gently flicking the bottom of your tube. Incubate the bacteria on ice for 5 min (*see* **Note 1**).
4. Heat shock bacteria in a 42 °C water bath or heat block for 30 s. Immediately transfer to ice and incubate for 2 min.
5. Add 970 μL of SOC media (without antibiotics) to transformed *E. coli* and shake on an orbital shaker at 220 rpm for 1 h at 37 °C.
6. Plate 100 μL of transformants onto prewarmed LB agar plate, perform a standard quadrant streak, and incubate the plates at 37 °C overnight.
7. Pick a single colony from LB agar plate and inoculate 5 mL of LB supplemented with appropriate antibiotic in a 10-mL culture tube. Incubate the culture at 37 °C on a shaker at 220 rpm overnight.
8. Add 2 mL of starting culture to 1 L of LB containing appropriate antibiotic in a 4-L flask (1:500 dilution with 75% headspace). Shake on an orbital shaker at 220 rpm, 37 °C. During inoculation, monitor the optical density (OD) of culture by spectrophotometry until OD_{600} reaches 0.6 (*see* **Note 2**).
9. Add 1 mL of 1000× (1 M) IPTG when OD_{600} 0.6 is achieved and continue shaking on orbital shaker at 220 rpm, 37 °C for 4 hr.
10. Pellet the induced bacterial culture by centrifuging at 3300× *g* for 15 min at 4 °C. Discard the supernatant. Cell pellets may be used immediately, or store at −80 °C for up to 1 month.

11. Add bacterial lysis buffer to cell pellet and supplement with 0.1 mg/mL DNAase and 2 mM $MgCl_2$. Vortex until fully resuspended.
12. Lyse the bacterial pellet by sonicating on ice (*see* **Note 3**). For Branson SFX250 sonicator, use the following parameters:

    Intervals of 1 s ON and 3 s OFF

    Total ON time: 30 s

    Amplitude: 20%
13. Preclarify the lysate by centrifuging 20 min at 16,000× *g*, 4 °C and transfer the supernatant into clean tube.
14. Filter the supernatant with a 0.22- or 0.45-μm PES membrane filter unit to remove cell debris and precipitates (*see* **Note 4**).
15. Load your samples on HisTrap HP column on the FPLC and run the following program. For NiNTA purification, set solvent A and solvent B as nickel wash buffer and nickel elution buffer, respectively. Set the flow rate of the buffers at an appropriate value for your column. Column volume (CV) varies depending on the size of your HisTrap HP column.
16. Wash the line with MQ water at 5 mL/min for 3 min.
17. Clean the column by running 100% solvent B for ~5 CV with the appropriate flow rate of the column.
18. Equilibrate the column by running 100% solvent A through the column with 7.5 CV same speed as last step.
19. Load the sample onto HisTrap HP column.
20. Wash the column by running 15 CV of solvent A.
21. Run gradient solution by mixing:

    5 CV from 100% solvent B to 95% solvent B.

    10 CV from 100% solvent A to 0% solvent A.
22. When the run is complete, check the chromatogram with UV absorbance reading to identify fractions that have protein and collect it (*see* **Note 5**).
23. If the total volume of your purified protein is >2.5 mL, concentrate to 2.5 mL or less using centrifugal unit (*see* **Note 6**).
24. Prepare a PD-10 column for desalting. Remove the top cap and discard the column storage buffer. Cut the bottom and remove the seal. Fill the column with 1× PBS, allow the buffer to flow-through gravity, and discard. Repeat three more times.
25. Add the sample to the column and discard the flow-through.
26. Elute the protein with 3.5 mL of protein storage buffer.

27. Determine the protein concentration via NanoDrop or Lowry assay (*see* **Note 7**).
28. Repeat protein concentration by centrifugal filtration if necessary.

3.2 Evaluate Heme Occupancy of APEX2-LPETG

APEX2 requires the presence of a heme cofactor for enzyme function. The presence of heme, which lends the APEX2-LPETG a reddish-brown color, can vary within different batches of protein expression. To supplement heme loading to appropriate levels it may be necessary to supplement *E. coli* cultures with 1 mM aminolevulinic acid concurrent with IPTG to improve heme occupancy. Heme occupancy may be measured using a benchtop spectrophotometer. We recommend using proteins with heme loading >60% for surfaceome labeling experiments to ensure high levels of APEX2 activity.

1. Set spectrophotometer (NanoDrop) to UV–Vis and measure absorbance at 405 nm and 280 nm.
2. Blank the instrument with 2 μL of 1× PBS, or whatever buffer the protein in question is stored in.
3. Measure the absorbance of 2 μL of the protein sample at 280 nm and 405 nm.
4. Heme occupancy may be calculated by $(Abs_{405}/Abs_{280})/2$, where 1 corresponds to 100% heme occupancy.

3.3 Preparation of GGGYC-PEG2000-Cholesterol

The tagged cholesterol reagent (Cholesterol-GGGYC) may be generated at high yield from a single step conjugation of commercially available Cholesterol-PEG2000-Maleimide and the short GGGYC peptide (Fig. 2a).

1. Solubilize 0.044 mmol GGGYC peptide in 133 μL DMSO to generate 330 mM solution.
2. Add 0.04 mmol (~113.8 mg) solid Cholesterol-PEG2000-Maleimide directly to peptide/DMSO solution.
3. Dilute reaction with 0.8 mL MOPS buffer to dissolve and stir reaction for 48 h at room temperature (*see* **Note 8**).
4. Collect reaction and dialyze over 3 days in 1× PBS at 4 °C using 1 kDa Micro Dispodialyzer units (*see* **Note 9**).

3.4 APEX2-LPETG Cell Surface Labeling

Cell surface labeling may be performed on a wide variety of live cells and cell culture formats, resulting in intact cells with CSPs covalently linked to biotin moieties for enrichment or detection via the strong streptavidin–biotin interaction.

1. Seed mammalian cell of interest in 24-well plate for microscopy analysis (*see* **Note 10**). Incubate the cells overnight at 37 °C in 5% CO_2. The cells should reach 90–95% confluency before proceeding with cell surface labeling (*see* **Note 11**).

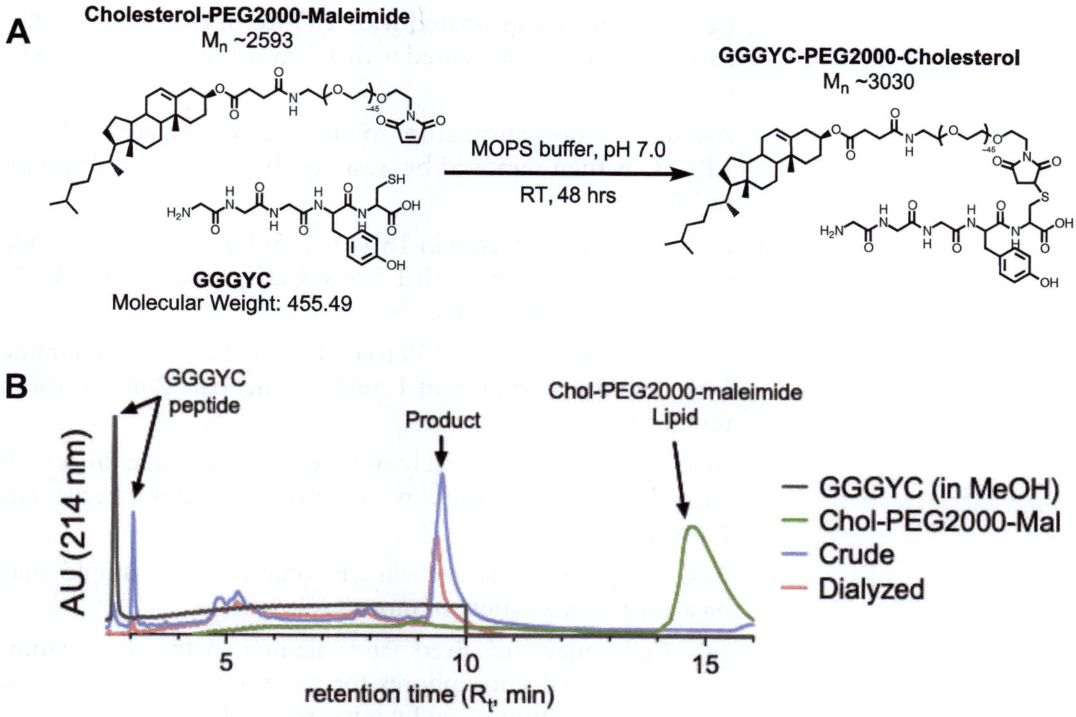

Fig. 2 Synthesis of Cholesterol-PEG2000-Maleimide. (**a**) Synthetic scheme demonstrating the conjugation of Cholesterol-PEG2000-Maleimide to the cysteine of the GGGYC peptide. (**b**) LC-MS analysis of starting reagents and reaction product, indicating the full conversion of Cholesterol-PEG2000-Maleimide to Cholesterol-PEG2000-GGGYC (product)

2. Aspirate the complete culture media (CM) from the culture dish and wash cells with 500 μL of PBS.
3. Remove PBS and wash cells with 500 μL of serum-free culture media (SFM). Aspirate SFM after washing.
4. Treat cells with 150 μL of 30 μM GGGYC-PEG2000-cholesterol. Incubate 1 h at 37 °C in 5% CO_2 for incorporation of cholesterol into cell membrane (*see* **Note 12**).
5. Prepare 15 μM APEX2-LPETG/15 μM eSrtA in SFM supplemented with 1.3 mM $CaCl_2$ (*see* **Note 13**).
6. Prepare 5.5 μL of 500 mM biotin-tyramide in DMSO and then dilute 100× into prewarmed SFM with 1.3 mM $CaCl_2$. Maintain in 37 °C before use (*see* **Note 14**).
7. Prepare fresh 1× quencher solution (*see* **Note 15**).
8. After 1 h of incubation, unincorporated cholesterol is removed by washing cells with 500 μL of PBS and followed 500 μL of SFM supplemented with 1.3 mM of $CaCl_2$.

9. Cells are then co-incubated with 150 μL of APEX2-LPETG/eSrtA in SFM supplemented with 1.3 mM $CaCl_2$ for 1 h at 37 °C in 5% CO_2.
10. Aspirate the protein-mixture solution. Unconjugated APEX2-LPETG is then removed by washing three times with 500 μL of PBS.
11. Cells are then incubated in 150 μL of 500 μM biotin-tyramide in SFM supplemented with 1.3 mM $CaCl_2$ for 30 min at 37 °C in 5% CO_2 (*see* **Note 16**).
12. Add 100 mM H_2O_2 in PBS to each well after biotin-tyramide for a final concentration of 1 mM and incubate 2 min at room temperature.
13. Aspirate the solution quickly and wash all wells three times with enough quencher solution to adequately cover cells (*see* **Note 17**).
14. At this stage cell surface labeling is completed and samples may be analyzed in a variety of modalities:
    (a) Cells may be fixed and stained with streptavidin-conjugated fluorophores for visualization of cell surface protein biotinylation by microscopy (Fig. 3).
    (b) Cells may be scraped and lysed for subsequent Western blot analysis or further prepared for proteomic analysis by mass spectrometry (Fig. 4).

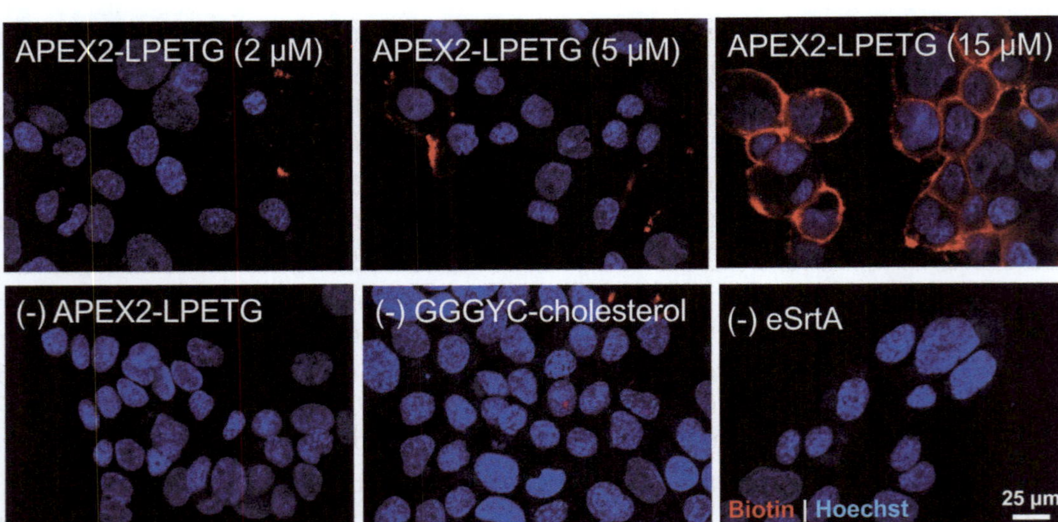

Fig. 3 Confocal fluorescence microscopy images of cells post-surfaceome labeling. Confocal microscopy of surfaceome-labeled HPAF-II cells stained with streptavidin-CY5 [25]. Surfaceome labeling generates strong, specific staining at the cell surface that is dose dependent and reliant on each reaction component

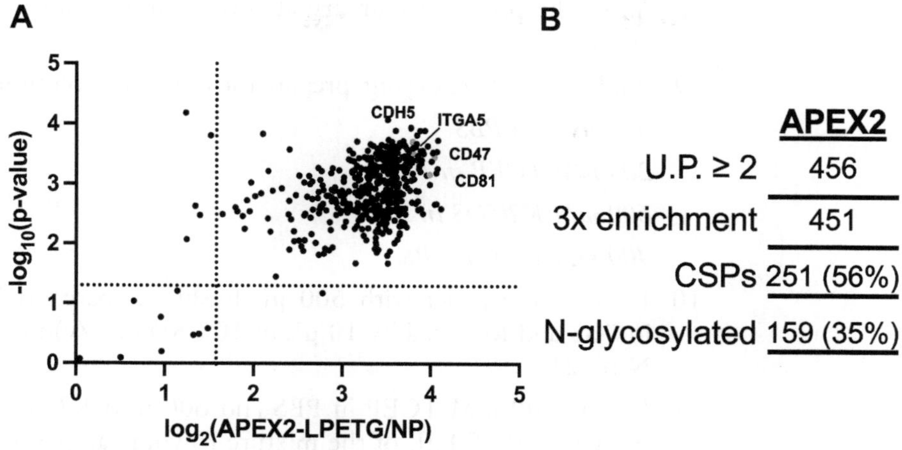

Fig. 4 Surfaceome-enabled proteomic analysis of JEG3 cells. Representative proteomic dataset of JEG3 cells after surfaceome labeling [25]. (**a**) Volcano plot depicting the enrichment of CSPs over an unlabeled "no protein" (NP) negative control. The vertical dashed line indicates threefold enrichment, while the horizontal dashed line indicates a p value of 0.05 obtained by multiple t-test. Example CSPs are highlighted. (**b**) Proteomic analysis of surfaceome-labeled JEG3 cells identified 456 proteins with at least 2 unique peptides (U.P.). 451 of these were significantly enriched over the negative control. Of these significantly enriched proteins, 251 (56%) were annotated as CSPs, and 35% were annotated as N-linked glycoproteins by UniProt

3.5 Cell Surface Labeling for Quantitative MS-Based Proteomics with Tandem Mass Tags

1. After cell surface labeling, scrape cells off plates and centrifuge at 500× *g* for 3 min at 4 °C.
2. Remove the supernatant and wash the pellet twice with cold PBS and centrifuge at 500× *g* for 3 min at 4 °C (*see* **Note 18**).
3. Resuspend the cells in 400 μL of cold PBS and lyse the cells by sonication. If Branson SFX250 sonicator is used, the following parameters can be used for sonication:

   Intervals of 15 ms ON and 40 ms OFF.

   Total ON time: 1 s.

   Amplitude: 15%.
4. Determine the protein concentration and normalize the concentration of every sample to 1.0–2.0 mg/mL by diluting with PBS to achieve a final volume of 500 μL in 15-mL centrifuge tubes (*see* **Note 19**).
5. Add 2 mL of LC/MS-grade MeOH to your samples and vortex quickly to mix. Precipitate the proteins overnight at −20 °C.
6. Pellet protein by centrifuging at 3260× *g*, 4 °C for 15 min.
7. Wash the pellet once by resuspending with cold LC/MS-grade MeOH and spin down with the same centrifuging condition (*see* **Note 20**).

8. Allow the protein to air dry at room temperature for about 10 min.
9. While samples are drying, prepare the following solutions fresh:

   6 M urea in PBS

   200 mM TCEP in PBS

   600 mM K2CO3 in PBS

   400 mM IAA in PBS.
10. Dissolve the pellet with 500 μL freshly prepared 6 M urea in PBS and followed by 10 μL of 10% SDS (w/v) in PBS (*see* **Note 21**).
11. Premix 200 mM TCEP in PBS and 600 mM K2CO3 in 1:1 ratio and add 50 μL of the mixture to your sample to reduce the proteins. Incubate 30 min at 37 °C on an orbital shaker at 180 rpm.
12. Alkylate free thiols by adding 70 μL of freshly prepared 400 mM IAA in PBS to each sample and incubate in dark at room temperature for 30 min (*see* **Note 22**).
13. To each solution, add 130 μL of 10% SDS in PBS and then dilute with 5.5 mL of PBS.
14. Prepare streptavidin agarose resin (100 μL/sample). Resuspend the beads in storage buffer by shaking. Do not vortex the bottle because the force can break the agarose beads. Cut ~5 mm off pipet tip and transfer the required amount of slurry to a 15-mL centrifuge tube.
15. Wash slurry with 10 CV of PBS. Spin down the beads at 500× *g*, room temperature for 2 min (*see* **Note 23**).
16. Transfer 200 μL of streptavidin agarose beads to each sample and rotate the tubes for 1.5 h at room temperature.
17. After incubation, pellet beads by centrifuging at 500× *g*, 4 °C for 2 min.
18. Remove supernatant and wash the beads with 0.2% SDS (w/v) in PBS, PBS, and freshly prepared 100 mM pH 8.5 TEAB once for each reagent. Centrifuge the beads between each wash and remove the washing buffer (*see* **Note 24**).
19. Pellet the beads and resuspend in 500 μL of 100 mM pH 8.5 TEAB and transfer to LoBind microcentrifuge tubes. Wash the tube once more with 500 μL of 100 mM TEAB and add to the LoBind tubes to ensure the complete transfer of your samples.
20. Centrifuge at 500× *g* for 2 min at room temperature and remove the supernatant.
21. Prepare a trypsin premix by adding 2 mL of 100 mM TEAB and 20 μL of CaCl2 to a vial of sequencing-grade porcine trypsin. Pipet or vortex to mix.

22. Resuspend the beads in 200 μL of trypsin premix (about 2 μg of trypsin per sample) and digest overnight on an orbital shaker at 37 °C.
23. Spin down the beads at 500× *g*, room temperature for 2 min, and transfer 200 μL of supernatant that contain the digested and released peptides into a new LoBind microcentrifuge tube.
24. Wash the beads once more with 100 μL of 100 mM TEAB. Spin down the beads and transfer the washing into the same LoBind tube as in the last step. The total volume of each sample should be approximately 300 μL.
25. Add 100% anhydrous acetonitrile (ACN) to your samples to achieve a final concentration of 30% ACN (~ 120 μL).
26. Add 100 μL of peptide mix to respective 20 μg/μL 10-plex TMT tag aliquoted per manufacturer's instructions (*see* **Note 25**). Pipette to recombine with the peptides in LoBind tube.
27. Incubate the peptide and TMT tag mixture at room temperature for an hour with occasional vortex.
28. Add 6 μL of freshly prepared 5% hydroxylamine to each sample to quench the reaction. Vortex and incubate at room temperature for 15 min.
29. Add 4 μL of LC/MS-grade formic acid to acidify the reaction.
30. Flash freeze samples with liquid nitrogen or a dry ice-methanol bath before drying by vacuum centrifugation.
31. Redissolve the first sample with 200 μL of Buffer A (*see* **Note 26**) by pipetting up and down, vortexing and microcentrifuge.
32. Transfer the solution to second tube, resuspend the TMT-labeled peptides, vortex, and microcentrifuge. Repeat until all samples are combined.
33. Wash the first tube with 100 μL of Buffer A and repeat the process of pipetting, vortexing and spinning. Transfer to next tube until all remaining peptides are collected.
34. Desalt the combined sample by Pierce™ C-18 desalting column (All centrifugations within the desalting step are done 1500× *g*, 1 min at room temperature). Place the C-18 column into a microcentrifuge tube to collect flow-through from washing.
35. Activate the resin with 200 μL of 50% ACN in H2O and spin down to remove the supernatant. Repeat twice and replace the microcentrifuge tube afterwards.
36. Equilibrate resin with 200 μL of Buffer A and centrifuge to discard the supernatant. Repeat three times.
37. Apply the sample to resin and spin down, collect the flow-through in a new microcentrifuge.

38. Add the flow-through back to resin and centrifuge once again. Repeat once more to make sure all your samples are bound to the resin.
39. Wash resin with Buffer A, spin down and remove the supernatant. Repeat two more times. Replace the microcentrifuge tube with a new LoBind tube before elution.
40. Elute the sample with 20 μL of 70% ACN in H2O three times and combine the eluents.
41. Flash freezes your sample and vacuum-centrifuge the sample until complete dryness (*see* **Note 27**).
42. Dissolve your sample in 20 μL of Buffer A and centrifuge at 12000× *g* for 10 min.
43. Transfer the supernatant to MS vial.
44. Analyze the sample by high-resolution mass spectrometer.
45. Process proteomics data in Proteome Discoverer 2.4 (Thermo Scientific). Peptide sequences are determined by matching fragmentation patterns to public databases using Sequest HT algorithm. Use the following settings: Precursor mass tolerance set to 10 ppm; fragment ion mass tolerance set to 0.6 Da; allow for 1 missed trypsin cleavage site; set carbamidomethyl and TMT-10 plex as a static modification; set oxidation as a variable modification; set target false discovery rate at 1%.
46. Filter the dataset for proteins with unique peptides (U.P.) ≥ 2.
47. Transform TMT values by log2(x).
48. Within GraphPad Prism, group transformed TMT values by treatment condition and calculate the enrichment ratio by dividing the TMT values of APEX-LPETG treated cells to those of the negative control (Fig. 4) (*see* **Note 28**).
49. Calculate statistical significance by performing t-test analysis with at least 3 technical replicates for experimental conditions, and at least 2 technical replicates for negative controls. Conditions were considered to be significantly different when p-values ≤0.05 (*see* **Note 29**).

4 Notes

1. About 10 pg–100 ng of DNA is required for successful transformation of plasmid. The volume of plasmid added should not more than 10% of the volume of competent cells to prevent the osmotic pressure from killing the cells. We recommended adding 1 μL of plasmid from 1:10 dilution of a 100 ng/μL stock solution.

2. The time taken for the culture to reach OD_{600} ~ 0.6 varies based on your construct and selection process. We recommend checking around 2.5 h and more frequently afterwards. Once the bacteria move past stationary phase into the exponential phase, the OD_{600} will double approximately every 20 min, which is the doubling time of *E. coli*. The protein expression efficiency will greatly decrease if you let the culture grow past OD_{600} 0.8 before you add IPTG for induction.
3. Keep sonicated material on ice during sonication to prevent heat generated from denaturing or degrading the proteins.
4. We recommend saving the supernatant after this filtration for SDS-PAGE analysis to check the purification efficiency by comparing it to the whole cell lysate.
5. We recommend that you collect and store the flow-through from the column wash steps in case your sample does not bind to column. The protein content of this wash fraction may be evaluated by SDS-PAGE analysis.
6. The molecular weight of eSrtA and APEX-LPETG are 18 and 29 kDa respectively. Thus, it is advised to use 3 or 10 kDa centrifugal unit to concentrate the proteins to prevent loss.
7. If NanoDrop is used, you can determine the extinction coefficient (ε) and molecular weight of the protein based on its amino acid sequence using online tools. (e.g., https://web.expasy.org/protparam/).
8. Reaction completion may be monitored by reverse phase HPLC (Fig. 2b). Consumption of starting materials is observed by the reduction of the Cholesterol-PEG2000-Maleimide and GGGYC peptide peaks, and the formation of a new peak representing the conjugate. Upon complete consumption of the limiting reagent Cholesterol-PEG2000-Maleimide, we presume 100% reaction completion and formation of 0.04 mmol product.
9. Upon completion, the reaction should appear to be light yellow and oily. If the reaction is too viscous to be easily handled, we recommend further diluting the reaction with 1× PBS prior to dialysis. If diluted further, it is necessary to recalculate the estimated Cholesterol-GGGYC concentration based on the new volume.
10. This APEX2-based surfaceome labeling technique has been tested with success in a variety of cell lines and cell types; however, its suitability cannot be universally guaranteed. We recommend treating new cell lines with titrations of surfaceome labeling reagents to ensure cell health during treatment and minimize nonspecific labeling.

11. Appropriate cell culture plate selection depends on downstream analysis. To minimize proteins and reagents consumption, 24-well plate is optimal for microscopy analysis. If mass-spectrometry-based proteomic analysis is performed, greater numbers of cell may be required, and labeling should be performed at the scale of a 10 or 15 cm dish to obtain enough protein for streptavidin beads enrichment.
12. The volume used for incubations during labeling is dependent on the surface area of the culture dish. We recommend using 750 μL for 6-well plate, 3 mL for 10 cm dish, and 8 mL for 15 cm dish.
13. While we recommend using surfaceome labeling reagents at the scale listed above (15 μM APEX2-LPETG, etc.), we have found that reducing the concentration of these reagents (down to 5 μM APEX2-LPETG, etc.) still yields detectable, although dose dependent signal at the cell surface in most modalities. This may be useful for small scale, less sensitive experiments to preserve reagents.
14. To prevent biotin-tyramide from crashing out in CM, it is important to aggressively pipette the stock solution into prewarmed and vortex immediately to facilitate the solubilization.
15. Sodium ascorbate and sodium azide may be prepared beforehand and stored in −20 °C for several months while Trolox should be prepared fresh before each experiment.
16. Given biotin-tyramide is relatively permeable to cell membrane, incubation time of biotin-tyramide may be optimized to improve cell surface protein tagging and prevent intracellular labeling. Ideal incubation times may be cell line dependent.
17. This step is essential for removing all excess biotin tyramide and hydrogen peroxide. Omitting this step or replacing the quencher solution with PBS results in nonspecific labeling.
18. Cell pellets may be stored in −80 °C for a week if necessary.
19. The ideal concentration for MS analysis is 1.5–2.0 mg/mL. If none of the samples can be adjusted to 1.0 mg/mL, normalize the concentration to the highest possible concentration. Concentrations less than 1.0 mg/ml may affect your proteomic result since less abundant proteins may not be tagged and quantified.
20. If the pellet does not dissociate, try sonication as in cell lysis step to fully dissolve.
21. When resuspending the pellet with 6 M urea, the pellet may stick in the pipette tips causing loss of samples. After addition of urea and 10% SDS, the mixture should clarify, either at this point or after incubation.

22. Dissolve IAA in PBS immediately before adding to your sample.
23. Since we will add 200 μL of agarose resin solution to each sample, resuspend the beads to a final volume of 2.2 mL with PBS to ensure an equal number of beads is added to samples. Agarose resin will occupy a significant volume in the tube. Therefore, do not directly add 2.2 mL PBS to the beads but slowly add PBS and bring it up to final volume.
24. 0.2% SDS in PBS can be prepared by 50-fold dilution of 10% SDS in PBS while 100 mM pH 8.5 TEAB can be prepared by diluting 1 part of 1 M TEAB in 9 parts of water.
25. Aliquot TMT tags in advance to 8 μL per tube and store in − 80 °C. It is important to remember the respective TMT channel you used for each sample for analysis.
26. Buffer A constitutes a final working concentration of 5% of ACN and 0.1% of formic acid in H_2O.
27. If the sample is not ready to be injected, store at −80 °C.
28. While we arbitrarily considered proteins with enrichment values ≥3 to be significantly enriched, this cutoff may be adjusted based on specific datasets.
29. According to previous mass spectrometry–based proteomics experiments, ~60% of the proteins labeled by our surfaceome labeling technique are annotated as cell surface proteins (an equivalent level of specificity as the sialic acid-based glycoprotein labeling strategy developed by the Paulson lab). We assigned "CSP" status to proteins using annotations provided by UniProt, however, some manual annotation may be necessary, as protein annotation often relies on the presences of canonical transmembrane regions, which may not be present in all CSPs. In addition to UniProt, we recommend the use of the SURFY database, maintained by the Wollscheid group as a stringent and robust collection of CSPs.

References

1. Haun RS, Quick CM, Siegel ER et al (2015) Bioorthogonal labeling cell-surface proteins expressed in pancreatic cancer cells to identify potential diagnostic/therapeutic biomarkers. Cancer Biol Ther 16(10):1557–1565. https://doi.org/10.1080/15384047.2015.1071740
2. Orentas RJ, Yang JJ, Wen X et al (2012) Identification of cell surface proteins as potential immunotherapy targets in 12 pediatric cancers. Front Oncol 2:194. https://doi.org/10.3389/fonc.2012.00194
3. da Cunha JP, Galante PA, de Souza JE et al (2009) Bioinformatics construction of the human cell surfaceome. Proc Natl Acad Sci USA 106(39):16752–16757. https://doi.org/10.1073/pnas.0907939106
4. Chen W, Mou KY, Solomon P et al (2021) Large remodeling of the Myc-induced cell surface proteome in B cells and prostate cells creates new opportunities for immunotherapy. Proc Natl Acad Sci USA 118(4). https://doi.org/10.1073/pnas.2018861118
5. Hopkins A, Groom C (2002) The druggable genome. Nat Rev Drug Discov 1:727–730

6. Wishart DS, Feunang YD, Guo AC et al (2018) DrugBank 5.0: a major update to the DrugBank database for 2018. Nucleic Acids Res 46 (D1):D1074–D1082. https://doi.org/10.1093/nar/gkx1037
7. Ferguson ID, Escobar BP, Tuomivaara ST et al (2021) Defining the cell surface proteomic landscape of multiple myeloma reveals immunotherapeutic strategies and biomarkers of drug resistance. BioRXIV. https://doi.org/10.1101/2021.01.17.427038
8. Leung KK, Wilson GM, Kirkemo LL et al (2020) Broad and thematic remodeling of the surfaceome and glycoproteome on isogenic cells transformed with driving proliferative oncogenes. Proc Natl Acad Sci USA 117(14): 7764–7775. https://doi.org/10.1073/pnas.1917947117
9. Ravenhill BJ, Soday L, Houghton J et al (2020) Comprehensive cell surface proteomics defines markers of classical, intermediate and non-classical monocytes. Sci Rep 10(1):4560. https://doi.org/10.1038/s41598-020-61356-w
10. Wallin E, Von Heline G (1998) Genome-wide analysis of integral membrane proteins from eubacterial, archaean, and eukaryotic organisms. Protein Sci 7:1029–1038
11. Weekes MP, Antrobus R, Lill JR et al (2010) Comparative analysis of techniques to purify plasma membrane proteins. J Biomol Tech 21:108–115
12. Helbig AO, Heck AJ and Slijper M (2010) Exploring the membrane proteome--challenges and analytical strategies. J Proteome 73(5):868–878 doi:https://doi.org/10.1016/j.jprot.2010.01.005
13. Wagner S, Baars L, Ytterberg AJ et al (2007) Consequences of membrane protein overexpression in Escherichia coli. Mol Cell Proteomics 6(9):1527–1550. https://doi.org/10.1074/mcp.M600431-MCP200
14. Mathieu K, Javed W, Vallet S et al (2019) Functionality of membrane proteins overexpressed and purified from *E. Coli* is highly dependent upon the strain. Sci Rep 9(1): 2654. https://doi.org/10.1038/s41598-019-39382-0
15. Zhou Y, Gao J, Zhu H et al (2018) Enhancing membrane protein identification using a simplified centrifugation and detergent-based membrane extraction approach. Anal Chem 90(4):2434–2439. https://doi.org/10.1021/acs.analchem.7b03710
16. Huang G (2012) Biotinylation of cell surface proteins. Bio-protocol 2(9):e170
17. Zeng Y, Ramya TN, Dirksen A et al (2009) High-efficiency labeling of sialylated glycoproteins on living cells. Nat Methods 6(3): 207–209. https://doi.org/10.1038/nmeth.1305
18. Li Y, Qin H, Ye M (2019) An overview on enrichment methods for cell surface proteome profiling. J Sep Sci 43:292–312
19. Griffin NM, Schnitzer JE (2011) Overcoming key technological challenges in using mass spectrometry for mapping cell surfaces in tissues. Mol Cell Proteomics 10(2): R110.000935. https://doi.org/10.1074/mcp.R110.000935
20. Kuhlmann L, Cummins E, Samudio I et al (2018) Cell-surface proteomics for the identification of novel therapeutic targets in cancer. Expert Rev Proteomics 15(3):259–275. https://doi.org/10.1080/14789450.2018.1429924
21. Woods EC, Yee NA, Shen J et al (2015) Glycocalyx engineering with a recycling Glycopolymer that increases cell survival in vivo. Angew Chem Int Ed Engl 54(52):15782–15788. https://doi.org/10.1002/anie.201508783
22. Lam SS, Martell JD, Kamer KJ et al (2015) Directed evolution of APEX2 for electron microscopy and proximity labeling. Nat Methods 12:51–54
23. Hung V, Udeshi ND, Lam SS et al (2016) Spatially resolved proteomic mapping in living cells with the engineered peroxidase APEX2. Nat Protoc 11(3):456–475. https://doi.org/10.1038/nprot.2016.018
24. Dorr BM, Ham HO, An C et al (2014) Reprogramming the specificity of sortase enzymes. Proc Natl Acad Sci USA 111(37): 13343–13348. https://doi.org/10.1073/pnas.1411179111
25. Vilen Z, Reeves AE, O'Leary TR et al (2022) Cell surface engineering enables Surfaceome profiling. ACS Chem Biol. https://doi.org/10.1021/acschembio.1c00865

Chapter 4

BioID-Based Proximity Mapping of Transmembrane Proteins in Human Airway Cell Models

Melissa Iazzi, Audrey Astori, Jonathan St-Germain, Sara Sadeghi, Brian Raught, and Gagan D. Gupta

Abstract

The cystic fibrosis transmembrane conductance regulator (CFTR), a chloride channel residing primarily at the apical membrane of epithelial cells, plays a major role in fluid secretion and the maintenance of epithelial surface hydration. Mutations in the *CFTR* gene lead to the fatal disease known as cystic fibrosis (CF). Drugs that improve mutant CFTR protein folding and channel function have dramatically improved CF patient outcomes. However, the current regimen only restores the function of the most common mutant, ΔF508, to ~62% of wildtype (WT). Notably, ~10% of patients harboring hundreds of less common *CFTR* mutations are not eligible or do not respond at all to treatment with current CFTR modulators. Better characterizing the WT and mutant CFTR protein interactomes could provide critical insight into how to treat patients with rarer mutations and thereby improve the druggability of this devastating disease. Here we describe how BioID (proximity-dependent biotin identification) can be used to map the CFTR interactome in a human airway model—bronchial epithelial cells grown at the air–liquid interface. Approximately 26% (>5500) of all human protein-coding genes are predicted to code for membrane proteins, which together account for ~30% of the druggable proteome. The methods described here could thus also be applied to improve our understanding of many additional respiratory, autoimmune, and metabolic diseases.

Key words Cystic fibrosis, Proteomics, Air, liquid interface, Human airway, BioID: proximity-dependent biotin identification, Transmembrane protein, Membrane protein, CFTR, ΔF508-CFTR

1 Introduction

Cell membranes are complex multicomponent structures, and the proteins that reside there play vital roles in a wide range of cellular processes (e.g., signaling, transport, and intercellular communication) [1, 2]. Membrane proteins (MPs) are encoded by approximately a quarter of the human protein-coding genes (26%), and ~ 30% of these proteins are druggable [3–6]. Indeed, multipass MPs with two or more transmembrane domains are of the highest interest for pharmaceutical development [6]. However, intrinsic properties of membrane-associated polypeptides, such as partial

Csaba Matta et al. (eds.), *The Surfaceome: Methods and Protocols*, Methods in Molecular Biology, vol. 2908, https://doi.org/10.1007/978-1-0716-4434-8_4,

protection by lipid membranes, generally low abundance, hydrophobicity, and low solubility in standard lysis buffers, pose significant experimental challenges for the characterization of MP protein–protein interaction (PPI) networks [1–3].

The cystic fibrosis transmembrane conductance regulator (CFTR) is an MP that resides at the apical membrane of epithelial cells and functions as an ion channel that mediates the flux of chloride and bicarbonate ions [7, 8]. Over 2000 CFTR coding region mutations have been identified in cystic fibrosis (CF) patients, the most common autosomal recessive lung disease in the Caucasian population [9, 10]. These disease-causing mutations most commonly result in a defective channel protein with lower surface abundance and disrupted function, leading to mucus stagnation in the lungs, airway obstruction, and a significantly reduced life expectancy [10, 11].

Modern CFTR-targeting small-molecule correctors (i.e., lumacaftor/VX-809, tezacaftor/VX-661, and elexacaftor/VX-445) can partially correct CFTR protein processing and trafficking defects [12, 13]. These CFTR modulators are often combined with ivacaftor (VX-770), a small molecule potentiator that increases channel open probability and restores channel activity at the cell surface [12–15], and has been approved for use in patients with gating defects (e.g., G551D-CFTR and 97 other mutations) [16, 17]. Together, these drugs have significantly improved survival outcomes of CF patients [9, 10, 18–20]. Importantly, however, the current treatment regimen restores ΔF508 (the most common CFTR mutant) channel function to only ~62% of WT CFTR, and ~ 10% of patients are not eligible or do not respond to treatment at all with current CFTR modulator compounds [21].

Several different proteomics approaches have been used to map the CFTR PPI network and to characterize how it is altered in response to mutations or therapeutic compounds [22–29]. Individual approaches have distinct advantages and disadvantages, with the ability to perform interactomics studies on an MP in a relevant disease model being the most noted limitation [30, 31]. Proximity-dependent biotin identification (BioID) is a relatively new method that can be used to identify physiologically relevant PPIs in a living cell [32–36] and has unique advantages for the characterization of MP PPIs [31, 32]. The classic version of the technique employs the fusion of an abortive *Escherichia coli* biotin ligase mutant (BirA R118G, or BirA*) to a protein of interest (or "bait"), allowing for the covalent biotinylation of endogenous proximal (within ~10 nm) polypeptides (or "preys") in mammalian cells [33, 37]. Since biotin is covalently bound to the prey proteins, harsh cell lysis conditions can be used to efficiently solubilize all membranes and organelles. This contrasts with previously reported methods, such as immunoprecipitation coupled with mass spectrometry (IP–MS), that use much more "gentle" lysis buffers to maintain PPIs. Limitations to IP–MS thus include inefficient solubilization of MPs and the inability to detect associations that are not preserved post-cell lysis.

Previous CFTR interactomes generated using BioID have been performed in the proteomics "workhorse" human embryonic kidney (HEK) 293 cell line, using a BirA*-CFTR fusion protein [23, 25]. While these experiments identified a number of novel CFTR interactors of interest, conducting BioID in a more relevant model system could reveal additional and/or unique tissue-specific PPIs critical for our understanding of CFTR function. To this end, here we describe BioID of the CFTR protein expressed in CF bronchial epithelial (CFBE) cells in an air–liquid interface (ALI) culture system.

Due to low transfection efficiency in human bronchial epithelial lines, a lentiviral BioID system (pSTV2) designed to accommodate larger open reading frames (e.g., CFTR: 4440 bp) was implemented (Fig. 1a, b) [37]. The BirA*-CFTR construct was transduced into the CFBE line, a suitable system for the study of a heterologously expressed CFTR protein, as it produces no detectable endogenous CFTR transcript [38]. BioID was performed on BirA*-CFTR-expressing CFBE cells grown both under normal 2D culture conditions and in an ALI culture system [39]. The ALI protocol induces further differentiation (a process that can be tracked by monitoring the zonula occludens-1 (ZO-1) protein at the tight junction and the development of transepithelial electrical resistance (TEER)) to produce a polarized, pseudostratified ciliated epithelial layer that reconstitutes many aspects of the in vivo airway [39].

Importantly, the ALI BioID approach described here, designed for characterizing the interactomes of human airway MPs, can potentially provide important insights relevant to developing personalized treatment strategies for CF and a broad spectrum of other respiratory diseases [40–42].

2 Materials

2.1 Cell Culture

1. Cystic fibrosis bronchial epithelial (CFBE) cells.
2. Human embryonic kidney (HEK) 293 Tv cells.
3. Cell culture dishes, 100 mm x 22 mm.
4. Cell culture dishes, 150 mm x 22 mm.
5. Dulbecco's modified Eagle's medium (DMEM) with high glucose, L-glutamine, and pyruvate. Storage at 4 °C.
6. Minimum essential medium (MEM), with Earle's salts and L-glutamine, 6 x 500 mL. Storage at 4 °C.
7. Albumin bovine serum.

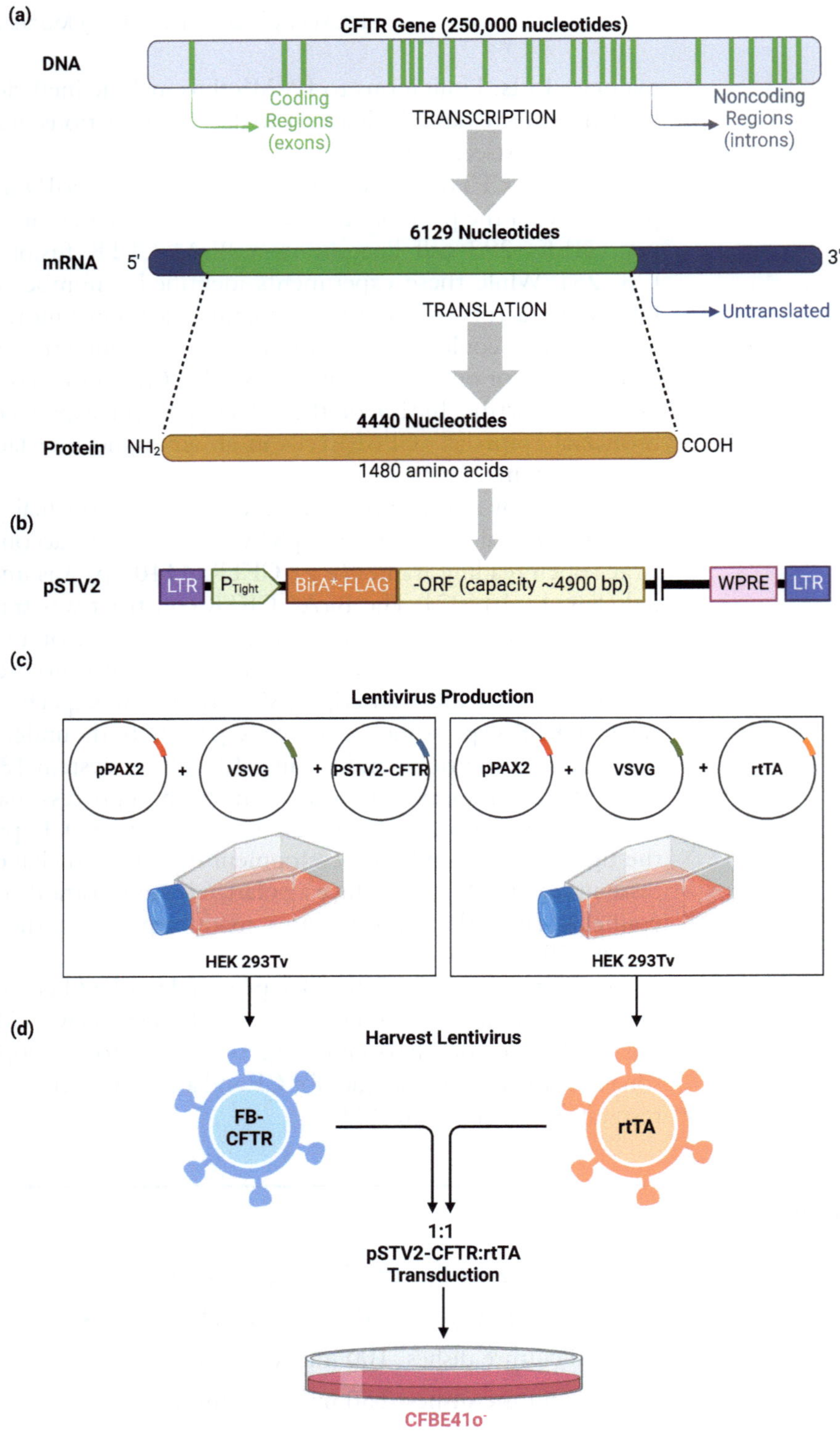

Fig. 1 Producing a pSTV2 lentivirus transfer vector to generate the inducible stable expression of FLAG-BirA*-CFTR in a human bronchial epithelial cell line. (**a**) The *CFTR* gene spans a significant region of DNA, 250,000 nucleotides, and is made up of 24 exons. The protein-coding portion of CFTR is 4440 nucleotides in length [43]. (**b**) A minimal packaging virus (pSTV2) was used to accommodate larger open reading frames (upper limit ~4900 nucleotides) [37]. (**c**) Production of a third-generation lentiviral packaging system. HEK293Tv cells are cotransfected with psPAX2 (packaging vector), VSV.G (envelope vector), and either pSTV2-FLAG-BirA*-CFTR or rtTA (transfer vectors). (**d**) The resulting packaged FLAG-BirA*-CFTR and rtTA lentiviruses are used to transduce CFBE cells at a 1:1 ratio to generate the inducible stable expression of FLAG-BirA*-CFTR

8. Collagen I, rat tail.
9. Fibronectin.
10. Fetal bovine serum (FBS). Long-term storage at −80 °C and short-term storage at −20 °C.
11. 100× penicillin/streptomycin (PS). Storage at −20 °C.
12. Sterile phosphate-buffered saline (PBS). Storage at room temperature.
13. 0.05% Trypsin-EDTA. Long-term storage at −20 °C and short-term storage at 4 °C.
14. Doxycycline stock solution: 10,000× stock. Store at 4°C.
15. Biotin stock solution: 1000× stock. Store at 4 °C.

2.2 Transfection and Transduction

1. Dulbecco's modified Eagle's medium (DMEM) with high glucose, L-glutamine, and pyruvate. Not supplemented with FBS or PS. Storage at 4 °C.
2. PolyJet transfection reagent (FroggaBio). Storage at 4 °C.
3. psPAX2 2nd/third generation lentiviral packaging plasmid [37].
4. VSV-G lentiviral envelope vector plasmid [37].
5. pSTV2 FLAG-BirA* N-terminally tagged CFTR expression vector [37].
6. Reverse tetracycline transactivator (rtTA) third-generation lentiviral vector [37].
7. T75 flasks for cell culture.

2.3 Immunoblotting

1. Nitrocellulose membranes.
2. Western blot transfer buffer: 0.025 M Tris–HCl, 0.192 M glycine, 20% methanol.
3. Tris-buffered saline (TBS; 10×): 1.5 M NaCl, 0.1 M Tris–HCl, pH 7.4.
4. TBST: TBS + 0.05% Tween-20, pH 7.6.
5. Blocking solution: 5% skim milk in TBST. Store at 4 °C.
6. Streptavidin-HRP (1:5000).
7. Mouse anti-GAPDH (1:1000).
8. ECL substrate.

2.4 Immunofluorescence

1. 100% methanol. Store at −20 °C.
2. Blocking/antibody dilution buffer: Prepare a 1× PBS/1% BSA/0.3% Triton™ X-100 buffer by adding 0.1 g BSA (#9998) and 30 μL Triton™ X-100 to 10 mL 1× PBS.
3. Streptavidin-488 (1:500).

4. Mouse anti-FLAGm2 (1:1000).
5. Rabbit anti-*ARL13B* (1:1000).
6. *ZO-1* (rabbit; 1:500).
7. Alexa Fluor 594 donkey anti-mouse.
8. Alexa Fluor 647 donkey anti-rabbit.
9. Microscope slides with beveled edge (1 end, 1 side frosted, precleaned), 75 × 25 mm.
10. 12 mm glass coverslips.
11. ProLong™ Gold antifade mountant.

3 Methods

Carry out all methods at room temperature and in a containment level 2 (CL2) lab with enhanced operational procedures unless otherwise instructed (*see* **Note 1**).

3.1 HEK293 Tv Transfection

3.1.1 Plate Cells

1. Day 1: Plate and maintain HEK293Tv cells in T75 tissue culture flask at a density of ~80% confluency. (*see* **Note 2,** *see* **Note 3**).

3.1.2 Transfect Cells

1. Day 2: Transfect cells—replace media with 10 mL of DMEM (+HI-FBS, +PS). Incubate for at least 30 min to 1 h prior to transfection.
2. Transfection mixture should be prepared in tissue culture hood where contamination risk is minimized. Prepare and label sterile 1.5 mL centrifuge tubes based on mix and transfection condition.
3. For "Mix 1," add 8 μg of the lentiviral expression vector (Transfection 1: pSTV2 FLAG-BirA* N-terminally tagged CFTR; Transfection 2: rtTA third-generation lentiviral vector (*see* **Note 4**)), 8 μg of psPAX2, 4 μg of VSV.G, and bring the volume in the tube to 720 μL with DMEM (no FBS, no PS) (*see* **Note 5**).
4. For "Mix 2," add 680 μL of DMEM (no FBS, no PS) and 40 μL of PolyJet. Vortex tubes for 2 s each to mix and centrifuge to bring all liquid to the bottom of tubes.
5. Add 720 μL of "Mix 2" into each tube of "Mix 1." Mix by gentle pipetting, being sure to avoid any bubbles. Incubate transfection mixture for 10 min at room temperature (*see* **Note 6**).
6. Add the transfection mixture directly (slowly and dropwise) onto the cells while gently swirling the media to mix (Fig. 1c) (*see* **Note 7**).

3.1.3 Stop Transfection

1. Day 3: Approximately 16–24 h post-transfection, aspirate the medium from the cells and add 10 mL of fresh DMEM (+FBS, +PS) to the cells (*see* **Note 8**).

3.2 CFBE Transduction

1. Day 4: Transduction—collect supernatant from the transfection flask with a syringe and pass through a 0.45 μm filter. Add 5 mL of pSTV2-FLAG-BirA*-CFTR virus and 5 mL of rtTA virus (1:1 ratio) directly onto 50% confluent CFBE cells in a 15 cm coated culture dish (*see* **Note 9**) (Fig. 1d).
2. Day 5: After 24 h, wash the transduced CFBE cells —four to five times with 10 mL of PBS. Change media with complete media (MEM, +FBS, +PS).
3. Day 6: Split cells into 10 cm culture dish and 6-well dishes for validation of construct expression.

3.3 Validation with Immunofluorescence

3.3.1 Cell Grown Under Normal Culturing Conditions

1. Seed transduced cells onto 12 mm glass coverslips coated with poly-d-lysine.
2. For induced condition samples, induce cells with 1 μg/mL doxycycline and 50 μM biotin for 24 h prior to fixing. For uninduced condition samples, change media.
3. Fix and permeabilize cells in 100% methanol at −20 °C for at least 20 min and store at −20 °C until staining.
4. Dilute primary antibodies in antibody diluent solution and incubate for 30 min at room temperature. Antibodies: anti-FLAGm2 (mouse; 1:500) and anti-*Na-ATPase* (rabbit; 1: 1000) (Fig. 2a).
5. Rinse cells twice, then incubate with secondary antibodies: Alexa Fluor 594 donkey anti-mouse (1:1500), Alexa Fluor 647 donkey anti-rabbit (1:1000), and Streptavidin-488 (1: 500) are diluted in antibody diluent solution and incubated for 30 min at room temperature.
6. Prepare and image cells using the same protocol outlined in Iazzi et al., 2023 [33] (*see* **Note 12**).

3.3.2 Cell Grown Under ALI Culturing Conditions

1. For validation of transduced cells grown under ALI culturing conditions: Seed into 6-well transwell insert culture dishes at the liquid–liquid interface (LLI) (*see* **Note 10**).
2. Allow cells to grow until 100% confluence and a transepithelial electrical resistance (TEER) output of 1000 Ω or greater (*see* **Note 11**).
3. Once cells reach a TEER >1000 Ω, remove the media in the transwell insert (leaving only media below inside the well) and allow cells to grow at the ALI for ~2 weeks (*see* **Note 11**).
4. Incubate cells for 24 h in complete media supplemented with 1 μg/mL doxycycline and 50 μM biotin.

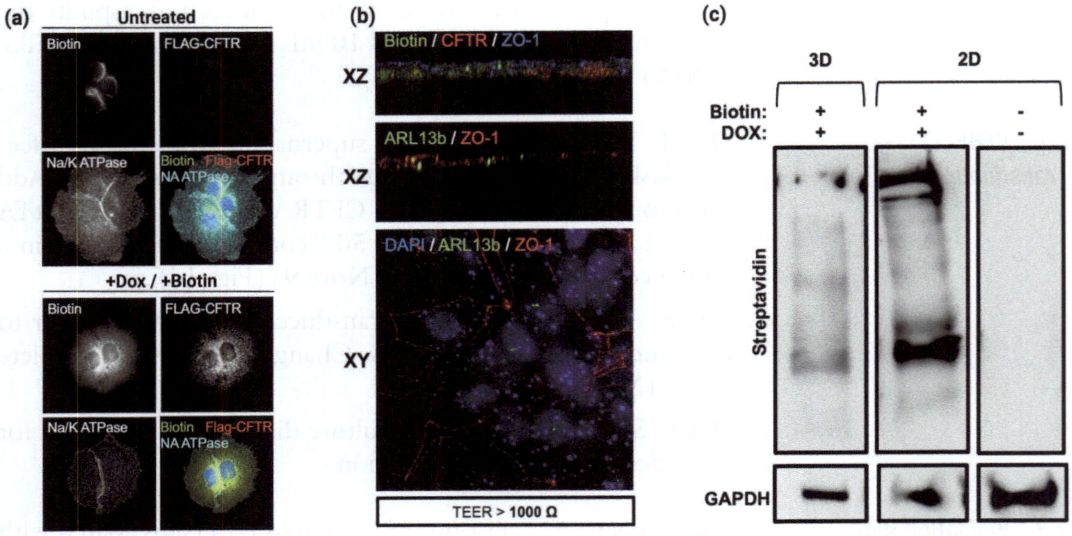

Fig. 2 Inducible expression of FLAG-BirA*-CFTR in CFBE cells grown as 2D and ALI cultures. (**a**) Immunofluorescence images of FLAG-BirA*-WT-CFTR grown in normal 2D culture conditions with or without 50 μM biotin and doxycycline for 24 h. Labelling was performed with fluorophore-conjugated streptavidin (green), anti-FLAG (red), and Na+/K+ ATPase (blue) as a PM marker. Scale bar = 50 μm. (**b**) FLAG-BirA*-CFTR in CFBE cells cultured at the air–liquid interface (ALI; transepithelial electric resistance (TEER) ≤ 1000 Ω). XZ (top): apical membrane/tight junction marker ZO-1 (blue), biotin (green), Flag-CFTR (red). XZ (bottom): ciliary marker ARL13B (green), ZO-1 (red). XY: DAPI (blue), ARL13B (green), ZO-1 (red). (**c**) Western blots showing levels of empty FLAG-BirA* biotinylation in varying conditions (normal culturing vs. ALI), supplemented with or without doxycycline and biotin for 24 h, with GAPDH as a loading control

5. Fix and permeabilize cells in 100% methanol at −20 °C for at least 20 min and store at −20 °C until staining.
6. Dilute primary antibodies in antibody diluent solution and incubate for 30 min at room temperature. Antibodies: FLAGm2 (mouse; 1:500), ARL13B (rabbit; 1:1000), and *ZO-1* (rabbit; 1:500) (Fig. 2b).
7. Rinse cells twice, then incubate with secondary antibodies: Alexa Fluor 594 donkey anti-mouse (1:1500), Alexa Fluor 647 donkey anti-rabbit (1:1000), and Streptavidin-488 (1: 500) are diluted in antibody diluent solution and incubated for 30 min at room temperature.
8. Prepare and image cells using the same protocol outlined in Iazzi et al., 2023 [33] (*see* **Note 12**).

3.4 Validation with Western Blotting

1. For confirmation of biotinylation in transduced cells grown under normal culturing conditions: Seed into 6-well tissue culture dishes. For induced condition samples, induce cells with 1 μg/mL doxycycline and 50 μM biotin for 24 h prior to fixing. For uninduced condition samples, change media.

2. For confirmation of biotinylation in transduced cells grown under ALI culturing conditions: Seed into 6-well transwell insert culture dishes at the LLI. Allow cells to grow until 100% confluence and a TEER output of 1000 Ω or greater (*see* **Note 11**).
3. Once cells reach a TEER >1000 Ω, remove the media in the transwell insert (leaving only media below inside the well) and allow cells to grow at the ALI for ~2 weeks (*see* **Note 11**).
4. Incubate cells for 24 h in complete media supplemented with 1 μg/mL doxycycline and 50 μM biotin.
5. Harvest, lyse, and process samples using the same protocol outlined in Iazzi et al. [33] (Fig. 2c).

3.5 BioID Sample Expansion and Harvest

1. For sample expansion of cells grown in normal culturing conditions, follow the steps outlined in Iazzi et al. [33].
2. For sample expansion of cells grown in ALI culturing conditions (Fig. 3): Expand the inducible FLAG-BirA*-CFTR cells into 30 transwell inserts per sample (5× 6-well transwell insert dishes) (*see* **Note 13**).
3. After 24 h, change media (*see* **Note 14**).
4. Once cells reach a TEER >1000 Ω, remove the media in the transwell insert (leaving only media below inside the well) and allow cells to grow at the ALI for ~2 weeks (*see* **Note 11**).
5. Treat cells with 1 μg/mL of tetracycline +50 μM biotin in full MEM (+FBS, +PS), 2 mL volume per well, for 24 h.
6. After 24 h incubation, scrape cells on ice and collect media into a 50 mL falcon tube. Pool all 30 inserts into one falcon tube per sample.
7. Pellet cells (4 °C; 1000 rpm; 5 min). Wash cells with 10 mL ice-cold PBS and transfer to a 15 mL falcon tube per sample (*see* **Note 15**).
8. Pellet cells (4 °C; 1000 rpm; 5 min). Freeze the pellets at −80 ° C (*see* **Note 16**).
9. For cell lysis, affinity purification, liquid chromatography–mass spectrometry (LC–MS), LC–MS data analysis, and data visualization in Cytoscape and ProHits-Viz, follow the steps outlined in Iazzi et al. [33, 44] (Fig. 4).

4 Notes

1. Review standard operating procedures for lentiviral/retroviral work specific to the institution in which the experiments are being performed.

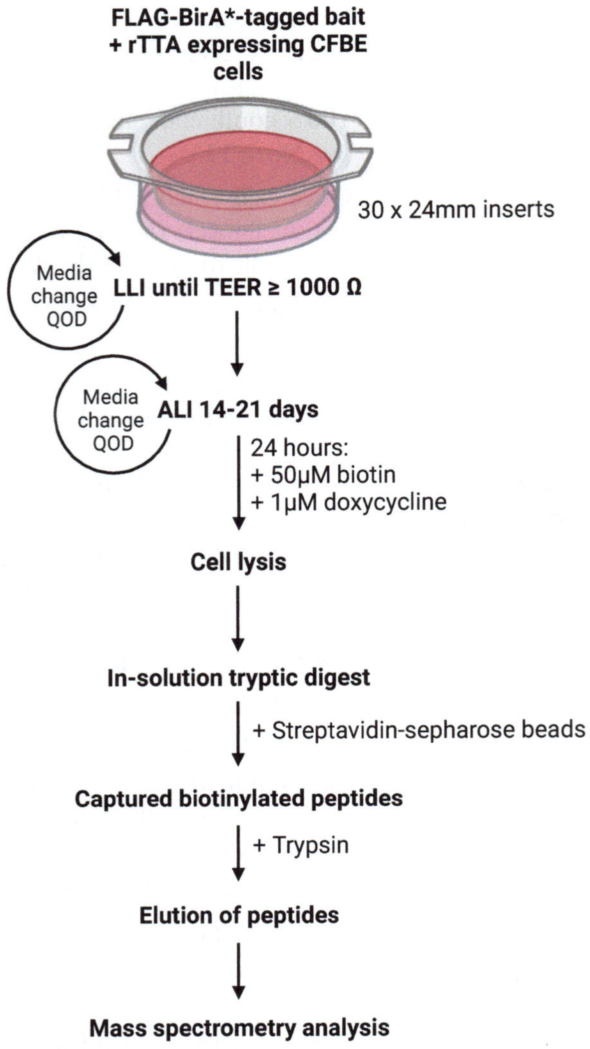

Fig. 3 Pipeline for generation of a BioID dataset on human bronchial epithelial cells grown at the ALI. Stable FLAG-BirA*-CFTR CFBE cells are expanded into 30× 6-well transwell inserts and grown at the LLI until a TEER >1000 Ω is achieved. Cells are then maintained for up to 3 weeks at the ALI before being supplemented with doxycycline and biotin for 24 h, and BioID conducted

2. Cells should be plated in a sterile tissue culture plate at a density of ~80% confluency and maintained in a humidified tissue culture incubator (5% CO2, 37 °C). Cultures should be free of contamination and grown in appropriate medium supplemented with heat-inactivated fetal bovine serum and all necessary growth factors.
3. Seed 1 flask per transfection that will be performed on Day 2.
4. pSTV2 is designed to accommodate large open reading frames (e.g., CFTR: 4440 bp) but does not harbor the rtTA

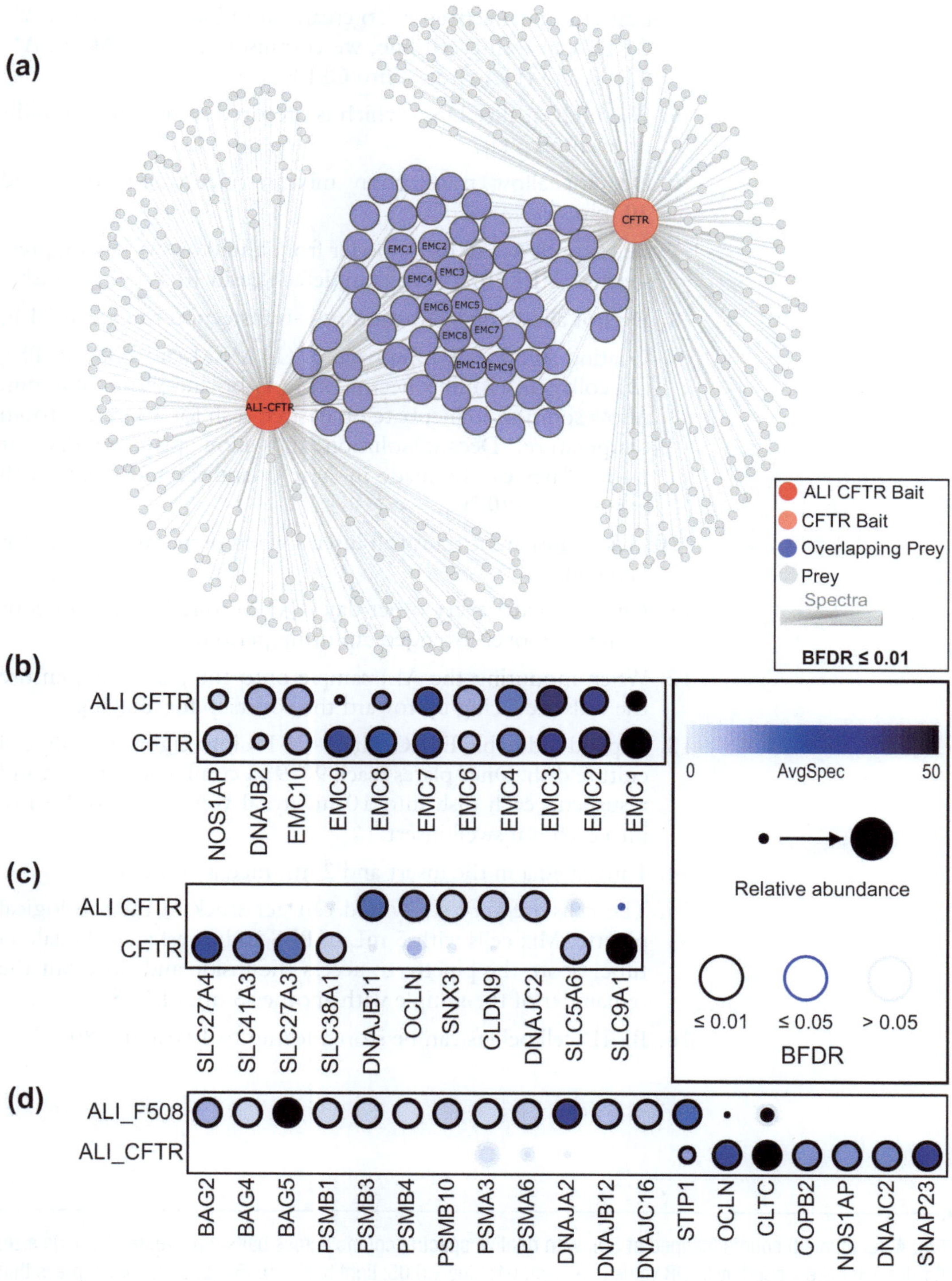

Fig. 4 Overlap of WT CFTR interactomes generated in CFBE cells in different culturing conditions. (**a**) The network was generated using Cytoscape79 version 3.8.0. Overlap of high confidence (BFDR ≤0.01) CFTR interactomes. Unsupervised, edge-weighted spring embedded view of FLAG-BirA*-CFTR in CFBE cells grown in normal and ALI conditions. All 10 subunits of the ER membrane protein complex (EMC) were identified as high confidence proximity interactors in both datasets [23, 45]. (**b, c, d**) Dot plots were generated using ProHits-viz78 [44]. Quantitation is encoded using the color gradient representing control-subtracted spectral

transcriptional activator. To create an inducible FLAG-BirA*-CFTR expressing cell line, we cotransduce the FLAG-BirA*-CFTR and rtTA viruses into CFBE cells.

5. Use high-quality DNA which is prepared using either a midi- or maxi-prep kit.
6. Do not allow transfection mixture incubation to exceed 20 min.
7. Be cautious not to drop too far from liquid or swirl too aggressively as HEK293Tv cells can detach easily from culture dish.
8. Do not allow the incubation post-transfection to exceed 24 h.
9. Coating solution is comprised of (1) MEM (no FBS, no PS), (2) collagen, (3) fibronectin, and (4) bovine serum albumin. Allow solution to incubate on desired dish for 10 min at room temperature. Decant solution and allow plate to dry for 2+ h. Plates can be made in large quantities and stored with parafilm at −20 °C.
10. Use transparent transwell inserts when preparing samples for imaging.
11. Change media every other day (QOD) to reduce risk of contamination over extended culturing period.
12. When mounting the ALI sample onto the glass slide, ensure the cells are facing up toward the coverslip for imaging.
13. Expand inducible FLAG-BirA*-CFTR into eight 100 mm cell culture dish. Once plates reach 90–95% confluence, trypsin and resuspend each dish into 10 mL total volume. Seed 2.5 mL into each transwell insert.
14. 1 mL media in the insert and 2 mL media in the well.
15. The cells are very sticky and can get stuck in the serological pipette. Mix cells with 5 mL of PBS and transfer to the falcon tube. Score the pipette to access the inside and rinse out the remainder of the sample with another 5 mL of PBS.
16. BioID cell pellets can be stored for up to 1 year at −80 °C.

Fig. 4 (continued) counts (capped at 50), with relative spectral counts across baits represented by node size. Border color is encoded by BFDR value (black ≤0.01; blue ≤0.05; light blue >0.05). (**b**) Snapshot of preys that exhibit a consistent response in the different culturing condition CFBE datasets as those reported in literature [23, 24, 46]. (**c**) Snapshot of preys that exhibit a unique response in the different culturing condition CFBE datasets [25, 47, 48]. (**d**) BioID dataset was generated of FLAG-BirA*-ΔF508-CFTR in CFBE cells grown at the ALI. Snapshot of preys behaving similarly to findings reported on WT- and ΔF508-CFTR datasets generated in HEK293 T-REx cells [23]

Acknowledgments

The authors acknowledge the gift of vectors from Dr. Anne Claude Gingras and the CFBE cell line from Dr. John P. Clancy.

References

1. Jelokhani-Niaraki M (2022) Membrane proteins: structure, function and motion. Int J Mol Sci 24:468
2. Rawlings AE (2016) Membrane proteins: always an insoluble problem? Biochem Soc Trans 44:790–795
3. Pankow S, Bamberger C, Calzolari D, Bamberger A, Yates JR (2016) Deep interactome profiling of membrane proteins by co-interacting protein identification technology. Nat Protoc 11:2515–2528
4. Schey KL, Grey AC, Nicklay JJ (2013) Mass spectrometry of membrane proteins: a focus on Aquaporins. Biochemistry 52:3807–3817
5. Fagerberg L, Jonasson K, von Heijne G, Uhlén M, Berglund L (2010) Prediction of the human membrane proteome. Proteomics 10:1141–1149
6. Gulezian E et al (2021) Membrane protein production and formulation for drug discovery. Trends Pharmacol Sci 42:657–674
7. Cottrill KA et al (2021) Sphingomyelinase decreases transepithelial anion secretion in airway epithelial cells in part by inhibiting CFTR-mediated apical conductance. Physiol Rep 9(15):e14928
8. Costantini C et al (2020) Selectively targeting key inflammatory pathways in cystic fibrosis. Eur J Med Chem 206:112717
9. Laselva O et al (2022) Small-molecule drugs for cystic fibrosis: where are we now? Pulm Pharmacol Ther 72:102098
10. Mergiotti M, Murabito A, Prono G, Ghigo A (2022) CFTR modulator therapy for rare CFTR mutants. J Resp 2:59–76
11. Lopes-Pacheco M (2016) CFTR modulators: shedding light on precision medicine for cystic fibrosis. Front Pharmacol 7:275
12. Iazzi M et al (2023) Synthesis and evaluation of Ivacaftor derivatives with reduced lipophilicity. ACS Omega. https://doi.org/10.1021/acsomega.3c05839
13. Van Goor F et al (2011) Correction of the F508del-CFTR protein processing defect in vitro by the investigational drug VX-809. Proc Natl Acad Sci 108:18843–18848
14. Bear CE (2020) A therapy for most with cystic fibrosis. *Cell* 180:211. Preprint at https://doi.org/10.1016/j.cell.2019.12.032
15. Cutting GR (2017) Treating specific variants causing cystic fibrosis. JAMA 318:2130
16. Laselva O et al (2021) Rescue of multiple class II CFTR mutations by elexacaftor+tezacaftor +ivacaftor mediated in part by the dual activities of elexacaftor as both corrector and potentiator. Eur Respir J 57:2002774
17. Terlizzi V et al (2021) Effectiveness of Elexacaftor/Tezacaftor/Ivacaftor therapy in three subjects with the cystic fibrosis genotype Phe508del/unknown and advanced lung disease. Genes (Basel) 12:1178
18. Wainwright CE et al (2015) Lumacaftor–ivacaftor in patients with cystic fibrosis homozygous for Phe508del *CFTR*. N Engl J Med 373:220–231
19. Phuan P-W et al (2019) Nanomolar-potency 'co-potentiator' therapy for cystic fibrosis caused by a defined subset of minimal function CFTR mutants. Sci Rep 9:17640
20. Vertex Pharmaceuticals Incorporated (2023) Who KALYDECO® Is For
21. Veit G et al (2020) Allosteric folding correction of F508del and rare CFTR mutants by elexacaftor-tezacaftor-ivacaftor (Trikafta) combination. JCI Insight 5(18):e139983
22. Iazzi M, Sadeghi S, Gupta GD (2023) A proteomic survey of the cystic fibrosis transmembrane conductance regulator Surfaceome. Int J Mol Sci 24:11457
23. Iazzi M, Astori A, St-Germain J, Raught B, Gupta GD (2022) Proximity profiling of the CFTR interaction landscape in response to Orkambi. Int J Mol Sci 23:2442
24. Pankow S et al (2015) ΔF508 CFTR interactome remodelling promotes rescue of cystic fibrosis. Nature 528:510–516
25. Chevalier B et al (2022) Differential CFTR-Interactome proximity labeling procedures identify enrichment in multiple SLC transporters. Int J Mol Sci 23(16):8937
26. Lim SH et al (2022) CFTR interactome mapping using the mammalian membrane two-hybrid high-throughput screening system. Mol Syst Biol 18(2):e10629

27. Hutt DM, Loguercio S, Campos AR, Balch WE (2018) A proteomic variant approach (ProVarA) for personalized medicine of inherited and somatic disease. J Mol Biol 430:2951–2973
28. McDonald EF, Sabusap CMP, Kim M, Plate L (2022) Distinct proteostasis states drive pharmacologic chaperone susceptibility for cystic fibrosis transmembrane conductance regulator misfolding mutants. Mol Biol Cell 33(7):ar62
29. Reilly R et al (2017) Targeting the PI3K/Akt/mTOR signalling pathway in cystic fibrosis. Sci Rep 7(1):7642
30. Cottrill KA, Farinha CM, McCarty NA (2020) The bidirectional relationship between CFTR and lipids. Commun Biol 3:179
31. Das N, Murray DT, Cross TA (2013) Lipid bilayer preparations of membrane proteins for oriented and magic-angle spinning solid-state NMR samples. Nat Protoc 8:2256–2270
32. Roux KJ, Kim DI, Burke B, May DG (2018) BioID: a screen for protein-protein interactions. Curr Protoc Protein Sci 91: 19.23.1–19.23.15
33. Iazzi M, St-Germain J, Acharya S, Raught B, Gupta GD (2024) Proximity mapping of ciliary proteins by BioID. Methods Mol Biol 2725: 181–198. https://doi.org/10.1007/978-1-0716-3507-0_11
34. Roux KJ, Kim DI, Raida M, Burke B (2012) A promiscuous biotin ligase fusion protein identifies proximal and interacting proteins in mammalian cells. J Cell Biol 196:801–810
35. Gingras AC, Abe KT, Raught B (2019) Getting to know the neighborhood: using proximity-dependent biotinylation to characterize protein complexes and map organelles. Curr Opin Chem Biol 48:44–54
36. Gupta GD et al (2015) A dynamic protein interaction landscape of the human centrosome-cilium Interface. Cell 163:1484–1499
37. Samavarchi-Tehrani P, Abdouni H, Samson R, Gingras A-C (2018) A versatile lentiviral delivery toolkit for proximity-dependent Biotinylation in diverse cell types. Mol Cell Proteomics 17:2256–2269
38. Ousingsawat J, Kongsuphol P, Schreiber R, Kunzelmann K (2011) CFTR and TMEM16A are separate but functionally related Cl− channels. Cell Physiol Biochem 28:715–724
39. Baldassi D, Gabold B, Merkel OM (2021) Air−liquid Interface cultures of the healthy and diseased human respiratory tract: promises, challenges, and future directions. Adv Nanobiomed Res 1:2000111
40. Enemark A, Jönsson ÅLM, Kronborg-White S, Bendstrup E (2021) Pulmonary alveolar Microlithiasis – a review. Yale J Biol Med 94: 637–644
41. Connolly B, Isaacs C, Cheng L, Asrani KH, Subramanian RR (2018) SERPINA1 mRNA as a treatment for Alpha-1 antitrypsin deficiency. J Nucleic Acids 2018:1–7
42. John A, Kizhakkedath P, Al-Gazali L, Ali BR (2015) Defective cellular trafficking of the bone morphogenetic protein receptor type II by mutations underlying familial pulmonary arterial hypertension. Gene 561:148–156
43. Clark DP, Pazdernik NJ (2016) Inherited defects and gene therapy. In: *Biotechnology*. Elsevier, pp 523–564. https://doi.org/10.1016/B978-0-12-385015-7.00017-X
44. Knight JDR et al (2017) ProHits-viz: a suite of web tools for visualizing interaction proteomics data. Nat Methods 14:645–646
45. Louie RJ et al (2012) A yeast phenomic model for the gene interaction network modulating CFTR-ΔF508 protein biogenesis. Genome Med 4:103
46. Shurtleff MJ et al (2018) The ER membrane protein complex interacts cotranslationally to enable biogenesis of multipass membrane proteins. elife 7:e37018
47. He L et al (2021) DNAJB12 and Hsp70 triage arrested intermediates of N1303K-CFTR for endoplasmic reticulum-associated autophagy. Mol Biol Cell 32:538–553
48. Ikpa PT et al (2020) Transcriptome analysis of the distal small intestine of Cftr null mice. Genomics 112:1139–1150

Chapter 5

Aminooxy Biotin-Based Characterization of the Surfaceome of Chondrogenic Cells

Patrik Kovács, David J. Boocock, Clare Coveney, Ali Mobasheri, and Csaba Matta

Abstract

This protocol outlines a comprehensive approach for characterizing the cell surface subproteome of chondroprogenitor cells, based on aminooxy biotinylation followed by mass spectrometry analysis. The first step involves the selective labeling of cell surface proteins with aminooxy biotin on living chondroprogenitor cells, ensuring the specific tagging of glycoproteins on the outer membrane. Subsequently, glycocapture technique is employed to enrich the glycosylated fraction of the cell surface proteins. Following multiple wash steps to reduce contamination with detergents and nonsurface proteins, shotgun mass spectrometry is applied for the quantitative and qualitative analysis of the enriched subproteome, allowing for the identification and characterization of surface proteins. The integration of these techniques offers a comprehensive and sensitive method for profiling the cell surface proteome during chondrogenesis, enabling a deeper understanding of the molecular composition of chondroprogenitor cells. This protocol holds promise for advancing our knowledge of chondrogenesis and may contribute to the identification of potential therapeutic targets for cartilage-related disorders.

Key words Glycocapture, Aminooxy biotin, Chondrogenesis, Chondroprogenitor cell, Surfaceome

1 Introduction

Cartilage, a highly specialized connective tissue type which provides support and flexibility to various structures in the body, is composed of chondrocytes and progenitor cells embedded in a specialized extracellular matrix [1]. Cartilage forms during the series of events termed chondrogenesis. This process is tightly regulated by specific signaling pathways [2]. During embryonic development, chondrogenesis is a crucial step in the formation of the early skeletal system. Chondrogenesis is also important in the

Supplementary Information The online version contains supplementary material available at https://doi.org/10.1007/978-1-0716-4434-8_5.

Csaba Matta et al. (eds.), *The Surfaceome: Methods and Protocols*, Methods in Molecular Biology, vol. 2908,
https://doi.org/10.1007/978-1-0716-4434-8_5,

context of tissue repair and regeneration upon cartilage defects and disorders. In mature articular cartilage, however, such intrinsic repair processes are usually incapable of a complete healing response. A better understanding of the cellular and molecular mechanisms involved in chondrogenesis is essential for advancing cartilage tissue engineering and regenerative medicine, with potential implications for treating conditions related to cartilage damage or degeneration such as osteoarthritis (OA) [3].

One such approach is a better characterization of chondrogenic cells undergoing chondrogenic differentiation; specifically, the chondroprogenitor cell surface proteome [4]. High-throughput surface subproteome analysis of chondrogenic cells helps understanding the intricate molecular processes underlying cartilage development and function. This method involves the comprehensive study of proteins expressed on the cell surface, shedding light on the dynamic interactions between differentiating cells and their developing microenvironment.

However, the dynamic nature of chondrogenic cells poses an important difficulty. These cells undergo various differentiation stages during cartilage formation [5], and their surface protein expression profiles change accordingly. Conventional static analyses may not capture these dynamic changes, necessitating the development of innovative methods to monitor surface proteome alterations in high temporal resolution. Furthermore, the limited sensitivity of detection technologies poses another significant challenge. Many surface proteins are expressed at low abundance, making their detection and quantification challenging. This limitation affects the depth of analysis, making the identification of rare or low-abundance proteins that may play critical roles in chondrogenesis rather difficult. Improving the sensitivity of detection methods, such as by using mass spectrometry, is crucial for unravelling the surface proteome of chondrogenic cells.

Cell surface proteome samples are complex mixtures of proteins, and the presence of high-abundance proteins can mask the signals of lower-abundance ones. Effective sample preparation techniques are needed to enrich surface proteins while minimizing contamination from intracellular proteins, enhancing the sensitivity of detection [6].

The vast majority of cell surface proteins are glycosylated, and these glycan modifications play crucial roles in various cellular functions, such as receptor-mediated signaling, adhesion, and transport processes [7]. Glycocapture, which refers to the specific enrichment of glycoproteins present on the cell surface, offers several advantages in cell surface proteomics compared to other methods. This specificity allows for the selective analysis of cell surface glycoproteins, providing a more focused and accurate representation of the cell surface proteome. Enriching glycoproteins can enhance the sensitivity of mass spectrometry-based proteomic analyses. This is particularly important for low-abundance cell surface proteins, as the enrichment step helps in detecting and identifying proteins that may be present in limited quantities [8].

In this paper, we provide a detailed protocol on the selective enrichment and mass spectrometry-based identification of the cell surface subproteome of chondroprogenitor cells in micromass cultures of limb bud-derived chondrogenic progenitor cells obtained from early-stage chicken embryos. To achieve specific enrichment of cell surface glycoproteins, aminooxy biotin labeling is performed. Aminooxy groups react with molecules containing aldehyde or ketone groups to form a stable oxime linkage. The reaction is fast and can be further accelerated using a catalyst such as aniline. Therefore, this approach provides a convenient and rapid way to label glycoproteins with a detectable tag, such as biotin, under mild conditions. The labeling process involves two steps, first introducing aldehyde groups into the glycoproteins by mild periodate oxidation, followed by treating the functionalized proteins with aminooxy biotin [9]. This method has been used to efficiently label cell surface sialic acid–containing glycans on living chondroprogenitor and mesenchymal stem cells [4] and equine chondrocytes [10].

2 Materials

For cell culture work, use sterile plasticware and sterile (or filter-sterilized) reagents. For protein labeling and mass spectrometry analysis, prepare all solutions using HPLC grade water and analytical grade reagents (unless otherwise stated). Prepare and store all reagents at room temperature (unless indicated otherwise).

2.1 General Lab Equipment and Disposables

1. Laminar flow biosafety cabinet.
2. Equipment for egg incubation (hatcher), candling, cracking, embryo retrieval, and limb bud collection (various sets of microforceps).
3. Stereomicroscope.
4. General CO_2 incubator for cell culturing.
5. General cell culturing plasticware (24-well plates, d = 15 cm plastic petri dishes, centrifuge tubes, Protein LoBind Eppendorf tubes).
6. Eppendorf pipet, pipet tips.
7. Centrifuge.
8. Vortex mixer.
9. Refrigerator.
10. Rocking platform.

11. Vacuum centrifuge.
12. Temperature-controlled mixing block.
13. High-recovery HPLC vials and caps.

2.2 Micromass Cell Cultures

1. Fresh chicken eggs (procured from poultry farm).
2. CMF-PBS (calcium and magnesium-free phosphate-buffered saline): 8 g NaCl, 0.3 g KCl, 0.184 g Na_2HPO_4 (× 12 H_2O), 0.02 g KH_2PO_4, 2 g glucose. Adjust pH to 7.2. Bring volume to 1000 mL with deionized water. Sterile filter (0.2 μm filter) in a laminar flow.
3. 0.25% trypsin-EDTA solution.
4. Ham's F12 culture medium, supplemented with 10% FBS (fetal bovine serum), 0.5 mM L-glutamine, 1% penicillin/streptomycin, and 50 mg/L vitamin C.
5. CMF-PBS, trypsin, cell culture media, and supplements should be sterile or filter sterilized (*see* **Note 1**).
6. Plastic petri dishes, cell culture treated (d = 150 mm).

2.3 Surface Protein Labeling Using Aminooxy Biotin

1. Prepare the following solutions using Milli-Q ultrapure water (prepared by purifying deionized water with a resistivity of at least 18.2 MΩ·cm at 25 °C) or HPLC grade water and analytical grade reagents.
2. FBS-free Ham's F12 medium.
3. CMF-PBS (same composition as above). Adjust pH to 7.4. Bring volume to 1000 mL with deionized water. Sterile filter (0.2 μm filter) in a laminar flow.
4. 0.25% type I collagenase in PBS. Adjust pH to 7.4.
5. Oxidation/biotinylation mix: 1 mM sodium meta-periodate, 150 mM aminooxy biotin and 10 mM aniline in ice-cold PBS, pH adjusted to 6.7. Aniline should be added prior to vortexing. Prepare fresh before use.
6. Glycerol (1 mM final concentration).
7. PBS with $CaCl_2$ and $MgCl_2$: 1 mM $CaCl_2$ and 0.5 mM $MgCl_2$, pH adjusted to 7.4.

2.4 Biotinylated Cell Surface Protein Isolation and Enrichment Using NeutrAvidin Beads

1. Prepare the following solutions using Milli-Q ultrapure water (prepared by purifying deionized water with a resistivity of at least 18.2 MΩ·cm at 25 °C) or HPLC grade water and analytical grade reagents.
2. Lysis buffer: 1% Triton X-100, 150 mM NaCl, 5 mM iodoacetamide, 0.1 mg/mL PMSF (phenylmethylsulfonyl fluoride), and 10 mM Tris–HCl, adjust pH to 7.6, add protease inhibitor cocktail.

3. NeutrAvidin agarose beads.
4. Snap Cap spin columns.
5. PBS, pH adjusted to 7.4.
6. PBS containing 0.5% (w/v) SDS (sodium dodecyl sulfate).
7. PBS containing 0.5% (w/v) SDS and 100 mM DTT (dithiothreitol). Prepare fresh before use.
8. UC buffer: 6 M urea, 100 mM Tris–HCl adjusted pH to 8.5.
9. UC buffer containing 50 mM iodoacetamide.
10. 5 M NaCl.
11. 100 mM Na_2CO_3.
12. HPLC grade/ultrapure water.
13. 50 mM NH_4HCO_3 and 5 μg Trypsin Gold.
14. 50 mM NH_4HCO_3.

2.5 Peptide Separation and Identification by UPLC-MS/MS

1. LCMS grade 0.1% formic acid (mobile phase A).
2. LCMS grade 0.1% formic acid in acetonitrile (mobile phase B).
3. LCMS grade 10% acetonitrile with 0.1% formic acid (dilute mobile phase B in mobile phase A).
4. Bioer mixing block/thermomixer.
5. Autosampler LC vials.
6. Waters M-Class HPLC.
7. Sciex 7600 ZenoTOF.
8. Phenomenex Kinetex XB-*C18* 2.6 μm 150 × 0.3 mm analytical column.
9. *Gallus gallus* SwissProt proteome .FASTA file.
10. DIA-NN software 1.8.1.

3 Methods

3.1 Chondrogenic Micromass Cultures from Embryonic Limb Buds

Chicken embryonic limb bud derived micromass cultures provide a valuable in vitro model to study the molecular and cellular background of hyaline cartilage development and maturation. They mimic the sequential processes of cellular differentiation, maturation, hypertrophy, calcification, and apoptosis seen in cartilage development in vivo. The micromass cultures undergo distinct chondrogenesis by culture day 3, forming large nodules with cartilage-like morphology and extracellular matrix. The distal parts of the forming limb buds of early-stage (Hamburger–Hamilton developmental stage 24; ~4.5 days old) chicken embryos are used to set up the chondrifying cultures [11].

1. Incubate eggs in a hatcher for 4.5 days (104 h) at 38 °C (*see* **Note 2**).
2. Carefully crack the eggs, retrieve the embryos, and using two pairs of fine microdissection forceps (a straight and a curved set), remove the distal parts of all four limb buds from each embryo (*see* **Note 3**). Pool collected limb buds in CMF–PBS.
3. Once all limb buds from all embryos have been gathered, carefully aspirate the CMF–PBS and transfer the limb buds to 0.25% trypsin-EDTA. Incubate for 55 min at 37 °C in a CO_2 incubator (*see* **Note 4**).
4. After the incubation time, carefully dissociate the limb buds. Terminate the enzymatic dissociation by adding equal volume of FBS, then centrifuge at 150 × *g* for 5 min at room temperature to pellet the cells.
5. Wash pelleted cells in complete Ham's F12 medium and centrifuge again at 150 × *g* for 10 min.
6. Filter the cell suspension through a 20 μm pore-size plastic filter, determine cell concentration, and adjust the volume with Ham's F12 medium to achieve the desired concentration of 1.5×10^7 cells/mL.
7. Plate micromass cultures by inoculating 100 μL droplets of the cell suspension into petri dish(es) (d = 150 mm) (*see* **Note 5**).
8. Flood cultures with medium after the 2-h incubation. Plating is considered day 0. Change the culture medium every other day of the experiment.

3.2 Surface Protein Labeling Using Aminooxy Biotin

In this step, sialic acid residues of cell surface proteins are first oxidized by sodium-meta-periodate and then biotinylated by aminooxy biotin (Fig. 1).

1. On designated culture days, depending on the required number of cultures (*see* **Note 6**), prepare one or two petri dishes with the differentiating micromass cultures. Prepare and filter sterilize 0.25% type I collagenase solution in PBS (pH 7.4) to digest the collagen-rich extracellular matrix (ECM) without affecting the surface proteins.
2. Wash the petri dish(es) twice with 15 mL sterile PBS (pH 7.4) and incubate in 15 mL serum-free Ham's F12 media for 1 h in a CO_2 incubator at 37 °C to help remove high-abundance proteins from FBS.
3. After the 1-h incubation, wash cells twice with 15 mL PBS (pH 7.4) and add 5 mL 0.25% type 1 collagenase in PBS (pH 7.4) to the micromass cultures. Replace the petri dish into the incubator and wait for 5–20 min until the ECM of cell cultures has been digested (*see* **Note 7**).

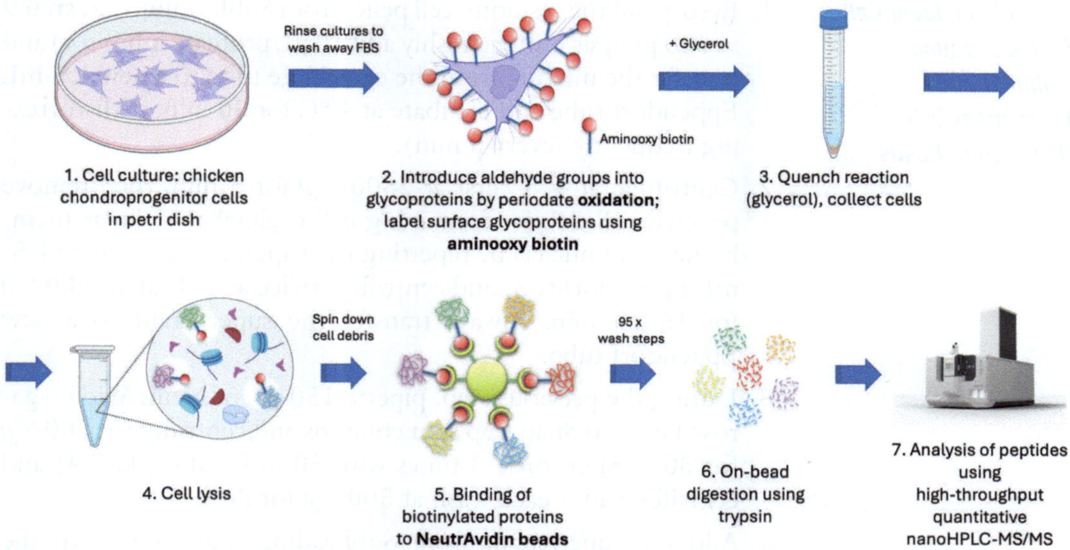

Fig. 1 Schematic representation of the workflow

4. Collect the liberated cells from the petri dish to a centrifuge tube using a 5-mL pipet, then wash the petri dish twice with 5 mL ice-cold PBS (pH 7.4) to collect the remaining cells. Pool these fractions into the same centrifuge tube. Centrifuge at 300 × *g* for 5 min at 4 °C to pellet the cells.

Perform the following steps on ice until Subheading 3.3, ***step 11****.*

5. Prepare the oxidation/biotinylation mix comprising 1 mM sodium meta-periodate, 100 mM aminooxy biotin, and 10 mM aniline in ice-cold PBS (pH 6.7). (4 mL of the mix is required for each 150-mm petri dish.)
6. Add the oxidation/biotinylation mix into the pelleted cells and transfer contents to a petri dish (60 mm) placed on a rocking platform in the dark at 4 °C. Incubate for 1 h with gentle rocking.
7. After 1 h, stop the reaction by the addition of glycerol at a final concentration of 1 mM (add 400 μL) and incubate the cells for 10 min in the dark at 4 °C.
8. Pipet the contents of the petri dish to a new 50-mL centrifuge tube. Wash the petri dish once using 10 mL PBS (pH 7.4) and add this to the centrifuge tube, then centrifuge at 300 × *g* for 5 min.
9. Resuspend the pellet in 10 mL PBS (pH 7.4) containing 1 mM $CaCl_2$ and 0.5 mM $MgCl_2$, then centrifuge again at 300 × *g* for 5 min.

3.3 Biotinylated Cell Surface Protein Isolation and Enrichment Using NeutrAvidin Beads

1. Resuspend the resulting cell pellet from Subheading 3.2, **step 9** in 500 μL lysis buffer (freshly add 10 μL protease inhibitor) and transfer the mixture from the centrifuge tube to a new 1.5-mL Eppendorf tube and incubate at 4 °C for 30 min, with vortexing frequently (every 5 min).
2. Centrifuge at 4 °C first at 2800 × *g* for 5 min, then remove pelleted cell debris (containing cell organelles, cellular membranes, and nuclei) by pipetting the supernatant to a new 1.5-mL Eppendorf tube and centrifuge twice at 4 °C at 16,000 × *g* for 15–15 min. Always transfer the supernatant to a new Eppendorf tube.
3. During the previous step, pipette 150 μL of NeutrAvidin agarose beads to Snap Cap spin columns and centrifuge at 500 × *g* for 30 s. Then rinse 3 times with 500 μL PBS (pH 7.4) and centrifuge after each wash at 500 × *g* for 30 s.
4. Add the supernatant from Subheading 3.3, **step 2,** to the previously prepared Snap Cap spin column (containing NeutrAvidin agarose beads) and incubate for 2 h at 4 °C with rocking continuously, with a rubber cap placed to the bottom of the columns. In this step, the biotin-labeled surface proteins will attach to the NeutrAvidin agarose beads by avidin-biotin binding, while the nonlabeled contaminant proteins remain unattached.
5. To eliminate nonspecifically bound contaminant proteins, perform multiple washing steps (15 times with 500 μL of the solutions specified below followed by centrifugation at 500 × *g* for 30 s). Remove the rubber cap before spinning.
6. Wash beads 15 times with lysis buffer, followed by 15 washes with 0.5% (w/v) SDS in PBS (pH 7.4).
7. Incubate the beads for 20 min at room temperature in 0.5% (w/v) SDS, 100 mM DTT in PBS (pH 7.4; prepare fresh), then centrifuge again at 500 × *g* for 30 s.
8. Perform further washing 15 times with UC buffer.
9. Alkylate for 20 min at room temperature with UC buffer containing 50 mM iodoacetamide.
10. Wash beads again with 500 μL of the following solutions (10 times per solution followed by centrifugation at 500 × *g* for 30 s after each step), as follows: UC buffer, 5 M NaCl, 100 mM Na_2CO_3, PBS (pH 7.4), and then HPLC grade water.
11. After the washing steps, replace a new rubber cap to the bottom of column and put it into a new 1.5 mL Protein LoBind Eppendorf tube.
12. Resuspend the beads in 400 μL HPLC water containing 50 mM NH_4HCO_3 and 5 μg Trypsin Gold or other mass

spectrometry grade trypsin. Allow biotinylated glycoproteins to be digested in solution (on-beads) overnight at 37 °C.

13. On the next day, remove the rubber cap from the bottom of column and collect the tryptic peptides via centrifugation at 500 × *g* for 1 min.
14. Transfer the Snap Cap spin column into a new 1.5-mL Protein LoBind Eppendorf tube and rinse the beads with 200 μL of 50 mM NH_4HCO_3 to collect the remaining tryptic fractions (500 × *g* for 1 min) and transfer the flow-through into the first Eppendorf tube.
15. Evaporate solvent in a vacuum centrifuge, then store samples at −80 °C until ready for analysis.

3.4 Peptide Separation and Identification by UPLC-MS/MS

1. To the tube of dried peptides from Subheading 3.3, **step 15,** add 150 μL of 10% acetonitrile 0.1% formic acid and shake on a thermomixer at 5 °C for 20 min, then add 150 μL of 0.1% formic acid and vortex thoroughly. This will result in a final concentration of 5% acetonitrile 0.1% formic acid (*see* **Notes 8** and **9**).
2. Remove any remaining debris or non-resuspended particulates by centrifuging at 12,000 RCF for 2 min (13,400 rpm in Eppendorf MiniSpin). Transfer to LC compatible vial, taking care not to disturb any pellet if visible.
3. Load LC vial into autosampler and inject an appropriate mass of peptides onto the column for MS (*see* **Note 10**).
4. Inject 3 μL of each reconstituted sample (*see* **Note 9**) from a Waters M-Class LC system connected to a Sciex 7600 Zeno-TOF via a Phenomenex Kinetex XB-C_{18} 2.6 μm 15 × 0.3 mm analytical column in Direct Inject mode with the column oven set at 30 °C.
5. Run the samples on an LC method with the following gradient: 3–35% mobile phase B over 12 min at 10 μL/min, increase flow rate to 12 μL/min to wash at 80% B at 13 min to 15 min and re-equilibrate to 3% at 16.5 min.
6. Run the MS in positive ion mode with Data-Independent Acquisition (zenoSWATH [12]) utilizing a 25 ms TOFMS scan followed by 65 variable SWATH windows each of 12 ms duration from m/z 400 to 750 giving to total cycle time of 1146 ms. The full methodological parameters are provided in Supplementary information (*MS Method Parameters.csv*). Alternatively, for a more qualitative analysis, the MS instrument can be run in IDA (Information/Data-Dependent Acquisition mode, DDA) mode. (Please see Chap. 6 in this volume).

7. The ZenoTOF 7600 generates three files for each acquisition with the same filename prefix, .wiff, .wiff2, and .wiff.scan (*see* **Note 11**).
8. Download the latest (or most appropriate) proteome FASTA file from UniProt (https://www.uniprot.org) for *Gallus gallus* (UniProt taxonomy [9031]) in uncompressed format, canonical plus isoforms.
9. Process the data files using the DIA-NN software [13] for label-free quantitation (LFQ) (*see* **Note 12**). Leave all settings at default and only change the following parameters: In the GUI, click "Raw" and select your .wiff files to be analyzed. Do not select a spectral library. Click "Add FASTA" and select the downloaded *Gallus gallus* proteome file from UniProt. Tick "FASTA digest for library-free search" and "Deep learning-based spectra." The full command line input is provided in the Supplementary information (*DIANN command line.txt*).
10. In the "Output" section, click "Main output" and select the folder and name for your output report. "Generate spectral library" will be ticked, click "Output library" and select the folder and name for the library file that will be generated.
11. Leave the Precursor FDR% at 1.0 and select the appropriate number of processing threads for your computer's CPU. Select log level 2 (*see* **Note 13**).
12. Leave the remaining settings as they are: mass accuracy 0.0, MS1 accuracy 0.0, scan window 0. Untick unrelated runs (*see* **Note 14**) and leave the rest in the "Algorithm" section ticked.
13. The remaining settings in this section should be as follows: Genes, Single-pass mode, Rbust LC (high precision), RT-dependent, Smart profiling, and Optimal results.
14. Click "Run" and check there are no errors in the log as it progresses.
15. The results will be output to the folder selected in Subheading 3.4, **step 10**. The folder contains several files. For initial analysis use the pg_matrix.tsv file which will list the proteins detected and a normalized LFQ value for each sample. For detailed explanation of the various output files, please read the manual for DIA-NN on the GitHub repository.
16. Export the protein list as an Excel file after analysis. Annotate proteins based on Gene Ontology (GO) information downloaded from the UniProt database (www.uniprot.org) (*see* **Note 15**). Infer cell surface/plasma membrane proteins according to the above annotation.

4 Notes

1. Sterilize solutions and equipment that get in contact with the cells to avoid bacterial/fungal infection of the chondrogenic cultures. Note that eggs are usually sourced from a poultry farm. Please check whether any decontamination procedures are in place (e.g., ozone treatment) to reduce the risk of infections (e.g., contagious poultry diseases which may be transmitted through eggs) of cell cultures. Use 70% (v/v) ethanol to wipe clean the surface of the eggs.
2. In our experience, embryos retrieved from approx. 100 eggs will yield a minimum of 6–7 mL of cell suspension with the required density (1.5×10^7 cells/mL). This volume is required for an experiment in which all main stages of in vitro chondrogenesis can be studied using the protocol described in this chapter. Pay attention to keep the interior of the egg hatcher humid and warm for the entire duration of the incubation.
3. Embryos must be at Hamburger–Hamilton developmental stage 23–24 [14]. At this developmental stage, chondrogenesis has not yet started in the limb buds, which appear club-shaped, without signs of future digits. It is important to remove only the apical/distal portions of limb buds, as the cellular composition and commitment of more proximal parts is not ideal for micromass culturing. See more details in Takacs et al., 2023 [11].
4. Limb bud collection is time sensitive: cells in the isolated limb buds quickly start losing vitality. It is strongly recommended to finish this step within 1 h. Do not allow the digestion to proceed longer than 1 h: excessive digestion will compromise the integrity of the cells and they will die.
5. Avoid air bubbles or irregularly shaped drops when plating the droplets of cells. This would interfere with cell culture morphology and chondrogenic differentiation. Make sure to handle the petri dishes containing the freshly inoculated droplets with care: inappropriate handling will result in misshapen cultures, which adversely affects chondrogenesis. Care must be taken when adding culture medium to freshly established micromasses: let the medium run down the sides of the petri dishes, rather than adding it directly onto the cultures with harsh pipetting, so as not to disturb the cells. If the medium is introduced too harshly, the cells will wash off the plastic surface and die.
6. Work out the number of micromass colonies/number of plated cells for each experimental day during pilot experiments. In our experience, approx. 25 droplets (100 μL each) are needed for

day 1, 20 droplets for day 3, 15 droplets for day 6, and 15–15 droplets for days 10 and 15 are required.

7. Note that the incubation time is dependent on the amount of collagen-containing extracellular matrix in the micromasses. In aged (10- or 15-day) cultures, digestion may require more time (up to 2 h).

8. The initial higher 10% acetonitrile concentration increases recovery of the dried protein from the tube. The final volume/concentration may be adjusted to suit instrument sensitivity and injection volume.

9. The concentration of digested peptides will need to be optimized for the mass spectrometer and LC method used, so a suitable injection volume and mass can be analyzed. For the ZenoTOF 7600 in zenoSWATH acquisition mode, 500–750 ng peptides is optimal on a 16.5 min LC run (11 min gradient), over 1 μg on column can saturate the detector.

10. Optimal mass of peptides injected on column will vary depending on mass spectrometer and LC method used, use the optimized amount for your system so you do not overload the column or the MS detector.

11. For downstream data processing, keep the prefix for all three data files identical, if you need to rename any samples then you must change all the related file prefixes to match. For third party software the .wiff and .wiff.scan pair must both be present in the same folder during analysis, while the .wiff file is usually the one to select for processing.

12. Read the installation instructions on the DIA-NN GitHub site in detail (https://github.com/vdemichev/DiaNN). To process Sciex .wiff data files you must first install extra files to allow them to be read by the software. For .wiff support, download and install ProteoWizard [15] (https://proteowizard.sourceforge.io/)—choose the version (64-bit) that supports "vendor files"). Then copy all files with 'Clearcore' or 'Sciex' in their name (these will be .dll files) from the ProteoWizard folder to the DIA-NN installation folder (the one which contains diann.exe, DIA-NN.exe, and several other files). If you do not do this DIA-NN will exit in the log when it tried to load the .wiff files.

13. The log level is not relevant to processing the data except to generate a text log output. Log level 1 does not generate enough information and the software will appear not to be doing anything. Level 2 gives enough granular feedback that you can see the progress, level 3 is necessary to send to the author for troubleshooting.

Table 1
A combination of the GO terms (cellular component, biological process) that are used to make inferences in terms of cell surface/plasma membrane proteins identified in the samples

GO term	Definition
[GO:0016020]	Membrane
[GO:0031012]	Extracellular matrix
[GO:0005576]	Extracellular region
[GO:0005615]	Extracellular space
[GO:0005886]	Plasma membrane
[GO:0009897]	External side of plasma membrane
[GO:0031234]	Extrinsic component of cytoplasmic side of plasma membrane
[GO:0045121]	Membrane raft
[GO:0009986]	Cell surface
[GO:0005890]	Sodium:Potassium-exchanging ATPase complex
[GO:0016324]	Apical plasma membrane
[GO:0016323]	Basolateral plasma membrane
[GO:0016328]	Lateral plasma membrane
[GO:0005905]	Clathrin-coated pit
[GO:0060170]	Ciliary membrane

14. Unrelated runs can optionally be selected which will optimize the mass accuracy and MS1 accuracy for each sample individually. This will take significant extra time (approximately 15 min per sample on a 32 thread 3.5 GHz AMD Threadripper system with 128 GB RAM) per sample. If all your samples were run in a single batch there is no need to tick this and the values will be generated from the first sample and used for all remaining ones.
15. Use a combination of the GO terms (cellular component, biological process) that are summarized in Table 1.

Acknowledgments

CM would like to express his gratitude to Dr. Susan Liddell (Exonate Ltd., UK) and Dr. Julia Smith (Coventry University, UK) for their invaluable help in introducing him into the world of proteomics and for their guidance in developing and adjusting this protocol. CM was supported by the Young Researcher Excellence Programme (grant number: FK-134304) of the National Research, Development and Innovation Office, Hungary. CM was also supported by the Bolyai János Research Fellowship of the Hungarian Academy of Sciences. This research was supported by the UNKP-23-5 and UNKP-23-5 New National Excellence Program of the Ministry for Culture and Innovation from the source of the National Research, Development and Innovation Fund, Hungary, awarded to CM (UNKP-23-5-DE-487) and PK (UNKP-23-3-II-DE-13). AM acknowledges financial support from the European Structural and Social Funds through the Research Council of Lithuania (Lietuvos Mokslo Taryba), according to the Program Attracting Foreign Researchers for Research Implementation (Grant No. 01.2.2-LMT-K-718-02-0022). AM also acknowledges funding from Horizon Europe (PROTO-Advanced PeRsOnalized Therapies for Osteoarthritis-TACKLING INFLAMMATION TO IMPROVE PATIENT OUTCOMES, Grant agreement ID: 101095635, https://cordis.europa.eu/project/id/101095635 and ENgineered CArtilage from Nose for the Treatment of Osteoarthritis [ENCANTO], Grant agreement ID: 101137315, https://cordis.europa.eu/project/id/101137315) and the Academy of Finland through the Profi6 336449 grant awarded to the University of Oulu. CM and AM also acknowledge financial support from the European Cooperation in Science and Technology (COST) Association Action CA21110—Building an open European Network on OsteoArthritis research (NetwOArk); https://www.cost.eu/actions/CA21110/).

Conflict of Interest The authors declare that they have no conflicts of interest. The authors wrote this paper within the scope of their academic and research positions. None of the authors has any relationships that could be construed as biased or inappropriate. Funding bodies were not involved in the study design, data collection, analysis, or interpretation. The decision to submit this paper for publication was not influenced by any funding bodies.

References

1. Jahr H, Matta C, Mobasheri A (2015) Physicochemical and biomechanical stimuli in cell-based articular cartilage repair. Curr Rheumatol Rep 17(3):22. https://doi.org/10.1007/s11926-014-0493-9
2. Takacs R et al (2023) The temporal transcriptomic signature of cartilage formation. Nucleic Acids Res 51(8):3590–3617. https://doi.org/10.1093/nar/gkad210
3. Fellows CR et al (2016) Adipose, bone marrow and synovial joint-derived mesenchymal stem cells for cartilage repair. Front Genet 7:213. https://doi.org/10.3389/fgene.2016.00213
4. Matta C et al (2019) Molecular phenotyping of the surfaceome of migratory chondroprogenitors and mesenchymal stem cells using biotinylation, glycocapture and quantitative LC-MS/MS proteomic analysis. Sci Rep 9(1):9018. https://doi.org/10.1038/s41598-019-44957-y
5. Richardson SM et al (2016) Mesenchymal stem cells in regenerative medicine: focus on articular cartilage and intervertebral disc regeneration. Methods 99:69–80. https://doi.org/10.1016/j.ymeth.2015.09.015
6. Pauwels J, Fijalkowska D, Eyckerman S, Gevaert K (2022) Mass spectrometry and the

cellular surfaceome. Mass Spectrom Rev 41(5): 804–841. https://doi.org/10.1002/mas.21690

7. Kissel T, Toes REM, Huizinga TWJ, Wuhrer M (2023) Glycobiology of rheumatic diseases. Nat Rev Rheumatol 19(1):28–43. https://doi.org/10.1038/s41584-022-00867-4
8. Li Y, Qin H, Ye M (2020) An overview on enrichment methods for cell surface proteome profiling. J Sep Sci 43(1):292–312. https://doi.org/10.1002/jssc.201900700
9. Weekes MP et al (2012) Proteomic plasma membrane profiling reveals an essential role for gp96 in the cell surface expression of LDLR family members, including the LDL receptor and LRP6. J Proteome Res 11(3): 1475–1484. https://doi.org/10.1021/pr201135e
10. Jeremiasse B et al (2020) Alterations in the chondrocyte surfaceome in response to pro-inflammatory cytokines. BMC Mol Cell Biol 21(1):47. https://doi.org/10.1186/s12860-020-00288-9
11. Takacs R et al (2023) Isolation and micromass culturing of primary chicken Chondroprogenitor cells for cartilage regeneration. Curr Protoc 3(7):e835. https://doi.org/10.1002/cpz1.835
12. Wang Z et al (2022) High-throughput proteomics of nanogram-scale samples with Zeno SWATH MS. elife 11. https://doi.org/10.7554/eLife.83947
13. Demichev V et al (2020) DIA-NN: neural networks and interference correction enable deep proteome coverage in high throughput. Nat Methods 17(1):41–44. https://doi.org/10.1038/s41592-019-0638-x
14. Hamburger V, Hamilton HL (1992) A series of normal stages in the development of the chick embryo. 1951. Dev Dyn 195(4):231–272. https://doi.org/10.1002/aja.1001950404
15. Chambers MC et al (2012) A cross-platform toolkit for mass spectrometry and proteomics. Nat Biotechnol 30(10):918–920. https://doi.org/10.1038/nbt.2377

Chapter 6

Isolation of the Surfaceome from Primary Human Epidermal Melanocytes Using Glycocapture and Shotgun Proteomics

Tibor Hajdú, Minh Ngoc Nguyen, Emese Zsigrai, Clare Coveney, David J. Boocock, and Csaba Matta

Abstract

Melanocytes play a crucial role in skin pigmentation and protection against UV radiation. Understanding the molecular mechanisms regulating melanocyte function is essential for elucidating skin pigmentation disorders and melanoma progression. This protocol details a comprehensive methodology for characterizing the cell surface subproteome of melanocytes and melanoma cell lines, utilizing aminooxy biotinylation followed by mass spectrometry analysis. The initial step involves selectively labeling cell surface proteins with aminooxy biotin on live melanocytes and melanoma cell lines to specifically tag glycoproteins. Labeled proteins are then captured on NeutrAvidin beads. Following multiple wash steps to minimize contamination with non-cell surface proteins, mass spectrometry is employed for quantitative and qualitative analysis of the enriched subproteome, facilitating the identification and characterization of surface proteins. This is a sensitive method for profiling the cell surface subproteome in melanocytes and melanoma cell lines in vitro, offering deeper insights into their molecular composition. This protocol holds promise for advancing our understanding of melanoma and may aid in identifying potential therapeutic targets for melanoma treatment.

Key words Glycocapture, Surfaceome, Aminooxy biotin, Melanocyte, Melanoma, Biomarker

1 Introduction

Cutaneous melanoma, a highly aggressive tumor, originates from the neoplastic transformation of epidermal melanocytes, the pigment-producing cells located primarily in the epidermis and hair follicles. Melanoma is the fifth most diagnosed tumor in both men and women in Europe, accounting for 4% of all new oncological cases and 1.3% of all malignancy-related deaths. The overall trends for cutaneous melanoma are increasing for both incidence

Supplementary Information The online version contains supplementary material available at https://doi.org/10.1007/978-1-0716-4434-8_6.

Csaba Matta et al. (eds.), *The Surfaceome: Methods and Protocols*, Methods in Molecular Biology, vol. 2908, https://doi.org/10.1007/978-1-0716-4434-8_6,

and mortality [1, 2]. It is estimated that there will be a significant increase in new cases and deaths caused by melanoma by 2040, emphasizing the importance of global prevention efforts [3]. The development of effective therapies against melanoma remains challenging due to its heterogeneity, resistance mechanisms, and propensity for metastasis [4]. To address these challenges, it is crucial to identify specific molecular markers and pathways involved in melanomagenesis, progression, and response to treatment.

Previous studies have demonstrated that alterations in cell surface proteins play significant roles in tumorigenesis and disease progression [5]. Cell surface proteins, which are collectively referred to as the *surfaceome*, play roles in melanoma progression by influencing cellular functions such as adhesion, signaling, and immune responses. The surfaceome, specifically in the context of melanoma, is important due to its potential as a source of diagnostic and prognostic biomarkers and druggable targets. However, comprehensive analyses of the surfaceome in melanocytes and melanoma are still limited, particularly at the level of individual protein identification and quantification.

Studies have shown that certain melanosomal glycoproteins, typically expressed in both melanoma and normal melanocytes, can serve as autoantigens and potential targets for therapy [6]. Some of these proteins are expressed on the cell surface of melanoma cells and can be recognized by antibodies and T lymphocytes, contributing to immune responses against tumors [6]. Alterations in cell surface protein expression have been correlated with the metastatic potential of melanoma cells. The metastatic ability of certain melanoma cell lines correlates with changes in cell surface protein profiles, potentially influencing tumor dissemination [7].

Melanocytes provide an effective and useful model for studying physiological cellular mechanisms, such as organelle transport and melanin production, and pathological processes, such as development of melanoma or vitiligo. Successful isolation and maintenance of melanocyte cultures is paramount to achieve these aims. Therefore, the first part of the following protocol describes the isolation and culturing procedure of human epidermal melanocytes from skin samples.

In melanoma research, understanding the qualitative and quantitative composition of the surfaceome of epidermal melanocytes and different grades of melanoma can help identify new therapeutic targets and markers of drug resistance, which are crucial for developing more effective treatment strategies. To achieve such objectives, we describe a combination of aminooxy biotin labeling and high-resolution mass spectrometry. Firstly, directly following conversion of cell surface sialic acid residues into aldehyde groups, aminooxy biotin labeling of live cells under mild conditions takes place to minimize internalization artefacts while maximizing membrane protein accessibility. By combining periodate oxidation with aniline-catalyzed oxime ligation, a bio-orthogonal approach is achieved, allowing for specific and efficient labeling of glycoproteins on cell surfaces [8]. Next, biotinylated cell surface proteins are isolated via NeutrAvidin binding followed by tryptic digestion and liquid chromatography tandem mass spectrometry (LC–MS/MS). Afterward, PEAKS Studio software is used to process raw data files and perform peptide and protein identifications based on the UniProt *Homo sapiens* database.

Profiling the surfaceome of epidermal melanocytes and melanoma cells is essential for identifying new biomarkers and druggable targets that can inform the development of more effective treatment strategies. Integrating proteomics data [9] with available transcriptome datasets [10–12] can also help rank potential standalone (immunotherapy) targets. The approach described in this protocol may provide valuable insights into the tumor microenvironment and help guide the design of targeted therapies to overcome drug resistance.

2 Materials

2.1 General Lab Equipment and Disposables

1. Laminar flow biosafety cabinet.
2. General CO_2 incubator for cell culturing.
3. General cell culturing plasticware (6-well plates, T-25, T-75, and T-175 plastic petri dishes, centrifuge tubes, Eppendorf tubes).
4. Eppendorf pipets and pipet tips.
5. Centrifuge.
6. Vortex mixer.
7. Refrigerator.
8. Rocking platform.
9. Vacuum centrifuge.

2.2 Primary Human Epidermal Melanocyte Isolation

2.2.1 Preparation

1. Skin sample (postoperative skin specimen or acquired by punch biopsy; minimum 5 mm diameter; 1 cm^2 size skin samples yield best amounts of cells).
2. Saline.
3. Betadine® (containing 10 mg/mL povidone iodine).
4. 70% alcohol.
5. Dulbecco's phosphate-buffered saline (DPBS) without calcium or magnesium.
6. 10 μg/mL gentamicin.

2.2.2 Separation of Epidermis and Dermis

1. 10 mg/mL dispase II in Hank's balanced salt solution.
2. Surgical equipment including anatomical (blunt) forceps, surgical scalpels.

2.2.3 Initiation and Maintenance of Melanocyte Culture

1. 500 μg/mL trypsin/EDTA solution: Dissolve 0.5 mL 10 × trypsin-EDTA in 4.5 mL 0.5 mM EDTA solution in DPBS without Ca or Mg. Store at −20 °C.
2. RPMI-1640 cell culture medium.
3. Hemocytometer or automated cell counter.
4. 40 μM TPA (12-O-tetradecanoylphorbol 13-acetate) in 0.1 mg/mL BSA in DPBS without Ca or Mg; filter-sterilized (*see* further details in **Note 6**).
5. 40 nM CT (cholera toxin) in 0.1 mg/ml BSA in DPBS without Ca or Mg; filter-sterilized (*see* further details in **Note 6**).
6. 5 μM ET1 (endothelin 1) in 0.1 mg/ml BSA in DPBS without Ca or Mg; filter-sterilized (*see* further details in **Note 6**).
7. 5 μg/mL SCF (human stem cell factor) in 0.1 mg/ml BSA in DPBS without Ca or Mg; filter-sterilized (*see* further details in **Note 6**).
8. 125 μg/mL trypsin/EDTA solution: Dilute 500 μg/mL trypsin/EDTA solution in 0.5 mM EDTA solution in 1:4 ratio.

2.2.4 Freezing and Thawing of Melanocytes

1. DMSO.
2. 100–300 μM PTU (phenylthiourea or phenylthiocarbamide) (*see* **Note 12**).
3. Fetal bovine serum (FBS).

2.3 Surface Protein Labeling Using Glycocapture

1. Prepare the following solutions using Milli-Q ultrapure water (prepared by purifying deionized water with a resistivity of at least 18.2 MΩ·cm at 25 °C) or HPLC grade water and analytical grade reagents.
2. FBS-free RPMI 1640 medium.
3. PBS: Adjust pH to 7.4. Bring volume to 1000 mL with deionized water. Sterile filter (0.2 μm filter) in a laminar flow.
4. Oxidation/biotinylation mix: 1 mM sodium meta-periodate, 150 mM aminooxy-biotin, and 10 mM aniline in ice-cold PBS, pH adjusted to 6.7. Aniline should be added prior to vortexing. Prepare fresh before use.
5. Glycerol (1 mM final concentration).
6. PBS with $CaCl_2$ and $MgCl_2$: 1 mM $CaCl_2$ and 0.5 mM $MgCl_2$, pH adjusted to 7.4.

2.4 Biotinylated Cell Surface Protein Isolation and Enrichment Using NeutrAvidin

1. Prepare the following solutions using Milli-Q ultrapure water (prepared by purifying deionized water with a resistivity of at least 18.2 MΩ·cm at 25 °C) or HPLC grade water and analytical grade reagents.
2. Lysis buffer: 1% Triton X-100, 150 mM NaCl, 5 mM iodoacetamide, 0.1 mg/mL PMSF (phenyl-methyl-sulfonyl fluoride), and 10 mM Tris–HCl, adjust pH to 7.6, add protease inhibitor cocktail.
3. NeutrAvidin agarose beads.
4. Snap Cap spin columns.
5. PBS, pH adjusted to 7.4.
6. PBS containing 0.5% (w/v) SDS (sodium dodecyl sulphate).
7. PBS containing 0.5% (w/v) SDS and 100 mM DTT (dithiothreitol). Prepare fresh before use.
8. UC buffer: 6 M urea, 100 mM Tris–HCl adjusted pH to 8.5.
9. UC buffer containing 50 mM iodoacetamide.
10. 5 M NaCl.
11. 100 mM Na_2CO_3.
12. HPLC grade/ultrapure water.
13. 50 mM NH_4HCO_3 and 5 μg Trypsin Gold.
14. 50 mM NH_4HCO_3.

2.5 Peptide Separation and Identification by NanoHPLC-MS/MS

1. LCMS grade 0.1% formic acid (mobile phase A).
2. LCMS grade 0.1% formic acid in acetonitrile (mobile phase B).
3. LCMS grade 10% acetonitrile with 0.1% formic acid (dilute mobile phase B in mobile phase A).
4. Bioer mixing block/thermomixer for 1.5 mL microtubes.
5. Autosampler LC vials.
6. Waters M-Class HPLC.
7. Sciex 7600 ZenoTOF.
8. Phenomenex Kinetex XB-C*18* 2.6 μm 15 × 0.3 mm analytical column.
9. Human SwissProt proteome .FASTA file.
10. Software to search Data-Dependent Acquisition MS data, e.g., PEAKS Studio 12 (*see* **Note 15**).

3 Methods

3.1 Isolation and Culturing of Primary Human Epidermal Melanocytes

This unit details the isolation and maintenance of primary melanocyte cultures from human skin. The original version of this protocol was first published by Godwin et al. [13], and then modified by our group [14]. The protocol is applicable on skin samples from around the body independently of the age of the donors. However, skin samples from much younger people are preferable and yield more cells compared to older donors. To cleave the basement membrane at the dermo-epidermal junction, we applied dispase instead of the conventionally used trypsin. Dispase is gentler regarding enzymatic functions, does not damage cell membranes, and yields primary cells in greater numbers. All materials and reagents included in this protocol must be sterilized or purchased sterile. All materials and reagents must be opened in a previously sterilized (if possible, UV-sterilized) laminar flow (*see* **Note 1**).

The protocol for the initiation of an epidermal melanocyte culture from human skin sample (as described in the steps under Subheadings 3.1.1, 3.1.2, and 3.1.3) is performed over 2 days. The first stage (Subheading 3.1.1) should take between 35–40 min, while the second stage (Subheadings 3.1.2 and 3.1.3) requires 1–2 h, depending on the number of samples.

3.1.1 Preparation

1. If the skin sample is too large, then before the initial washing steps remove subcutaneous fat and/or cut the sample into smaller pieces (~1 cm wide) with a surgical scalpel. A small piece (1 cm^2) of skin is sufficient to yield a T-25 culture flask with cells.
2. Place the skin sample in a Petri dish and wash it in saline for a few times to remove remaining blood.
3. Then change washing saline to saline containing povidone iodine (3–4 droplets of Betadine®, dilution ratio 1:10) and submerge the sample in it for 5 min. After removal of the saline-iodine solution, repeat this step again.
4. After the iodine treatment, put the sample into a 6-well plate where the first well contains washing saline, while the second and third wells contain 70% ethanol and DPBS without Ca or Mg, respectively. Keep the sample in each well for a couple of seconds. After this, place the sample into the fourth and fifth well for 10 min each, both wells containing 10 μg/mL gentamicin in DPBS.

3.1.2 Separation of Dermis and Epidermis

1. Dissolve 10 mg/mL dispase II in 5 mL Hank's solution and place it into a Petri dish. Carefully submerge the skin sample with the epidermis-side-down and incubate at 4 °C overnight (minimum 18 h) (*see* **Note 2**).

2. On the following day, with the help of two sets of anatomical forceps, carefully separate the epidermis and the dermis and transfer the epidermis to 5 mL of 500 μg/mL trypsin/EDTA solution in a centrifuge tube. The epidermis is a browner and less lustrous tissue layer compared to the dermis. While peeling away the epidermis, one pair of forceps is used to hold the dermis and the other one to remove the epidermis. Avoid changing the two forceps during the process as this may significantly improve chances of dermal fibroblast contamination (*see* **Note 3**).
3. If necessary, use a sterile surgical scalpel to cut larger sized epidermal tissues into smaller pieces.
4. Avoid forceful removal of the epidermis in case it is attached firmly to the dermis, as this may risk fibroblast contamination from the dermis (*see* **Note 3**).

3.1.3 Initiation of Melanocyte Culture

1. Make sure the epidermis is not attached to the internal surface of the centrifuge tube but floats in the 500 μg/mL trypsin/EDTA solution.
2. Place the centrifuge tube in a 37 °C water bath and incubate the tissue for approximately 10–15 min (and not more than 15 min as longer incubation increases the risk of undesired cell death).
3. After incubation, resuspend the solution with the epidermis (pieces) in it by carefully pipetting up and down at a slow and steady speed until the tissue is totally dispersed into a cell suspension.
4. Add 5 mL of RPMI growth medium to the suspension and centrifuge the mixture at 450× *g*, 4 °C, for 15 min.
5. After centrifugation discard the supernatant and add 5 mL of fresh RPMI medium to resuspend the pellet.
6. To determine cellular density in the suspension, use a hemocytometer (or an automated cell counter of your choice).
7. Add 5 μL/mL 40 μM TPA, 5 μL/mL 40 nM CT, 2 μL/mL 5 μM ET1, and 2 μL/mL 5 μg/mL SCF to fresh RPMI medium (final concentrations are 200 nM TPA, 100 pM CT, 10 nM ET1, and 10 ng/mL SCF) and pipette it into a T-25 or T-75 flask. Recommended optimal plating density is at 1.5×10^5 cells per mL (*see* **Notes 4**, **5**, and **6**).
8. Place the initial culture in a humidified (90% humidity) incubator in the presence of 95% air and 5% CO_2 atmosphere, at 37 °C.

3.1.4 Maintaining Melanocyte Cultures

1. Change medium every 4–5 days and add mitogens in the same concentrations as in Subheading 3.1.3, **step** 7 (*see* **Note** 7).
2. When the culture reaches 80% confluence, remove the RPMI medium and wash cells in 2 mL of DPBS without Ca or Mg. Tilt the flask gently around to rinse away remnants from the medium. Repeat this once.
3. Add 2 mL of 125 μg/mL trypsin/EDTA solution onto the culture. Tilt the flask again.
4. To minimize trypsin exposure, check the detachment of melanocytes under light microscope. Cells detach relatively quickly, but slight lateral physical exertion (gentle knocking) on the side of the flask makes the process quicker.
5. When cells are detached, add 4 mL BSA supplemented fresh RPMI medium to the flask (this is four times the amount of the trypsin-EDTA solution), rinse, and collect the suspension in a centrifuge tube. Repeat this step to remove all cells. Instead of 8 mL BSA supplemented fresh RPMI medium, 2 mL BSA stock can also be used (1:1 ratio with trypsin-EDTA solution). Irrespective of the size of the flask (T-25 or T-75), the volume of solutions used for subculturing is the same.
6. Centrifuge the mixture at 115× *g*, 4 °C, for 10 min. After centrifugation, remove the supernatant and add 100 μL of fresh RPMI medium to resuspend the pellet.
7. Melanocytes can rapidly start to attach at room temperature, especially to plastic surface. Attachment can be inhibited at low temperatures. Therefore, the cell suspension is advised to be kept on ice (for up to approximately 1 h) to prevent loss of melanocytes.
8. Determine cell count with a hemocytometer.
9. Resuspend the suspension and plate melanocytes at the desired density in fresh RPMI medium supplemented with BSA and growth factors at the same concentration as in Subheading 3.1.3, **step** 7. According to their growth rate, $3–6 \times 10^4$ cells/mL is recommended as optimal plating density.
10. Place the new culture in a humidified (90% humidity) incubator in the presence of 95% air and 5% CO_2 atmosphere, at 37 °C (*see* **Notes 8**, **9**, and **10**).

3.1.5 Freezing and Thawing of Melanocytes

1. Freezing cells can take place at any convenient time after a conventional passage.
2. Trypsinize melanocytes as described in Subheading 3.1.4 and centrifuge at 115× *g*, 4 °C, for 10 min.

3. Make a mixture of 7.5% (v/v) DMSO in 50% RPMI medium supplemented with 42.5% FBS. Label cryotubes while centrifugation, to minimize exposure to DMSO.
4. Resuspend pellet in the cryogenic mixture (*see* **Note 11**).
5. Pipette 1 mL aliquots into cryotubes and gradually freeze at −80 °C for at least 24 h, before placing into liquid nitrogen. Do not keep cryotubes at −80 °C for longer periods.
6. If melanocytes are well pigmented, add 100–300 μM PTU for a whole passage before freezing. PTU inhibits melanin synthesis through inhibition of tyrosinase enzyme. Addition of PTU improves viability, as DMSO promotes release of toxic melanin intermediates into the cytoplasm (*see* **Note 12**).
7. When thawing cells, quickly place the cryotube into a 37 °C water bath.
8. Resuspend and transfer the mixture to a 50 mL centrifuge tube. Firstly, pipette 1 mL of warm RPMI medium dropwise into the cell suspension while carefully whirling around over 30 s, then pipette another 1 mL RPMI medium dropwise over 15 s, and finally add 18 mL RPMI medium dropwise over 30 s.
9. After this, centrifuge the cell suspension at 115× *g*, 4 °C, for 10 min, and then resuspend pellet in fresh RPMI medium containing growth factors and mitogens as described in Subheading 3.1.3, **step** 7.
10. Also supplement heavily pigmented melanocytes with 100–300 μM PTU for a couple of days after thawing.

3.2 Cell Surface Protein Isolation from Epidermal Melanocytes

In the following protocol, cell surface sialylated membrane glycoproteins are enriched in a two-step process. Firstly, aldehyde groups are introduced to sialic acid residues through oxidization by sodium-meta-periodate, and then the functionalized proteins are labeled by aminooxy biotin as the aldehyde groups covalently bind to biotin.

3.2.1 Surface Protein Labeling Using Aminooxy Biotin

1. Culture melanocytes in T-175 flask(s) until approx. 90% confluence (*see* **Note 13**). When the culture is ready for the surface protein labeling process, wash twice with 20 mL sterile PBS (pH 7.4), and then incubate the culture in 20 mL serum-free RPMI medium for 1 h at 37 °C in a CO_2 incubator to help the removal of serum proteins from the culture medium.
2. After incubation, remove the serum-free media and wash the flask(s) thrice with 10 mL ice-cold sterile PBS (pH 7.4). Perform all the following steps on ice.
3. Prepare the oxidation/biotinylation mixture in ice-cold PBS (pH 6.7) containing 1 mM sodium meta-periodate, 100 mM aminooxy biotin, and 10 mM aniline and add the mix into the

T-175 flask (4 mL of the mixture is required for one T-175 flask).

4. After this, place the flask(s) on a rocking platform and incubate the cultures while gentle rocking in the dark for 1 h at 4 °C.
5. After 1 h, quench the oxidation mixture by the addition of glycerol at a final concentration of 1 mM (400 μL to 4 mL oxidation/biotinylation mix). Incubate melanocyte cultures in the dark for 10 min at 4 °C.
6. Wash cells twice in 10 mL PBS (pH 7.4), then gently scrape off the surface of flask(s) by using a cell scraper. Collect sample (s) into centrifuge tube(s) and centrifuge at 1000 × *g* for 5 min.
7. After centrifugation, resuspend pellets in 10 mL PBS (pH 7.4) containing 1 mM $CaCl_2$ and 0.5 mM $MgCl_2$, then centrifuge again at 1000 × *g*, 4 °C for 5 min.

3.2.2 Biotinylated Cell Surface Protein Isolation and Enrichment Using NeutrAvidin Beads

1. After the second centrifugation in Subheading 3.2.1, **step** 7, resuspend the resulting cell pellets in 500 μL lysis buffer (containing 1% Triton X-100, 150 mM NaCl, 1× protease inhibitor, 5 mM iodoacetamide, 0.1 mg/mL PMSF, and 10 mM Tris–HCl; pH 7.6). After this, transfer the mixture from the centrifuge tube to a new 1.5-mL Eppendorf tube and incubate at 4 ° C for 30 min, vortexing in every 5 min.
2. In the following step, perform new rounds of centrifugation to help remove pelleted cell debris (containing cell organelles, cellular membranes, and nuclei). First, centrifuge at 2800× *g*, 4 °C, for 10 min, and then centrifuge twice, both at 16,000× *g*, 4 °C, for 15–15 min. Always pipette the supernatant to a new 1.5-mL Eppendorf tube and discard the pellet.
3. Meanwhile, add 150 μL aliquots of NeutrAvidin agarose beads to Snap Cap spin columns and centrifuge at 500× *g* for 30 s. Then rinse 3 times with 500 μL PBS (pH 7.4) and centrifuge after each wash at 500× *g* for 30 s.
4. To isolate labeled glycoproteins, pipette the supernatant from Subheading 3.2.2, **step 2**, to the previously prepared Snap Cap spin column (containing NeutrAvidin agarose beads) and incubate with the supernatant for 2 h at 4 °C with continuous rocking, with a rubber cap placed to the bottom of the columns (*see* **Note 14**).
5. To eliminate nonspecifically bound contaminant proteins, wash multiple (15) times with 600 μL of the solutions specified below followed by centrifugation at 1000× *g* for 1 min. Remove the rubber cap before spinning.
6. Wash beads 15 times with lysis buffer, followed by 15 washes with 0.5% (w/v) SDS in PBS (pH 7.4).

7. Incubate the beads for 20 min at room temperature in 0.5% (w/v) SDS, 100 mM dithiothreitol in PBS (pH 7.4; prepare fresh), then centrifuge again at 1000× *g* for 1 min.
8. Perform further washing 20 times with UC buffer, containing 6 M urea, 100 mM Tris–HCl (pH 8.5).
9. This is followed by alkylation for 20 min at room temperature with UC buffer containing 50 mM iodoacetamide.
10. Wash beads again 20 times with 500 μL of the following solutions: UC buffer, 5 M NaCl, 100 mM Na_2CO_3, PBS (pH 7.4) and then HPLC grade water and centrifuge 500× *g* for 30 s after each step.
11. After the washing steps, plug a new rubber cap to the bottom of the column and place it into a new 1.5-mL Eppendorf tube.
12. Resuspend the beads in 400 μL HPLC water containing 50 mM NH_4HCO_3 and 5 μg Trypsin Gold. This step allows biotinylated cell surface proteins to be digested on-beads overnight at 37 °C.
13. On the following day, remove the rubber cap from the bottom of Snap Cap spin column(s) and collect tryptic peptides via centrifugation at 1000× *g* for 1 min.
14. Place the Snap Cap spin column into a new 1.5-mL Eppendorf tube and rinse the beads with 200 μL of 50 mM NH_4HCO_3 (AMBIC) to pool the remaining tryptic fractions (1000× *g* for 1 min). Transfer the flow-through into the first (collecting) Eppendorf tube.
15. Evaporate solvents in a vacuum centrifuge in approximately 2 h, then store samples at −80 °C.

3.3 Analysis of Peptides by Mass Spectrometry

1. To the tube of dried peptides from Subheading 3.2.2, **step 15**, add 150 μL of 10% acetonitrile 0.1% formic acid, and shake on a thermomixer at 5 °C for 20 min, then add 150 μL of 0.1% formic acid and vortex thoroughly. This will result in a final concentration of 5% acetonitrile 0.1% formic acid (**Notes 16** and **17**).
2. Remove any remaining debris or non-resuspended particulates by centrifuging at 12,000 RCF for 2 min (13,400 rpm, max speed, in Eppendorf MiniSpin). Transfer to LC compatible vial, taking care not to disturb any pellet if visible.
3. Load LC vial into autosampler and inject an appropriate mass of peptides onto the column for MS (*see* **Note 17**).
4. Inject 3 μL of each reconstituted sample (*see* **Note 17**) from a Waters M-Class LC system connected to a Sciex 7600 Zeno-TOF Optiflow source via a Phenomenex Kinetex XB-C_{18} 2.6 μm 150 × 0.3 mm analytical column in Direct Inject

mode with the column oven set at 30 °C. Different instrument sources and LC systems will necessitate different column dimensions and flow rates, optimize for your particular system.

5. For the 150 × 0.3 mm column, an appropriate method would be to run the samples with the following gradient: 3–35% mobile phase B over 21 min at 10 μL/min, increase flow rate to 12 μL/min to wash at 80% B at 22.5 min to 24.5 min and re-equilibrate to 3% at 25 min for a total run time of 27 min.
6. For a qualitative analysis, run the MS in positive ion mode with Information-Dependent Acquisition (IDA/DDA) mode with CID fragmentation utilizing a 100 ms TOFMS scan followed by the Top 45 ions in MSMS each of 10 ms duration from m/z 100 to 2000, zeno pulsing on giving to total cycle time of 795 ms. The full methodological parameters are provided in Supplementary information (*MS DDA Method Parameters.txt*). Alternatively for a quantitative analysis, a Data-Independent Acquisition method (DIA/SWATH or equivalent) could be applied here (*see* Chap. 5 in this volume).
7. The ZenoTOF 7600 generates three "raw" files for each acquisition with the same filename prefix, .wiff, .wiff2, and .wiff.scan (*see* **Note 18**).
8. Download the latest (or most appropriate) proteome FASTA file from UniProt (https://www.uniprot.org) for Human SwissProt .fasta proteome file [taxonomy 9606] in uncompressed format, canonical plus isoforms.
9. Process the data from the mass spectrometer using an appropriate search engine for Data-Dependent Acquisition (DDA) data files, e.g., PEAKS Studio 12 (Bioinformatics solutions, Waterloo, Canada [15]) using the PEAKS DB (in depth de novo assisted search) with the appropriate error tolerance for the MS instrument (e.g., for the ZenoTOF 7600 10 ppm MS tolerance and 0.05 Da MSMS). Select the .fasta proteome downloaded in **step 8**. As the samples were alkylated, set a fixed modification of carbamidomethylation [+57.02] and variable modification for oxidation of methionine [+15.99]. At this step, you can also include any particular variable posttranslational modifications (PTMs) you are interested in. PEAKS Studio can also carry out a PTM search for 300+ additional modifications, select this "find unspecified PTMs with PEAKS PTM" as an additional step in the setup. Full method parameters can be found in the supplementary file "PEAKS Studio search parameters.txt" (*see* **Note 15**).
10. Export the protein list as a csv file after analysis. Annotate proteins based on Gene Ontology (GO) information downloaded from the UniProt database (www.uniprot.org) (*see* **Note 15**). Infer cell surface/plasma membrane proteins according to the above annotation.

4 Notes

1. Follow strictly laboratory safety rules and wear lab safety clothes, gloves, and hairnet. Spray 70% ethanol or other sterilizing reagent on the gloves after each step of the protocol that requires the user to move out of the laminar flow. Make sure that all tools (scissors, scalpels, forceps) are sterilized before use by dry heat sterilizer.
2. Dispase, extracted from a bacterium called *Paenibacillus polymyxa*, is a neutral protease specifically acting through the cleavage of fibronectin and type VI collagen [16]. Compared to trypsin, dispase is a more useful enzyme for the dermo-epidermal separation as it is characterized by low cytotoxicity and is independent from skin sample thickness [17].
3. The use of two pairs of forceps during dermo-epidermal separation avoids fibroblast contamination during melanocyte culture initiation. Do not force the removal of the epidermis as fibroblasts from the dermis can attach to the separated tissue. Also, the forceps gripping the epidermis should be only used to transfer the epidermis to the trypsin solution, as the other tool may cause contamination.
4. The following growth factors are used for culturing human epidermal melanocytes: 12-O-tetradecanoylphorbol 13-acetate (TPA), also known as phorbol 12-myristate 13-acetate (PMA); cholera toxin (CT); human stem cell factor (SCF); and endothelin 1 (ET1). TPA and CT were identified as useful mitogens in 1982, when the first reproducible melanocyte culture was documented [18]. TPA is a phorbol ester that activates PKC signaling, resulting in the phosphorylation of several transcription factors, and eventually leading to proliferation and differentiation of cultured melanocytes [19]. CT mimics the effects of melanocyte stimulating hormone (MSH) resulting in specific activation of adenylate cyclase and increase of intracellular cyclic AMP concentrations, enhancing proliferation, differentiation, and pigment synthesis [20].
5. ET1 and SCF synergistically stimulate melanocyte (and melanoblast) proliferation if a cyclic AMP inducer, such as CT, is present in the culture medium. This finding was observed in melanocyte-keratinocyte co-cultures (a close representation of the epidermis), as keratinocytes (feeder cells) produce tumor necrosis factor (TNF)-α and interleukin (IL)-1α, that promotes ET1 and SCF release by melanocytes [21]. Without feeder cells or in the absence of ET1 and SCF, melanocytes show less proliferative activity and increased apoptotic ratio [22, 23].
6. Recipes for mitogens:

- CT:

  According to the manufacturer's instructions, dissolve 0.5 mg CT (Sigma) in 500 μL HPLC water to make a 20 μM stock solution. Store at 4 °C. Dilute with 0.1 mg/ml BSA (in DPBS without Ca or Mg; filter-sterilized) to make a 40 nM working solution (1:500 ratio). Store aliquots at 4 °C.

- TPA:

  To generate a 2 mM stock solution, dissolve TPA (Sigma) in pure ethanol. Aliquots are recommended to be stored at −20 °C. Dilute with 0.1 mg/mL BSA (in DPBS without Ca or Mg; filter-sterilized) to make a 40 μM working solution. Store in aliquots at −20 °C for an extended period, or for 2 weeks at 4 °C. TPA is a gel and is difficult to see, dissolve it carefully. TPA is light sensitive. Prepare stock solutions in a dim environment and cover aliquot vials in aluminum foil.

- ET1:

  Make a 50 μM stock solution of ET1 (Bachem) by dissolving in 0.1 mg/mL BSA (in DPBS without Ca or Mg; filter-sterilized). Store stock at −20 °C. When making a 5 μM working solution, dilute stock solution in 0.1 mg/mL BSA (in DPBS without Ca or Mg; filter-sterilized). Store aliquots at −20 °C or up to 2 weeks at 4 °C. Do not refreeze a third time.

- SCF:

  Prepare stock solution of human SCF (Invitrogen) according to the manufacturer's instructions. Dilute to 5 μg/mL with 0.1 mg/mL BSA (in DPBS without Ca or Mg; filter-sterilized). Store stock and aliquots at −80 °C or for 2 weeks at 4 °C after first use of aliquots. Do not refreeze a third time.

7. At the beginning, keratinocytes are alive in the culture but fail to grow in RPMI-1640 medium supplemented with mitogens. Keratinocytes detach from the bottom of the flask prior to the need of subculturing, and any adherent cells that remain will not proliferate upon further passaging.

8. The number of cells may decrease in the early stages of culture growth. Theoretically, reducing the medium volume by 50% concentrates autocrine growth factors produced by melanocytes and normalizes cell density. Medium volume can be increased to normal once melanocyte density is satisfactory.

9. The culture will become confluent after approximately 2 weeks. Based on optimal plating density, the subsequent passages will take between 1.5–2 weeks to become confluent. Population

doubling depends on the age of the skin donor, but as time progresses the frequency of passages decreases for each individual culture. Melanocytes close to senescence may require 6 weeks between passages. Based on our experience, a primary melanocyte culture needs 6–12 months to reach full senescence (this may vary according to donor age and other parameters).

10. If present, in subsequent passages fibroblasts grow rapidly and can take over a culture. To remove fibroblasts, add 75/150 μM geneticin to the RPMI-1640 medium for the period of at least two passages. In this concentration, geneticin is not toxic to melanocytes. To aid the effects of geneticin, exclude the use of ET1 as it can promote the proliferation of fibroblasts.
11. The recommended total density of melanocytes is $0.5–1 \times 10^6$ cells/mL in a single-tube cryogenic mixture.
12. Recipe for making PTU solution: Make a 100 mM stock solution by dissolving PTU in pure ethanol. Store at −20 °C or for a month at 4 °C.
13. The protocol described in Subheading 3.2 is also applicable to melanoma cell lines. Various types of melanoma cell lines are available for biotin labeling and avidin-bead based separation of cell surface proteins. We culture melanoma cells in RPMI-1640 culture medium supplemented with 10% FBS, 4.1 g/L glucose, 2 mmol/L L-glutamine, penicillin (100 units/mL), and streptomycin (100 μg/mL). Melanoma cells are maintained at 37 °C, in a humidified (90% humidity) incubator in the presence of 95% air and 5% CO_2 atmosphere [24, 25].
14. This step is based on the specificity of the avidin-biotin binding, meaning that the biotin-labeled surface proteins attach to the NeutrAvidin agarose beads, while the non-labeled contaminant proteins remain unattached.
15. Depending on what is available in your laboratory there are many different software packages for searching DDA bottom-up proteomics data. The example here uses PEAKS Studio, a commercial software but there are numerous free and commercial software packages utilizing different search engines available. The fundamental steps for analysis remain the same: Specify the raw mass spectrometry data files (and if necessary convert them to a universal .mzml format using a tool such as msconvert [26]), select an appropriate proteome file (e.g., human from Uniprot.org), specify appropriate instrument specific parameters based on mass accuracy, fragmentation type, and posttranslational modifications (mass shift).
16. The initial higher 10% acetonitrile concentration increases recovery of the dried protein from the tube. The final volume/concentration may be adjusted to suit instrument sensitivity and injection volume.

17. The concentration of digested peptides will need to be optimized for the mass spectrometer and LC method used, so a suitable injection volume and mass can be analyzed. For the ZenoTOF 7600 in IDA acquisition mode (with zenoSWATH selected), 500–1500 ng peptides is optimal on a 27 min LC run (21 min gradient). For an analysis in zenoSWATH acquisition mode (DIA) >1000 ng of peptides injected may cause the detector to be overloaded, optimize injection mass accordingly if this is the case.
18. For downstream data processing, keep the prefix for all three datafiles identical, if you need to rename any samples then you must change all the related file prefixes to match. For third party software the .wiff (or .wiff2) and .wiff.scan pair must both be present in the same folder during analysis, while the .wiff file is usually the one to select for processing. Some software packages may not be able to read Sciex .wiff or .wiff2 files natively and may need installation of additional software to read them. Usually, an alternative is to convert the raw Sciex files to .mzml format (*see* **Note 15**).

Acknowledgments

CM was supported by the Young Researcher Excellence Programme (grant number: FK-134304) of the National Research, Development and Innovation Office, Hungary. CM was also supported by the Bolyai János Research Fellowship of the Hungarian Academy of Sciences. This research was supported by the UNKP-23-5-DE-487 New National Excellence Program of the Ministry for Culture and Innovation from the source of the National Research, Development and Innovation Fund, Hungary. CM also acknowledge support from the European Cooperation in Science and Technology (COST) Association Action CA21110—Building an open European Network on OsteoArthritis research (NetwOArk); https://www.cost.eu/actions/CA21110/).

Conflict of Interest The authors declare that they have no conflicts of interest. The authors wrote this paper within the scope of their academic and research positions. None of the authors has any relationships that could be construed as biased or inappropriate. Funding bodies were not involved in the study design, data collection, analysis, or interpretation. The decision to submit this paper for publication was not influenced by any funding bodies.

References

1. Forsea AM (2020) Melanoma epidemiology and early detection in Europe: diversity and disparities. Dermatol Pract Concept 10(3): e2020033. https://doi.org/10.5826/dpc.1003a33
2. Koczkodaj P et al (2023) Melanoma mortality trends in 28 European countries: a retrospective analysis for the years 1960-2020. Cancers (Basel) 15(5). https://doi.org/10.3390/cancers15051514
3. Arnold M et al (2022) Global burden of cutaneous melanoma in 2020 and projections to 2040. JAMA Dermatol 158(5):495–503. https://doi.org/10.1001/jamadermatol.2022.0160
4. Boutros A et al (2024) The treatment of advanced melanoma: current approaches and new challenges. Crit Rev Oncol Hematol 196: 104276. https://doi.org/10.1016/j.critrevonc.2024.104276
5. Pessoa J et al (2022) Editorial: altered expression of proteins in cancer: function and potential therapeutic targets. Front Oncol 12: 949139. https://doi.org/10.3389/fonc.2022.949139
6. Takechi Y et al (1996) A melanosomal membrane protein is a cell surface target for melanoma therapy. Clin Cancer Res 2(11): 1837–1842
7. Qiu H, Wang Y (2008) Quantitative analysis of surface plasma membrane proteins of primary and metastatic melanoma cells. J Proteome Res 7(5):1904–1915. https://doi.org/10.1021/pr700651b
8. Zeng Y et al (2009) High-efficiency labeling of sialylated glycoproteins on living cells. Nat Methods 6(3):207–209. https://doi.org/10.1038/nmeth.1305
9. Betancourt LH et al (2021) The human melanoma proteome atlas-complementing the melanoma transcriptome. Clin Transl Med 11(7): e451. https://doi.org/10.1002/ctm2.451
10. Marie KL et al (2020) Melanoblast transcriptome analysis reveals pathways promoting melanoma metastasis. Nat Commun 11(1):333. https://doi.org/10.1038/s41467-019-14085-2
11. Reemann P et al (2014) Melanocytes in the skin–comparative whole transcriptome analysis of main skin cell types. PLoS One 9(12): e115717. https://doi.org/10.1371/journal.pone.0115717
12. Xie J et al (2023) Transcriptome analysis reveals tumor antigen and immune subtypes of melanoma. Oncol Res 31(3):389–403. https://doi.org/10.32604/or.2023.029274
13. Godwin LS et al (2014) Isolation, culture, and transfection of melanocytes. Curr Protoc Cell Biol 63:1.8.1–1.820. https://doi.org/10.1002/0471143030.cb0108s63
14. Hajdu T et al (2021) Pituitary adenylate cyclase activating polypeptide has inhibitory effects on melanoma cell proliferation and migration in vitro. Front Oncol 11:681603. https://doi.org/10.3389/fonc.2021.681603
15. Tran NH et al (2019) Deep learning enables de novo peptide sequencing from data-independent-acquisition mass spectrometry. Nat Methods 16(1):63–66. https://doi.org/10.1038/s41592-018-0260-3
16. Stenn KS et al (1989) Dispase, a neutral protease from bacillus polymyxa, is a powerful fibronectinase and type IV collagenase. J Invest Dermatol 93(2):287–290. https://doi.org/10.1111/1523-1747.ep12277593
17. Kitano Y, Okada N (1983) Separation of the epidermal sheet by dispase. Br J Dermatol 108(5):555–560. https://doi.org/10.1111/j.1365-2133.1983.tb01056.x
18. Eisinger M, Marko O (1982) Selective proliferation of normal human melanocytes in vitro in the presence of phorbol ester and cholera toxin. Proc Natl Acad Sci USA 79(6): 2018–2022. https://doi.org/10.1073/pnas.79.6.2018
19. Yamasaki T et al (2009) Phosphorylation of activation transcription Factor-2 at serine 121 by protein kinase C controls c-Jun-mediated activation of transcription. J Biol Chem 284(13):8567–8581. https://doi.org/10.1074/jbc.M808719200
20. O'Keefe E, Cuatrecasas P (1974) Cholera toxin mimics melanocyte stimulating hormone in inducing differentiation in melanoma cells. Proc Natl Acad Sci USA 71(6):2500–2504. https://doi.org/10.1073/pnas.71.6.2500
21. Hachiya A et al (2004) Biphasic expression of two paracrine melanogenic cytokines, stem cell factor and endothelin-1, in ultraviolet B-induced human melanogenesis. Am J Pathol 165(6):2099–2109. https://doi.org/10.1016/S0002-9440(10)63260-9
22. Hirobe T, Shinpo T, Higuchi K, Sano T (2010) Life cycle of human melanocytes is regulated by endothelin-1 and stem cell factor in synergy with cyclic AMP and basic fibroblast growth factor. J Dermatol Sci 57(2):123–131. https://doi.org/10.1016/j.jdermsci.2009.11.006

23. Sviderskaya EV et al (2003) p16/cyclin-dependent kinase inhibitor 2A deficiency in human melanocyte senescence, apoptosis, and immortalization: possible implications for melanoma progression. J Natl Cancer Inst 95(10): 723–732. https://doi.org/10.1093/jnci/95.10.723
24. Hajdu T et al (2018) NR1 and NR3B composed Intranuclear N-methyl-d-aspartate receptor complexes in human melanoma cells. Int J Mol Sci 19(7). https://doi.org/10.3390/ijms19071929
25. Katona E et al (2016) PP2B and ERK1/2 regulate hyaluronan synthesis of HT168 and WM35 human melanoma cell lines. Int J Oncol 48(3):983–997. https://doi.org/10.3892/ijo.2015.3313
26. Adusumilli R, Mallick P (2017) Data conversion with ProteoWizard msConvert. Methods Mol Biol 1550:339–368. https://doi.org/10.1007/978-1-4939-6747-6_23

Chapter 7

Dissecting the Heterogeneity of Senescent Cells Through CITE-seq Integration

Kotb Abdelmohsen, Krystyna Mazan-Mamczarz, Dimitrios Tsitsipatis, Martina Rossi, Rachel B. Munk, and Myriam Gorospe

Abstract

Cellular senescence, a state of persistent growth arrest following cell damage, is associated with aging and age-related diseases. Understanding cell heterogeneity within senescent populations is crucial for developing therapies to mitigate senescence-associated pathologies. The protocol described here outlines an integrated approach to exploit the presence of cell surface proteins on subsets of senescent cells to study their heterogeneity at the single-cell level. After identifying senescence-associated surface proteins by mass spectrometry (MS) and then performing cellular indexing of transcriptomes and epitopes sequencing (CITE-seq) single-cell analysis, we were able to identify unique transcriptomic programs associated with specific surface protein markers expressed in some senescent cells but not in others. We illustrate the utility of this approach by investigating the complex heterogeneity of senescent cell populations. However, this methodology can be applied to other biological scenarios where cells with unique transcriptomic profiles can be studied individually, thanks to the presence of specific cell surface proteins that distinguish them from other cells within the same population.

Key words Single-cell transcriptome, Cell cycle, Senescence, CITE-seq, Surface proteins, Transcriptome, Proteome, Multiomics

1 Introduction

Senescence is a state of persistent growth arrest in cells that have encountered damage that is not repaired but is insufficient to cause cell death [1]. Senescent cells are characterized by the secretion of a variety of pro-inflammatory cytokines, chemokines, and matrix metalloproteinases; this trait is known as the senescence-associated secretory phenotype (SASP) [2]. Although the transient generation of senescent cells is beneficial during developmental processes and wound repair, senescent cells progressively accumulate in the body

Kotb Abdelmohsen and Krystyna Mazan-Mamczarz authors contributed equally to this work.

Csaba Matta et al. (eds.), *The Surfaceome: Methods and Protocols*, Methods in Molecular Biology, vol. 2908,
https://doi.org/10.1007/978-1-0716-4434-8_7,

during aging, with a detrimental effect, as their removal was found to improve health and extend lifespan in animal models [3]. Senescent cells can be classified into different subtypes based on the damaging stimulus, the tissue of origin, the length of time elapsed since the damage, and the SASP profile, among other factors [4]. This heterogeneity has hampered efforts to identify markers to selectively eliminate senescent cells, which is critical for developing targeted therapies for age-related diseases.

To better understand the molecular mechanisms underlying senescence and its heterogeneity, this protocol outlines an integrated approach that combines mapping of senescence-associated surface proteins [5] followed by cellular indexing of transcriptomes and epitopes by sequencing (CITE-seq) analysis [6]. Surface proteins are pivotal markers of cell identity and function whose expression can be affected by senescence and potentially targeted by senotherapeutics, a growing catalog of interventions that selectively eliminate senescent cells [7]. CITE-seq enables the simultaneous profiling of the transcriptomes and surface protein markers in individual senescent cells, providing a comprehensive view of the molecular changes associated with senescence (Figs. 1

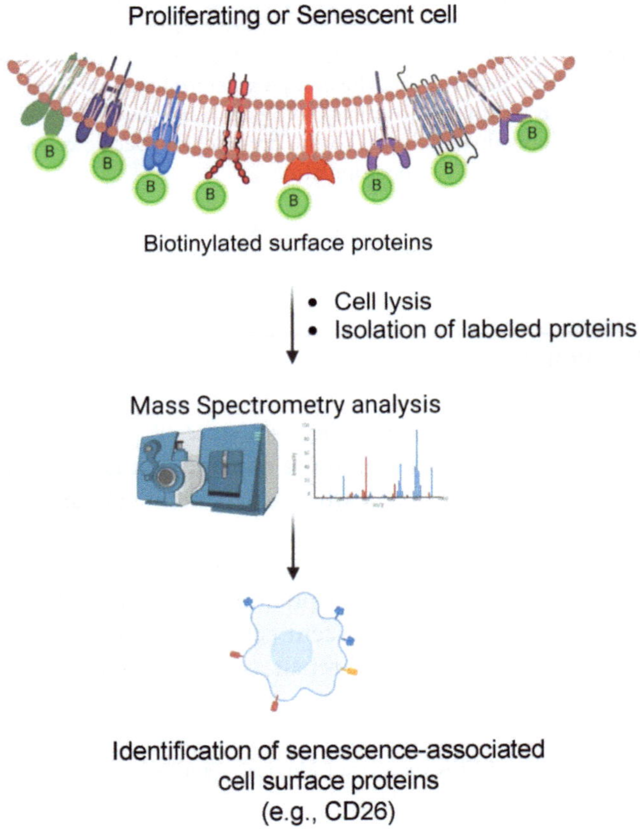

Fig. 1 Schematic representation of the workflow for identifying senescence-associated surface proteins

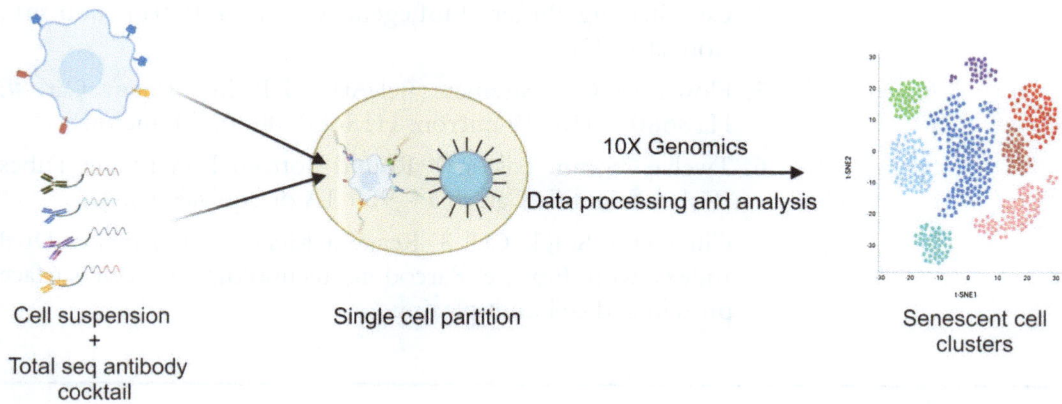

Fig. 2 Workflow of cellular indexing of transcriptomes and epitopes by sequencing (CITE-seq)

and 2). Furthermore, CITE-seq can be used to identify specific subpopulations of senescent cells to investigate the interactions between senescent cells and their microenvironment. Jointly analyzing the RNAs and cell surface protein markers of specific senescent cells offers unique insight into the heterogeneity of senescent cell subpopulations, enabling the development of strategies to target specific subclasses of senescent cells for therapeutic benefit.

2 Materials

2.1 Cell Culture

1. Primary human WI-38 fibroblasts, derived from fetal lung, are used. One to two million cells (proliferating or senescent) are required. Senescent cells can be prepared in different ways; here we used ionizing radiation (IR)-induced senescence.
2. Dulbecco's modified Eagle medium (DMEM).
3. 20% Fetal Bovine Serum (FBS).
4. 1% antibiotics.
5. 1% nonessential amino acids.

2.2 Materials for Cell Surface Protein Isolation and CITE-seq

1. Pierce Cell Surface Biotinylation and Isolation Kit (Thermo Scientific) which includes Sulfo-NHS-SS-Biotin, Tris Buffered Saline (TBS), Phosphate Buffered Saline (PBS), NeutrAvidin Agarose, Dithiothreitol (DTT), Lysis Buffer, Wash Buffer, Elution Buffer, and spin columns.
2. TotalSeq™-B antibodies (BioLegend), barcoded antibodies targeting surface proteins highly abundant in senescent cells identified by the MS analysis.
3. Human TruStain FcX (Fc Receptor Blocking Solution) (BioLegend Cat. #: 422301) or TruStain FcX™ PLUS (anti-mouse CD 16/32) (BioLegend Cat. #: 156603/156604).

4. Cell Staining Buffer (BioLegend Cat. #: 420201) (500 mL, store at 4 °C).
5. Flowmi™ Cell Strainer (Bel-Art, H-B Instrument Cat. #: H13680–0040, 40 micron; H13680–0070, 70 micron).
6. Twelve 75-mm Falcon™ Round-Bottom Polystyrene Tubes (Fisher Scientific Cat. #: 14–959-1A or equivalent).
7. Chromium Single Cell 3' Reagent Kits (v3.1 Chemistry Dual Index) with Feature Barcoding technology for cell surface protein and cell multiplexing.

3 Methods

This protocol enables the simultaneous detection of surface proteins (Fig. 1) and transcriptomes present in each individual cell, offering a comprehensive view of the cellular state (Fig. 2). Generally, this approach is completed by whole-cell transcriptomic analysis using TotalSeq™ and 10× Genomics).

3.1 Senescence Induction

Cellular senescence can be induced through various methods depending on the specific research question, cell type, and model of senescence. Below are common ways to induce senescence:

1. Replicative senescence (RS): repeatedly culturing cells until they reach the Hayflick limit, a high population doubling level (PDL) beyond which they cease to divide, is a well-established way to render cells senescent [8].
2. Nutrient depletion: removing essential growth factors from the culture medium can induce senescence in certain cell types, especially in primary cultures [9].
3. Exposure to ionizing radiation (IR): exposing cells to X-rays or gamma rays, causing DNA double-strand breaks, is a common way to trigger senescence [10].
4. Oncogene-induced senescence (OIS): expressing activated oncogenes (e.g., RAS, BRAF, AKT, E2F1, or cyclin E) or inactivated tumor suppressor genes (e.g., PTEN and NF1) can induce senescence [11, 12].
5. Senescence triggered by exposure to genotoxic drugs: treating cells with sublethal doses of chemotherapeutic drugs (e.g., doxorubicin, etoposide, camptothecin) induces DNA damage leading to senescence [13].
6. Exposure to oxidants: treating cells with hydrogen peroxide or other agents generating reactive oxygen species (ROS) can

damage cellular components and DNA, triggering senescence [14, 15].

7. Treatment with chromatin-altering drugs: exposure to DNA methyltransferase inhibitors (e.g., 5-azacytidine) or inhibitors of histone deacetylases (HDACs; e.g., trichostatin A) alters chromatin structure and epigenetic marks, changing transcriptional patterns that can lead to senescence [13].

In this protocol, we used both replicative exhaustion and ionizing radiation to induce senescence. For replicative exhaustion, we serially passaged the cells from population doubling level (PDL 20–24) until they reached PDL 50–56. This method of senescence induction is based on the fact that normal human fibroblasts in culture have a limited proliferation capacity, and they eventually become replicatively senescent (RS). For IR-induced senescence, we exposed the cells to 10 Gy to induce DNA double-strand breaks and cultured the cells for an additional 10 days until they reached senescence.

3.2 Validation of Senescence

There are several methods to confirm a senescent phenotype, including the assessment of general senescence-associated traits.

1. The activity of the lysosomal enzyme senescence-associated β-galactosidase (SA-β-gal) is a widely used marker of senescence, as it is activated in senescent cells and can be detected by staining with a tractable substrate such as X-gal [16].
2. The absence of proliferation can be measured by monitoring de novo DNA synthesis (e.g., by measuring BrdU incorporation) or levels of Ki-67, a marker of proliferation. Measuring the levels of general senescence-associated markers, such as the cyclin-dependent kinase (CDK) inhibitors p21 (CDKN1A) and p16 (CDKN2A), can also be used to validate senescence [17]. These senescence markers can be detected by monitoring mRNA levels (e.g., by reverse transcription followed by quantitative polymerase chain reaction (RT-qPCR) analysis) or protein levels (e.g., by western blot or immunofluorescence analyses).
3. The levels of pro-inflammatory SASP factors, such as interleukin (IL)-6 (IL6) and IL8, as well as matrix metalloproteinases, can be assessed by measuring the corresponding proteins or mRNAs by RT-qPCR, western blot analysis, or enzyme linked immunosorbent assay (ELISA) [18].

3.3 Cell Surface Protein Biotinylation

1. For each sample, prepare at least two 15-cm dishes of >85% confluent cells or 9×10^6 to 10^7 cells.
2. Remove the media and wash cells with 20 mL of PBS per plate.

3. Dissolve the contents of each vial of Sulfo-NHS-SS-Biotin in 24 mL of PBS. Add 10 mL of this solution to the cells.
4. Incubate cells for 10 min at room temperature with gentle agitation, ensuring that the biotin solution completely covers the cells.
5. Remove labeling solution from cells and wash cells twice with 20 mL ice-cold TBS.
6. Add 10 mL ice-cold TBS to the cells and scrape the cells into the TBS solution using the flat side of the scraper. Transfer the scraped cells into a 50-mL conical tube on ice.
7. Centrifuge cells at 500× *g* for 3 min at 4 °C and discard supernatant.
8. To lyse the cells, add protease and phosphatase inhibitors to 500 μL of ice-cold Lysis Buffer (Thermo Scientific, A44390), and resuspend the cell pellet.
9. Pipette up and down 20 times to suspend the cells.
10. Transfer cells in the lysis solution to a 1.5-mL microcentrifuge tube.
11. Incubate cells 30 min on ice. Vortex for 5 s at the beginning and end of the incubation.
12. Centrifuge cell lysate at 15,000× *g* for 5 min at 4 °C.
13. Transfer clarified supernatant to a new 1.5-mL tube.
14. To isolate labeled proteins, first twist off the bottom tip from one column and insert column into a collection tube.
15. Gently swirl the bottle of NeutrAvidin Agarose to obtain an even suspension. Using scissors, cut off the end of a p1000 pipet tip to widen the borehole. Add 250 μL of the NeutrAvidin Agarose slurry to the column and screw on the top cap, slightly loosened.
16. Centrifuge 1 min at 1000× *g* and discard storage solution.
17. Apply the white bottom plug to the column, add the clarified cell lysate to the resin, and then apply top cap to column.
18. Incubate for 30 min at room temperature with end-over-end mixing on a rotator.
19. Loosen the top cap and then remove the bottom plug from the column. Place the column in the collection tube.
20. Centrifuge the column for 1 min at 1000× *g* and discard flow-through.
21. Return the column to the collection tube and add 500 μL Wash Buffer (Thermo Scientific, A44390). Cap the column and mix by inverting two to three times. Centrifuge for 1 min at 1000× *g*. Discard the wash and remove top cap. Repeat this step for three to four times.

22. Replace bottom cap on column.
23. Mix 225 μL Elution Buffer (Thermo Scientific, A44390) and 25 μL DTT stock solution. Add 200 μL of Elution Buffer to the resin and cap the column. Incubate the reaction for 30 min at room temperature on a rotator.
24. Loosen the top cap, remove the bottom cap, and place into a collection tube.
25. Centrifuge column for 2 min at 1000× *g*.
26. The eluate can be assessed by western blot analysis to confirm the presence of surface proteins (*see* **Note 1**) or directly used for mass spectrometry (MS) analysis.

3.4 Labeling Cells with Barcode Oligonucleotide-Conjugated Antibodies for CITE-seq Analysis

1. Start with a single-cell suspension from tissues or cell culture. It is critical to obtain a high-quality, viable, single-cell suspension free of clumps.
2. Resuspend 1–2 million cells in 50 μL Cell Staining Buffer with 10–15% FBS.
3. Add 5 μL of Human TruStain FcX™ Fc Blocking reagent or 0.5 μL of TruStain FcX™ PLUS (anti-mouse CD 16/32) blocking reagent and incubate for 10 min at 4 °C.
4. Prepare an antibody pool using 1 μg of each TotalSeq™ antibody. Centrifuge the antibody pool at 14,000× *g* at 4 °C for 10 min before adding to the cells (*see* **Note 2**).
5. After centrifugation, carefully pipette out the liquid containing the antibody cocktail, avoiding contact with the bottom of the tube, and add the TotalSeq™ antibody cocktail to the cell suspension.
6. Incubate for 30 min at 4 °C.
7. Wash with 1 mL of Cell Staining Buffer, spin at 400× *g* for 5 min, 4 °C.
8. Resuspend cells in 1 mL of PBS containing 5 mM EDTA for a final concentration of 1×10^6 cells/mL.
9. Filter cells through a 40 or 70 μm Flowmi™ cell strainer or these can be sorted directly.
10. Stain cells with 2 μg/mL of propidium iodide for 10 min at 25 °C in the dark.
11. Sort 16,000 viable cells using WOLF G2 (NanoCellect Biomedical, Inc.) at a flow rate of 500–800 events/s.
12. Spin down sorted cells at 500× *g* for 15 min.
13. Remove supernatant until reaching a final volume of 16.5 μl.

    Critical point: It is crucial to keep the precise volume as it can affect the following steps.

3.5 Library Preparation and Sequencing

1. Encapsulation and barcoding. Load cells in single-cell suspension onto the 10× Genomics Chromium Controller following the manufacturer's detailed protocol. This instrument facilitates the encapsulation of individual cells into nanoliter-scale gel beads, a process that generates Gel Beads-in-Emulsion (GEMs). Each GEM encloses a single cell within a distinct oil droplet, ensuring isolated processing during reverse transcription.
2. cDNA synthesis and antibody capture. Inside the GEMs, cells are lysed, allowing mRNA to be reverse-transcribed to generate barcoded cDNA. Simultaneously, barcoded antibodies bound to cell surface proteins are also captured. These uniquely tagged antibodies enable protein expression profiling alongside gene expression, providing a comprehensive single-cell information that integrates both transcriptomic and proteomic data.
3. Library preparation. Following the breakdown of the GEMs and recovery of the barcoded material (cDNA and antibody tags), the library preparation proceeds through several key steps:
   - cDNA and antibody-tag amplification: This step amplifies the combined genetic material to prepare it for sequencing.
   - Fragmentation, end repair, and A-tailing: These steps prepare the ends of the amplified products for subsequent adapter ligation.
   - Adapter ligation: This step ligates sample index and Illumina read sequence adapters to the prepared fragments.
   - PCR amplification: This step enriches the adapter-ligated fragments through PCR, making them suitable for high-throughput sequencing.

3.6 Computational Analysis of CITE-seq Data

The computational analysis of CITE-seq data shares many steps used for processing single-cell RNA sequencing (scRNA-seq) data [19]. The analysis is designed to identify cell barcodes and demultiplex sequencing reads, then align them to appropriate references and create matrices with counts for each feature in each cell. The count matrices are further filtered using quality control (QC) parameters so as to retain for analysis only a high-quality cell population. Then, the selected dataset is normalized, and the top highly variable features are employed for dimensionality reduction to create cell clusters for biological interpretation.

In this workflow, we use the 10× Genomics Cell Ranger software [20] with Feature Barcode pipeline, followed by downstream analysis in Seurat package [21]. Other computational tools available for analyzing CITE-seq data include Scanpy, CiteFuse, and totalVI [22–24].

3.7 Data Preprocessing to Create RNA and ADT Count Matrices

Libraries sequenced for both the RNA and the cell surface protein antibody-derived tags (ADT) from the same cell are preprocessed in Cell Ranger using the Feature Barcode approach (10× Genomics) that quantifies each feature in each cell. The analysis includes the following steps:

1. Demultiplexing sequencing reads from each library (RNA and ADT) to generate two types of FASTQ files for each sample, one with reads for the cellular RNA and one with the reads captured by ADTs.
2. Aligning the obtained FASTQ files with RNA expression reads to the reference genome (in this study, 10× Genomics human GRCh38) and the FASTQ files containing reads for ADT-captured cell surface proteins to the feature barcode reference created specifically for the antibodies used in the study.
3. Creating the two count matrices, one for the RNA and one for the ADT that are used in the downstream analysis (*see* **Note 3**).

3.8 Cell Quality Control

QC of both RNA and ADT data is a crucial step before proceeding to downstream single-cell data analysis. This step aims to exclude cells with features that markedly differ from the majority of cells in the dataset. Specifically, QC assessment removes from downstream analysis cells with the following:

1. Low numbers of RNA corresponding to few genes and high proportion of mitochondrial RNAs, as these cells may be apoptotic or damaged, with leakage of cytoplasmic RNA.
2. Markedly large numbers of abundant RNAs corresponding to large numbers of genes, which may indicate capture of multiple cells within the same droplet.
3. Extremely high ADT counts that may be caused by antibody aggregation.

The threshold for each QC metric can vary substantially depending on the studied cells or experimental conditions. Different cell states and cell types may exhibit distinct cellular characteristics; for example, high counts may characterize large cells, low transcript numbers may indicate quiescent cells, and high proportion of mitochondrial transcripts is a feature of cardiomyocytes [25]. In sum, there are no standard thresholds of the QC metrics that can be applied for each dataset and the filtering criteria should be adjusted specifically to the type of data analyzed (*see* **Note 4**).

3.9 Data Normalization and Variable Gene Selection

Filtered raw counts of RNA and ADT matrices are normalized to correct for technical variation and make counts comparable across cells. The RNA data are normalized using the global scaling normalization method, while the ADT data are normalized with the centered log ratio (CLR) approach. Then, a subset of 2000 genes with the highest cell-to-cell variation in RNA expression is used for data integration, dimensional reduction, and cell clustering.

3.10 Dimensional Reduction and Clustering Cells Based of the RNA Data

The high-dimensional complexity of CITE-seq data requires dimensional reduction and clustering of cells to enable data visualization and biological interpretation. To this end, the following steps must be taken:

1. Reduce the dimensions of the data with the principal component analysis (PCA) method followed by uniform manifold approximation and projection (UMAP) or t-distributed stochastic neighbor embedding (t-SNE).
2. Identify cell clusters based on the RNA data by calculating k-nearest neighbors (KNN) for all cells, constructing a shared nearest neighbor SNN graph, and applying Louvain community-detection algorithm.
3. Build the UMAP or t-SNE plot that arranges the cell clusters in two-dimensional space, based on the similarity of their transcriptional profiles.
4. Find differentially expressed marker genes that characterize each cluster.
5. Visualize the side-by-side expression of RNA or protein (ADT) molecules in the dataset.

4 Notes

1. Performing Western blot analysis is important to confirm successful labeling and isolation before MS analysis. Positive control surface proteins that are highly expressed in senescence include DPP4 (CD26), CAV1, and EGFR [5].
2. When utilizing various antibodies in CITE-seq, adjustments and optimization may be required, particularly when using custom-designed antibodies to ensure effective labeling and sequencing [18]. If the antibody cocktail volume is less than 50 μL, add Cell Staining Buffer to bring the volume up to 50 μL.
3. For CITE-seq datasets employing a larger panel of antibodies, the analysis will benefit from the application of weighted nearest neighbors (WNN) approach [21], which enables clustering of cells based on a weighted combination of both RNA and ADT datasets.
4. Regarding the cell QC steps, plotting the distribution of cells for each filtering metric before and after applying the filtering criteria is good practice, as it allows the user to reassess the selected QC strategy and set the optimal criteria necessary to address the specificity of the dataset studied.

Acknowledgments

This work was supported in its entirety by the NIA IRP, NIH.

References

1. Herranz N, Gil J (2018) Mechanisms and functions of cellular senescence. J Clin Invest 128(4):1238–1246
2. Yue Z et al (2022) Senescence-associated secretory phenotype and its impact on oral immune homeostasis. Front Immunol 13:1019313
3. Chaib S, Tchkonia T, Kirkland JL (2022) Cellular senescence and senolytics: the path to the clinic. Nat Med 28(8):1556–1568
4. Sun Y, Li Q, Kirkland JL (2022) Targeting senescent cells for a healthier longevity: the roadmap for an era of global aging. Life Med 1(2):103–119
5. Kim KM et al (2017) Identification of senescent cell surface targetable protein DPP4. Genes Dev 31(15):1529–1534
6. Lakkis J et al (2022) A multi-use deep learning method for CITE-seq and single-cell RNA-seq data integration with cell surface protein prediction and imputation. Nat Mach Intell 4(11):940–952
7. Rossi M, Abdelmohsen K (2021) The emergence of senescent surface biomarkers as Senotherapeutic targets. Cells 10(7):1740
8. Marcotte R, Wang E (2002) Replicative senescence revisited. J Gerontol A Biol Sci Med Sci 57(7):B257–B269
9. Carroll B, Korolchuk VI (2018) Nutrient sensing, growth and senescence. FEBS J 285(11):1948–1958
10. Ibragimova M et al (2024) The molecular mechanisms in senescent cells induced by natural aging and ionizing radiation. Cells 13(6):550
11. Rattanavirotkul N, Kirschner K, Chandra T (2021) Induction and transmission of oncogene-induced senescence. Cell Mol Life Sci 78(3):843–852
12. Liu XL, Ding J, Meng LH (2018) Oncogene-induced senescence: a double-edged sword in cancer. Acta Pharmacol Sin 39(10):1553–1558
13. Kumari R, Jat P (2021) Mechanisms of cellular senescence: cell cycle arrest and senescence associated secretory phenotype. Front Cell Dev Biol 9:645593
14. Wang Z, Wei D, Xiao H (2013) Methods of cellular senescence induction using oxidative stress. Methods Mol Biol 1048:135–144
15. Chen JH, Ozanne SE, Hales CN (2007) Methods of cellular senescence induction using oxidative stress. Methods Mol Biol 371:179–189
16. Rossi M et al (2023) Pleiotropic effects of BAFF on the senescence-associated secretome and growth arrest. elife 12:e84238
17. Kohli J et al (2021) Algorithmic assessment of cellular senescence in experimental and clinical specimens. Nat Protoc 16(5):2471–2498
18. Herman AB et al (2023) DPP4 inhibition impairs senohemostasis to improve plaque stability in atherosclerotic mice. J Clin Invest 133(12):e165933
19. Mazan-Mamczarz K et al (2022) Single-cell analysis of the transcriptome and Epigenome. Methods Mol Biol 2399:21–60
20. Zheng GX et al (2017) Massively parallel digital transcriptional profiling of single cells. Nat Commun 8:14049
21. Hao Y et al (2021) Integrated analysis of multimodal single-cell data. Cell 184(13):3573–3587
22. Wolf FA, Angerer P, Theis FJ (2018) SCANPY: large-scale single-cell gene expression data analysis. Genome Biol 19(1):15
23. Kim HJ et al (2020) CiteFuse enables multimodal analysis of CITE-seq data. Bioinformatics 36(14):4137–4143
24. Gayoso A et al (2021) Joint probabilistic modeling of single-cell multi-omic data with totalVI. Nat Methods 18(3):272–282
25. Galow AM et al (2021) Quality control in scRNA-Seq can discriminate pacemaker cells: the mtRNA bias. Cell Mol Life Sci 78(19–20):6585–6592

Chapter 8

Metal Organic Layers–Based Sample Preparation Method for Membrane Proteome Analysis

Jiang Qianqian, Jiang Bo, Zhang Yukui, and Zhang Lihua

Abstract

As a central platform for exchange of signaling communication and substance between cells and within cells, mapping of membrane proteome can provide new insights into the pathophysiology of cancer and other diseases, explore potential drug targets and biomarkers for prognosis or diagnosis, and ultimately contribute to the development of precision medicine. However, mass spectrometry–based identification of membrane proteins poses many challenges due to the intrinsic properties of membrane proteins, especially high hydrophobicity and low abundance. Herein, we described a metal organic layers (MOL)-based sample preparation method for membrane proteome analysis in living cells, involving the cell surface engineering process through the exquisite interaction between MOL and phospholipid bilayers on the membrane and the following biomembrane fusion process, providing a simple and robust tool for dissection of membrane proteome with high specificity. This chapter details the preparation of such metal organic layers and enrichment workflow of HeLa membrane proteins.

Key words Membrane proteome, Metal organic layers, Enrichment, Mass spectrometry, based identification

1 Introduction

Membrane proteins, which account for one-third of total cellular proteins, are the executors of many important functions of biomembranes and play a variety of roles in substance transport, receptors, enzymes, immunity, and interactions with extracellular matrix. The study of the overall level of membrane proteins has gradually emerged and developed into an important component of proteomics research on biological functions [1, 2]. Although proteomics has made great progress in recent years [3], the research progress of membrane proteins, especially integrated membrane proteins, lags far behind soluble proteins, which is mainly due to the low abundance and strong hydrophobicity, leading to the underrepresentation of the hydrophobic membrane proteins in the total cell lysate [4, 5]. According to the neXtProt (2020) [6],

Csaba Matta et al. (eds.), *The Surfaceome: Methods and Protocols*, Methods in Molecular Biology, vol. 2908, https://doi.org/10.1007/978-1-0716-4434-8_8,

over 30% of the missing proteins that lack high-stringency evidence (for example, mass spectrometry identification) to confirm their existence are membrane proteins such as olfactory receptors, GPCRs, solute carrier proteins. In addition to the advance on high-precision mass spectrometry and high efficiency search engines [7, 8], improved protocols for sample preparation in membrane proteomics show great potential to increase the overall sensitivity of detection.

Two significant challenges must be faced in the sample processing of comprehensive and in-depth research of the membrane proteome [9, 10]: (1) isolating membrane proteins with maximum specificity and minimal sample loss and (2) separating membrane proteins in their native states to increase quantitative precision. Differential centrifugation, as a powerful tool for membrane separation, has been used in numerous studies of membrane proteomics [11, 12]. The membrane specificity can be further facilitated using high pH washing (carbonate washing) or high salt washing, chaotropes extraction or two-phase partition [13, 14]. However, there are still limitations in practice such as time-consuming steps, large amounts of cells, and incomplete coverage. A method that can identify and quantify membrane proteins on a large scale and with high specificity is in urgent demand.

Here, we described the development of a new method for comprehensive membrane enrichment with high selectivity and extensive coverage based on the in situ wrapping of living cells with functional metal organic layers (MOL-DpaZn) and following biomembrane fusion process [15]. MOL with advantages of sheet structure, abundant accessible active sites, and tunable biocompatibility is capable of cell membrane targeting [16, 17]. Therefore, cell membrane of living cells is first modified with the nanosheets, while inner organelle membranes are then adsorbed to them after mild cell lysis, enabling the separation of MOL-DpaZn/biomembrane system and subsequent extraction of membrane proteins. The method exhibits excellent qualitative and quantitative reproducibility and enables the plotting of membrane proteome of different cells and extracellular vesicles. During the analysis of HeLa cells, 1526 MPs with a specificity of 77% were identified, and the deep coverage of the HeLa membrane proteome (5087 MPs) was further realized. Herein, we describe a detailed sample preparation protocol for the enrichment of membrane proteomes using MOL-DpaZn.

2 Materials

2.1 MOL-DpaZn Material Synthesis

1. N,N-dimethylformamide (DMF).
2. Formic acid (FA).
3. ddH_2O.

4. Zirconium(IV) chloride.
5. 1,3,5-Tris(4-carboxyphenyl)benzene (H_3BTB).
6. Ethanol.
7. 9,10-bis(chloromethyl)anthracene.
8. 2,2′-dipicolylamine.
9. Potassium carbonate (K_2CO_3).
10. Potassium iodide (KI).
11. 1 M HCl: dilute 8.32 mL of 37% (w/v) HCl to a final volume of 100 mL with ddH_2O.
12. Ethyl acetate (AcOEt).
13. 4 M NaOH: dissolve 16.0 g of NaOH in 100 mL of ddH_2O.
14. Tetrahydrofuran (THF).
15. Brine.
16. Magnesium sulfate ($MgSO_4$).
17. $Zn(NO_3)_2{\cdot}6H_2O$.
18. Methanol (MeOH).
19. Heavy-wall pressure vessel (capacity 15 mL), round-bottom flask (capacity 100 mL), constant pressure dropping funnel, and separatory funnel.
20. Devices: microcentrifuge, ultrasonic water bath, vacuum oven, rotary evaporator, magnetic stirrer.

2.2 Enrichment of the MOL-DpaZn/Biomembrane System

1. Human HeLa cells.
2. Cell culture medium: MEM with 10% (v/v) FBS and penicillin (100 U/mL)/streptomycin (100 μg/mL).
3. MOL-DpaZn incubation medium: MEM without FBS.
4. 150 mm petri dish.
5. MOL-DpaZn material stock: 10 mg/mL in ddH_2O (*see* **Note 1**).
6. Ice-cold phosphate-buffered saline (PBS) without calcium and magnesium.
7. Ice-cold hypotonic buffer: 10 mM NaCl, 10 mM Tris–HCl, 1.5 mM $MgCl_2$, 1% cocktail (v/v) in ddH_2O. Adjust the pH to 7.4 on ice with 1 M HCl (*see* **Note 2**).
8. Cell scraper, 2 mL low-bind Eppendorf tubes (*see* **Note 3**).
9. Devices: CO_2 incubators, microcentrifuge, pH meter.

2.3 Extraction of the Membrane Proteins

1. 500 mM and 50 mM ammonium bicarbonate aqueous solution (ABC). Prepare fresh.
2. Washing buffer 1: 1 M NaCl: dissolve 5.85 g of NaOH in 100 mL of 50 mM ABC.

3. Washing buffer 2: 0.1 M Na_2CO_3 (pH 11.5): dissolve 1.06 g of Na_2CO_3 in 100 mL of 50 mM ABC.
4. Extracting buffer: ionic liquid extraction buffer (1-dodecyl-3-methylimidazolium chloride (C12Im-Cl) buffer [18] (*see* **Note 4**).
5. 1.5 mL and 2 mL low-bind Eppendorf tubes (*see* **Note 3**).
6. Devices: microcentrifuge, vortex mixer, ultrasonic water bath with temperature control.

2.4 Mass Spectrometry–Based Proteomic Analysis

1. Dithiothreitol (DTT): 100 mM DTT in 50 mM ABC.
2. Iodoacetamide (IAA): 25 mM IAA in 50 mM ABC (*see* **Note 5**).
3. Washing solution: 50 mM ABC.
4. 10 kDa molecular weight cut off FASP filter.
5. Trypsin: before use, dissolve the lyophilized trypsin with trypsin resuspension dilution buffer to reach a concentration of 1 μg/μL (*see* **Note 6**).
6. Devices: microcentrifuge, vortex mixer, water bath with temperature control, SpeedVac concentrator.

2.5 LC-MS/MS and Data Analysis

1. 0.1% FA: 99 μL water (ddH_2O) with 1 μL FA (MS grade).
2. Nanodrop spectral photometer.
3. HPLC solvent A: 0.1% formic acid/2% ACN in ddH_2O.
4. HPLC solvent B: 0.1% formic acid/98% ACN in ddH_2O.
5. High-resolution mass spectrometry system with online nano-UPLC system.
6. Proteome Discoverer (PD) software (for these studies we used PD version 2.5.0.400, Thermo Scientific, just for an example).
7. Mascot search engine (Matrix Science).

3 Methods

The key steps of the protocol are summarized in Fig. 1. We developed the cell surface engineering coupling biomembrane fusion method to map the whole membrane proteome from the plasma membrane to various organelle membranes taking advantage of the exquisite interaction between two-dimensional MOL and phospholipid bilayers on the membrane.

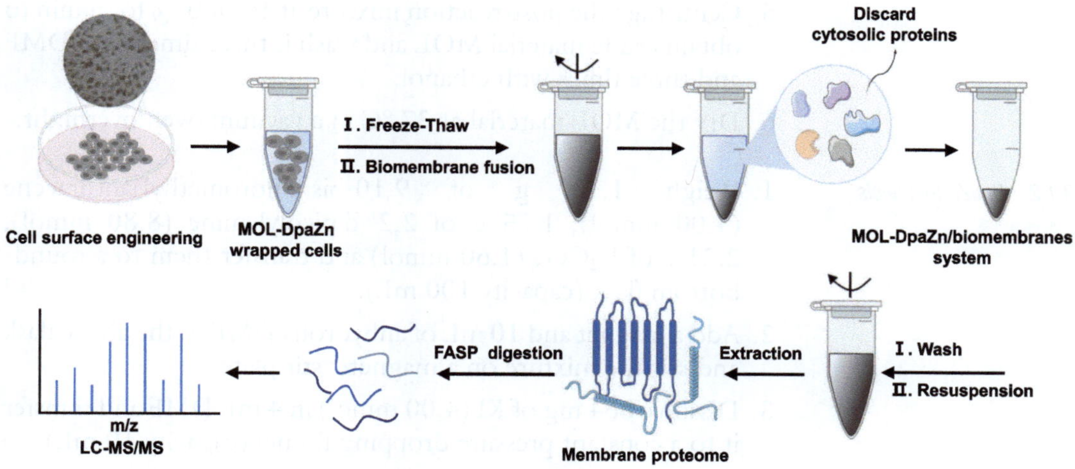

Fig. 1 Schematic representation of the workflow

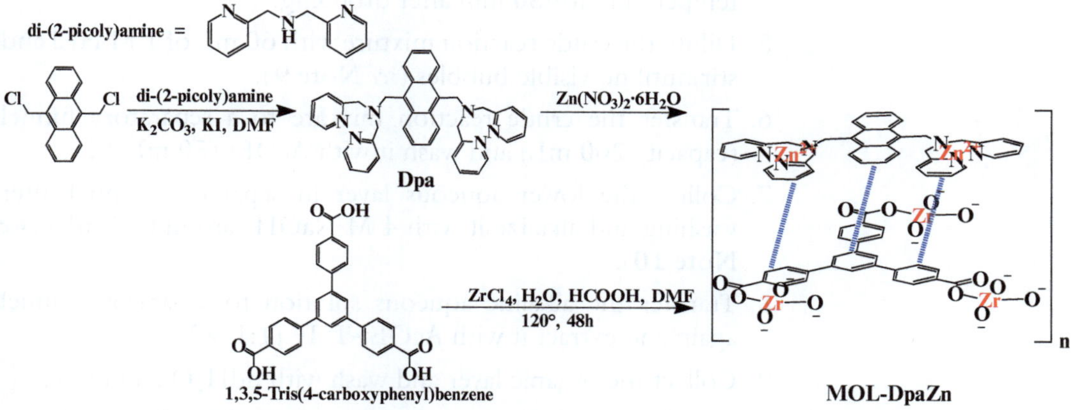

Fig. 2 Synthetic route of MOL-DpaZn

3.1 MOL-DpaZn Material Synthesis

The synthetic Route of MOL-DpaZn is shown in Fig. 2.

3.1.1 MOL Synthesis

1. Weigh 14.0 mg of $ZrCl_4$ and 12.5 mg of H_3BTB, and transfer them to a heavy-wall pressure vessel (capacity 15 mL).
2. Add 2.05 mL of DMF, 0.88 mL of formic acid, and 0.30 mL of ddH_2O to the above heavy-wall pressure vessel (*see* **Note** 7).
3. Dissolve the reaction mixture by treating the mixture containing vessel to sonication in a water bath for around 2 min.
4. Transfer the mixture containing vessel to an oven at 120 °C for reaction for 48 h.

5. Centrifuge the post-reaction mixture at 16,000× *g* for 5 min to obtain crude material MOL and wash it three times with DMF and three times with ethanol.
6. Dry the MOL material at 37 °C in a vacuum oven overnight.

3.1.2 DpaZn Synthesis

1. Weigh 1.10 g of 9,10-bis(chloromethyl)anthracene (4.00 mmol), 1.75 g of 2,2′-dipicolylamine (8.80 mmol), 2.21 g of K_2CO_3 (1.60 mmol) and transfer them to a round-bottom flask (capacity 100 mL).
2. Add a magnet and 10 mL of anhydrous DMF to the above flask and stir the mixture on a magnetic stir plate.
3. Dissolve 664 mg of KI (4.00 mmol) in 4 mL DMF and transfer it to a constant pressure dropping funnel (capacity 10 mL).
4. Drop KI solution through constant pressure dropping funnel to the above mixture containing round-bottom flask while stirring (*see* **Note 8**) and stir the reaction mixture at room temperature for 30 min after dropping.
5. Dilute the crude reaction mixture with 60 mL of 1 M HCl and stir until no visible bubbles (*see* **Note 9**).
6. Transfer the crude reaction mixture to a separatory funnel (capacity 200 mL) and wash it with AcOEt (50 mL, ×2).
7. Collect the lower aqueous layer in separatory funnel after washing and alkalize it with 4 M NaOH (around 15 mL) (*see* **Note 10**).
8. Transfer the alkaline aqueous solution to separatory funnel again and extract it with AcOEt-THF (1:1, ×2).
9. Collect the organic layer and wash with ddH_2O and brine.
10. Dry the organic layer over anhydrous $MgSO_4$, filter, and concentrate under reduced pressure using the rotary evaporator.
11. Recrystallize the residue from THF-AcOEt to give Dpa product.
12. Weigh 1.00 g of Dpa (1.66 mmol) and dissolve it with 80 mL MeOH-THF (10: 1) in a round-bottom flask (capacity 150 mL).
13. Weigh excess Zn $(NO_3)_2 \cdot 6H_2O$ (over 3.32 mmol) and dissolve it with 5 mL MeOH-THF (10: 1).
14. Drop the Zn $(NO_3)_2 \cdot 6H_2O$ solution to Dpa solution slowly and stir the mixture over 2 h at room temperature (*see* **Note 11**).
15. Collect the precipitate DpaZn, wash three times with MeOH-THF (10: 1), dry at 37 °C in a vacuum oven overnight.

3.1.3 MOL-DpaZn Synthesis

1. Disperse 30 mg of MOL in 4 mL ddH_2O and dissolve 10 mg of DpaZn in 1 mL of ddH_2O.
2. Add DpaZn solution to the MOL dispersion and react at room temperature overnight with end-over-end mixing (*see* **Note 12**).
3. Collect the yellow precipitate MOL-DpaZn and re-disperse it in ddH_2O for further usage (10 mg/mL) after washing repeatedly by centrifugation in ddH_2O (16,000× *g* for 10 min).

3.2 Enrichment of the MOL-DpaZn/Biomembrane System

The analysis of the membrane proteome can be performed on a wide range of cells and extracellular vesicles. Here, we exhibit the detailed steps towards HeLa cells.

1. Grow HeLa cells in 150 mm Petri dishes until they reach a confluency of about 80%.
2. Remove the medium and carefully wash the cells with 10 mL PBS to remove cellular debris and non-adherent cells.
3. Add 20 mL of serum-free culture medium (*see* **Note 13**) containing 0.1 mg/mL MOL-DpaZn carefully to the Petri dishes containing HeLa cells and cultivate for another 3 h in a humidified 5% CO_2 incubator at 37 °C.
4. Remove the medium and carefully wash the cells three times with 20 mL ice-cold PBS to remove excess MOL-DpaZn material.
5. Harvest the cells with cell scrapers using ice-cold PBS and collect them in a 2 mL Eppendorf tube, resuspend, and wash the cells gently three times using ice-cold PBS at 500× *g* for 5 min at 4 °C.
6. Resuspend the cells with 1.6 mL ice-cold hypotonic buffer and swell on ice for 5 min.
7. Repeated freezing and thawing for three times for cell mild disruption (*see* **Note 14**).
8. Collect the MOL-DpaZn/biomembrane precipitate after centrifugation (1000–2000× *g*, 3 min, 4 °C) of cell lysate and collect the supernatant into a new 1.5 mL low-bind Eppendorf tube for another twice centrifugation (1000–2000× *g*, 3 min, 4 °C) to increase the recovery.
9. Combine the MOL-DpaZn/biomembrane precipitate carefully using ice-cold PBS to a 1.5 mL low-bind Eppendorf tube and centrifuge to remove the PBS supernatant (16,000× *g*, 2 min, 4 °C).
10. Store at −80 °C for further usage.

3.3 Extraction of the Membrane Proteins

1. Wash the MOL-DpaZn/biomembrane precipitate three times with 1 mL of 1 M NaCl. Rotate the dispersion of MOL-DpaZn/biomembrane in washing buffer strongly for around 1 min and pellet the MOL-DpaZn/biomembrane for 1 min at 16,000× *g* at 4 °C, discard the supernatant (*see* **Notes 15** and **16**).
2. Wash the MOL-DpaZn/biomembrane precipitate three times with 1 mL of 0.1 M Na_2CO_3. Specific washing steps are the same as the previous (*see* **Notes 15** and **16**).
3. Wash the MOL-DpaZn/biomembrane precipitate once with 1 mL of 500 mM ABC and once with 1 mL of 50 mM ABC, respectively. Specific washing steps are the same as the previous.
4. Resuspend the washed MOL-DpaZn/biomembrane precipitate in 100 μL of the ionic liquid extraction buffer, sonicate for 10–20 min in a water bath at 4 °C (working for 30 s and pausing for 30 s), and then centrifuge at 16,000× *g* for 5 min at 4 °C.
5. Repeat the operation of **step 4** twice and obtain 300 μL of membrane protein solution (*see* **Note 17**).

3.4 Mass Spectrometry–Based Proteomic Analysis

Carry out filter-aided sample preparation (FASP) proteolytic digestion to obtain tryptic peptides from the enriched proteins solubilized with ionic liquid for LC-MS/MS analysis.

1. Determine the protein concentration using BCA assay and adjust the concentration to around 1 μg/μL using ionic liquid.
2. Reduce disulfide bonds by adding 1 M DTT to the protein lysate solution (50–100 μg, ~1 μg/μL) to achieve a final DTT concentration of 100 mM and incubate at 90 °C for 10 min (*see* **Note 18**). Cool the sample to room temperature.
3. Activate the 10 kDa molecular weight cut off FASP filter with 150 μL of ddH_2O and centrifuge for around 20 min at 14,000× *g* (*see* **Note 19**).
4. Transfer the cooled membrane protein solution to activated filter and centrifuge around 40 min at 14,000× *g*.
5. Wash filter with 200 μL of 50 mM ABC to remove ionic liquid and DTT by centrifuging 30 min at 14,000× *g*.
6. Alkylate free cysteines by adding 150 μL of 20 mM IAA and incubate at room temperature for 30 min, in the dark.
7. Remove IAA through centrifuging 20 min at 14,000× *g*.
8. Wash filter with 200 μL of 50 mM ABC to remove ionic liquid by centrifuging 30 min at 14,000× *g*.
9. Repeat this previous step over two times until the ionic liquid is removed thoroughly (*see* **Note 20**).
10. Replace with a new collection tube under the filter.

11. Add 100 μL of 10 mM ABC containing 3 μg of proteomics-grade trypsin to the filter and incubate overnight (at least 12 h) at 37 °C in a water bath (*see* **Note 21**).
12. Elute peptides from the filter by centrifuging for 20 min at 14,000× *g* and 50 μL of ddH_2O was used for another two elution.
13. Dry peptides in a SpeedVac concentrator or similar apparatus and store peptides at −80 °C until ready for mass spectrometry analysis.

3.5 LC-MS/MS and Data Analysis

3.5.1 LC-MS/MS Analysis

Various LC separation and MS acquisition methods can be used for membrane protein analysis. Standard LC-MS/MS setups have been described previously (*see* examples in [15]) and we only provide a brief outline here.

1. Resuspend peptides in 20 μL of 0.1% formic acid and vortex the tube rigorously for 15–20 s for complete dissolution.
2. Determine peptide concentration using nanodrop spectral photometer and adjust to 0.5–1 μg/μL with 0.1% formic acid.
3. Centrifuge peptide solution at 20,000× *g* over 1 h, carefully transfer the supernatant to the sample bottle without disturbing the debris pellet.
4. Load about 1 μg of peptides per sample onto the capillary column, separate through nano reversed-phase liquid chromatography (RPLC), and analyze using a high-resolution mass spectrometer. Operate the mass spectrometer in positive ion mode and acquire all MS/MS spectra in data-dependent mode.

3.5.2 Peptide and Protein Identification

Analyze MS data using the search engine of your choice. The following is intended to provide guidance for commonly used Proteome Discoverer (PD) software (Version2.5.0.400, Thermo Fisher Scientific).

1. Use the Mascot as search engine and a reviewed human SwissProt database (www.uniprot.org).
2. Specify trypsin as cleavage enzyme with a maximum of two missed cleavage sites.
3. Set oxidation (M) and acetyl (Protein N-term) as the variable modification, set carbamidomethyl (C) as the static modification.
4. Set the precursor ion mass tolerance for all searches to 10 ppm and the product ion mass tolerance to 0.05 Da (*see* **Note 22**).
5. Filter the data at a false discovery rate (FDR) of <1% on the peptide and protein levels.
6. Other settings were the same default as the conventional search.

3.5.3 Data Analysis

The data analysis process of membrane proteins is detailed here as a reference.

1. Export the protein list as an Excel file after analysis by Proteome Discoverer 2.5, filter and exclude keratins according to proteins names prior to calculating the number of identified proteins.
2. Annotate proteins based on Gene Ontology (GO) information from the UniProt database (www.uniprot.org) (*see* **Note 23**) and TMHMM-2.0 algorithm (services.healthtech.dtu.dk/services/TMHMM-2.0) (*see* **Note 24**).
3. Characterize membrane proteins according to the above annotation. Specifically, a protein is recognized as "membrane protein" when it is annotated as "membrane" in Gene Ontology cellular components or predicted to possess at least one transmembrane helix domain according to TMHMM.
4. Assign membrane proteins to different organelle membranes according to GO cellular component annotation. The example of the assigned membrane proteins is shown in Fig. 3.

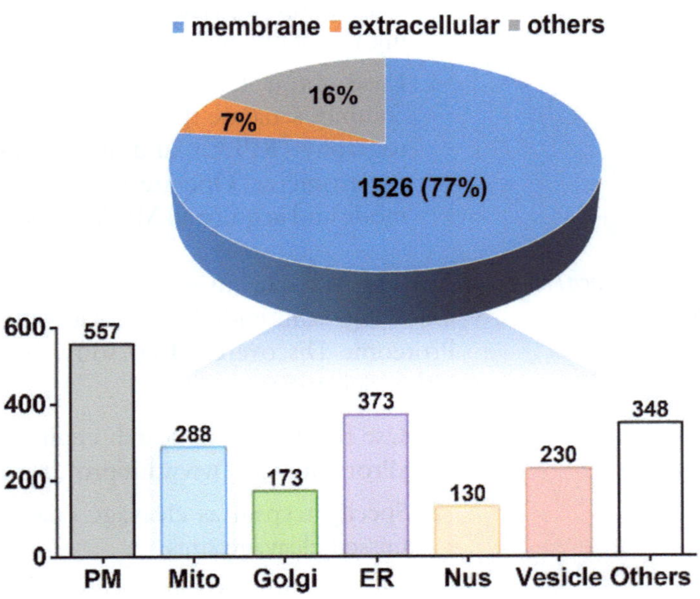

Fig. 3 The membrane proteome mapped in 1E7 HeLa cells using the nano reversed-phase liquid chromatography (RPLC) on a 15 cm C18 capillary column packed in-house with ReproSil-Pur C18-AQ particles (1.9 μm, 120 Å) and analyzed using the LTQ Orbitrap Velos mass spectrometer coupled on-line to the Dionex Ultimate 3000 RSLC nano-LC system. The detailed parameter settings are shown in [15]

5. Perform functional categories and pathway analysis using DAVID Bioinformatics Resources (david.ncifcrf.gov).
6. Create protein–protein interaction network through STRING database (string-db.org) and visualize using Cytoscape software.

4 Notes

1. MOL-DpaZn can be uniformly dispersed in water. To prevent concentration changes, the MOL-DpaZn dispersion is kept at 4 °C until further use. The dispersion should be sonicated in a water bath over 10 min before usage.
2. Adjust the pH on ice since the pH of Tris–HCl buffer is affected by temperature.
3. Low-bind Eppendorf tubes (or similar) should be used to minimize sample loss through adsorption to the vessel.
4. According to downstream experimental applications, other suitable solution can be selected for extraction.
5. IAA is sensitive to light, so use a fresh solution of IAA in each experiment to minimize exposure to light. To maintain the IAA activity, transport the fresh solution in an Eppendorf tube wrapped in tin foil. Rapid pipetting when using.
6. Trypsin can be stored in acid resuspension buffer at −80 °C for a long time. Avoid freeze-thaw cycles wherever possible as protease activity decreases with each cycle.
7. The addition of three solvents has no strict sequence requirements.
8. The dropping time should be controlled at about 1 h. After dropping, the yellow reaction solution becomes turbid.
9. The addition of HCl leads to production of large amounts of carbon dioxide gas, so the HCl should be added slowly.
10. With alkalization, the solution becomes turbid.
11. With reaction, the solution becomes turbid, and the yellow product precipitation (DpaZn) is generated.
12. Before reacting with DpaZn, MOL dispersion should be sonicated in a water bath over 10 min to prevent material aggregation.
13. The use of serum-free medium is to prevent the adsorption of protein in serum on MOL-DpaZn material and causing contamination.
14. Freeze at −80 °C or in liquid nitrogen, dissolve at room temperature.

15. Avoid prolonged high-speed centrifugation, otherwise the MOL-DpaZn/biomembrane precipitate will be difficult to disperse.
16. The washing frequency can be reduced to one time to improve the recovery.
17. Membrane protein recovery can be improved through multiple extractions.
18. High-concentration proteins are prone to precipitation during the heating at 90 °C.
19. Pay attention to the amount of residual liquid on the filter membrane, leaving the dry membrane will cause the next centrifugation difficult to carry out.
20. The ionic liquid is removed completely if there is no obvious foam appears when the effluent is repeatedly pipetted in and out.
21. The mass ratio of trypsin to protein is approximately 1:25.
22. The parameters should be set according to the different mass spectrometer.
23. GO (Gene Ontology) is set up by the Gene Ontology Consortium to create a common database that is applicable to a variety of species, to define and describe the function of genes and proteins. There are three main categories of GO database, namely biological processes (BP), cellular localization (CC), and molecular functions (MF).
24. TMHMM is an online analysis tool for protein transmembrane helices based on Hidden Markov models.

Acknowledgments

This work was supported by the National Natural Science Foundation (21834006, 22074140, 21725506, 32088101), National Key R&D Program of China (2020YFE0202200), and Talent innovation support program of Dalian (2019CT07).

References

1. Mermelekas G, Zoidakis J (2014) Mass spectrometry-based membrane proteomics in cancer biomarker discovery. Expert Rev Mol Diagn 14(5):549–563. https://doi.org/10.1586/14737159.2014.917965
2. Groen AJ, Lilley KS (2010) Proteomics of total membranes and subcellular membranes. Expert Rev Proteomics 7(6):867–878. https://doi.org/10.1586/epr.10.85
3. Savas JN, Stein BD, Wu CC, Yates JR (2011) Mass spectrometry accelerates membrane protein analysis. Trends Biochem Sci 36(7): 388–396. https://doi.org/10.1016/j.tibs.2011.04.005
4. Helbig AO, Heck AJR, Slijper M (2010) Exploring the membrane proteome—challenges and analytical strategies. J Proteome 73(5):868–878. https://doi.org/10.1016/j.jprot.2010.01.005

5. Vuckovic D, Dagley LF, Purcell AW, Emili A (2013) Membrane proteomics by high performance liquid chromatography–tandem mass spectrometry: analytical approaches and challenges. Proteomics 13(3–4):404–423. https://doi.org/10.1002/pmic.201200340
6. Adhikari S et al (2020) A high-stringency blueprint of the human proteome. Nat Commun 11(1):5301. https://doi.org/10.1038/s41467-020-19045-9
7. Wu S et al (2020) Open-pFind verified four missing proteins from multi-tissues. J Proteome Res 19(12):4808–4814. https://doi.org/10.1021/acs.jproteome.0c00370
8. Baker MS et al (2017) Accelerating the search for the missing proteins in the human proteome. Nat Commun 8(1):14271. https://doi.org/10.1038/ncomms14271
9. Wang X, Liang S (2012) Sample preparation for the analysis of membrane proteomes by mass spectrometry. Protein Cell 3(9):661–668. https://doi.org/10.1007/s13238-012-2062-1
10. Sun J et al (2021) Nanoparticles and photochemistry for native-like transmembrane protein footprinting. Nat Commun 12(1):7270. https://doi.org/10.1038/s41467-021-27588-8
11. Kongpracha P et al (2022) Simple but efficacious enrichment of integral membrane proteins and their interactions for in-depth membrane proteomics. Mol Cell Proteomics 21(5):100206. https://doi.org/10.1016/j.mcpro.2022.100206
12. Wu CC, MacCoss MJ, Howell KE, Yates JR (2003) A method for the comprehensive proteomic analysis of membrane proteins. Nat Biotechnol 21(5):532–538. https://doi.org/10.1038/nbt819
13. Zhao Q et al (2017) In-depth proteome coverage by improving efficiency for membrane proteome analysis. Anal Chem 89(10):5179–5185. https://doi.org/10.1021/acs.analchem.6b04232
14. Zhou Y et al (2018) Enhancing membrane protein identification using a simplified centrifugation and detergent-based membrane extraction approach. Anal Chem 90(4):2434–2439. https://doi.org/10.1021/acs.analchem.7b03710
15. Jiang Q et al (2023) Metal organic layers enabled cell surface engineering coupling biomembrane fusion for dynamic membrane proteome profiling. Chem Sci 14(42):11727–11736. https://doi.org/10.1039/D3SC03725H
16. Cao L et al (2016) Self-supporting metal–organic layers as single-site solid catalysts. Angew Chem Int Ed 55(16):4962–4966. https://doi.org/10.1002/anie.201512054
17. Ji Z et al (2018) Cytoprotective metal-organic frameworks for anaerobic bacteria. Proc Natl Acad Sci 115(42):10582–10587. https://doi.org/10.1073/pnas.1808829115
18. Fang F et al (2020) Molecular dynamics simulation-assisted ionic liquid screening for deep coverage proteome analysis. Mol Cell Proteomics 19(10):1724–1737. https://doi.org/10.1074/mcp.TIR119.001827

Chapter 9

Updated Methods for Elucidating Bacterial Surfaceomes

Veronica M. Jarocki, Maria C. Rezcallah, Pauline M. L. Coulon, Matthew P. Padula, and Steven P. Djordjevic

Abstract

The bacterial surfaceome, representing the full complement of surface-expressed proteins, is integral to interactions between bacteria, their hosts, and the environment, influencing processes such as adhesion, invasion, immune evasion, and biofilm formation. These proteins, due to their accessibility and vital roles in bacterial physiology, are prime targets for therapeutic intervention. This protocol describes a detailed methodology for mapping bacterial surfaceomes, including enzymatic surface shaving, cell surface biotinylation, solid-phase extraction, and liquid chromatography-tandem mass spectrometry.

Key words Surfaceome, Mycoplasma, Proteomics, LC-MS/MS, Host, Pathogen interactions, Therapeutic targets

1 Introduction

Surfaceomes, which encompass the entirety of proteins expressed on the cell surface, play critical roles in host–pathogen interactions, including adhesion, invasion, and immune evasion. These surface-associated proteins are prime candidates for therapeutic targets due to their accessibility and essential functions in bacterial physiology. Our understanding of surfaceomes have been significantly advanced by studies on *Mycoplasma* species, unusual for their ability to form host-specific, chronic infections in humans, animals, including all major food production species, and diverse, agriculturally significant plants.

Mycoplasmas (also termed Mesomycoplasmas and Metamycoplasmas) evolved by a process of degenerative evolution and are consequently minimalistic bacteria lacking cell walls and many anabolic metabolic pathways. Despite their small genomes, Mycoplasmas possess a diverse array of surface proteins involved in crucial

Veronica M. Jarocki and Maria C. Rezcallah have equally contributed to this chapter.

Csaba Matta et al. (eds.), *The Surfaceome: Methods and Protocols*, Methods in Molecular Biology, vol. 2908,
https://doi.org/10.1007/978-1-0716-4434-8_9,

biological processes. Systematic mapping of the *Mycoplasma* surfaceomes using advanced proteomic techniques has unveiled proteins that are vital for host colonization and pathogenicity including an unprecedented diversity of posttranslational processing events, generating a complex proteome that far exceeds the proteome predicted by in silico methods [1–3]. Proteolytic processing involves the cleavage of precursor proteins to generate multiple, surface accessible functional forms with novel binding characteristics, thereby increasing the functional repertoire of the surface proteome [1–5]. This process allows Mycoplasmas to adapt dynamically to host environments and evade immune responses. Surfaceome studies conducted in low G + C Firmicutes species such as *Staphylococcus aureus* indicate that proteolytic processing is not confined to the Mollicutes [6, 7]. *Mycoplasma* surfaceome studies have led to the discovery and characterization of several moonlighting proteins [8–12]. Moonlighting proteins are multifunctional proteins that perform more than one distinct biological function, and the detection of a protein in unexpected locations (e.g., proteins with intracellular canonical functions detected on the cell surface) is often indicative of moonlighting functions. Moonlighting is particularly advantageous for bacteria with limited genomic resources, enabling them to maximize their functional proteome.

The protocol detailed in this chapter includes enzymatic surface shaving, cell surface biotinylation (modified from [13]), solid phase extraction, liquid chromatography coupled with tandem mass spectrometry (LC-MS/MS), and data analysis. These methodologies ensure the accurate representation of a bacterial surfaceome, and the resulting data provides a robust framework for understanding host–pathogen interactions and developing novel vaccines and therapeutic agents. The protocols here are tailored for *Mycoplasma* species, their lack of a cell wall forced the development of gentle surface-labeling protocols, minimizing artefactual labeling of intracellular proteins. Nonetheless, the same protocol has been used successfully for other bacterial species displaying Gram-negative and Gram-positive cell wall compositions, including *Staphylococcus*, *Klebsiella*, *Pseudomonas*, and *Proteus*.

2 Materials

2.1 Enzymatic Cell Surface Shaving

1. Personal protective equipment (PPE): safety goggles, gloves, laboratory coat, enclosed shoes.
2. Benchtop centrifuge.
3. 15 mL falcon tubes.
4. 50 mL falcon tubes.
5. 2 mL Eppendorf tubes.

6. Phosphate-buffered saline (PBS).
7. Scale.
8. Pipettes and pipette tips.
9. Trypsin solution 1: 5 mg/mL MS Grade Trypsin in PBS.
10. Water or dry bath.
11. Ice.
12. Heat block.
13. Reduction stock (100×): 500 mM tris(2-carboxyethyl) phosphine (TCEP) in dH_2O. TCEP must first be adjusted to pH >8 with NaOH.
14. Alkylation stock (100×): 1000 mM acrylamide in dH_2O.
15. Dithiothreitol (DTT) stock (100×): 1000 mM DTT in dH_2O.
16. Trypsin solution 2: 1 μg/μL Trypsin Gold (MS grade) in 50 mM acetic acid.

2.2 Cell Surface Biotinylation

1. PPE.
2. Benchtop centrifuge.
3. 15 mL falcon tubes.
4. 50 mL falcon tubes.
5. 2 mL Eppendorf tubes.
6. PBS.
7. Scale.
8. Pipettes and pipette tips.
9. EZ-Link Sulfo-NHS-LC-Biotin [Pierce].
10. Ice.
11. 1 M Tris–HCl.
12. 1% sodium deoxycholate (SDC) in 4-(2-hydroxyethyl)-1-piperazineethanesulfonic acid (HEPES).
13. Heat block.
14. Reduction stock (100×): 500 mM TCEP in dH_2O. TCEP must first be adjusted to pH >8 with NaOH.
15. Alkylation stock (100×): 1000 mM acrylamide in dH_2O.
16. DTT stock (50×): 1000 mM DTT in dH_2O.
17. Dynabeads™ MyOne™ Streptavidin T1 [Pierce].
18. PBS + 0.1% bovine serum albumin (BSA) or PBS + 0.1% Tween.
19. Rotating mixer.
20. Vortex.
21. 30% acrylonitrile (ACN) + 0.4% trifluoroacetic acid (TFA).

2.3 Making STAGE-Tips

1. PPE.
2. 2 mL Eppendorf tubes.
3. Styrenedivinylbenzene-reversed phase polySulfonate (SDB-RPS) disk (Empore—3 M).
4. Flat ended 14G flat end needle (chromatography injection valve) and syringe plunger.
5. Pipette tips (200 μL if using 14G needle).
6. Petri dish.
7. Rotary tool with pink grinder tip attached or 3D-printed tube inserts for holding tips.
8. Microcentrifuge.
9. Bench minifuge.
10. Permanent marker.
11. Autosampler vial inserts.

2.4 Solid Phase Extraction (SPE)

1. SpeedVac concentrator.
2. 2 mL Eppendorf tubes.
3. Benchtop centrifuge.
4. SPE load buffer: 90% acetonitrile, 1.0% trifluoroacetic acid.
5. SPE wash buffer: 10% acetonitrile, 0.1% trifluoroacetic acid.
6. SPE elution buffer: 71 μL 1 M NH_4OH, 800 μL of 100% acetonitrile, 129 μL dH_2O (*see* **Note 1**).
7. Buffer A*: 0.1% formic acid in water.

2.5 LC-MS/MS Analysis

1. Acquity M-class nanoLC system [Waters, USA].
2. nanoEase Symmetry C18 trapping column (180 μm × 20 mm).
3. PicoFrit column [75 μm ID × 100 mm; New Objective, Woburn, MA].
4. SP-120-1.7-ODS-BIO resin [1.7 μm, Osaka Soda Co, Japan].
5. Q Exactive Plus mass spectrometer [Thermo Scientific].

2.6 Data Analysis of MS/MS Spectra

1. Software options: Peaks Studio 11 or FragPipe (v22).
2. Databases: Appropriate organism database (*see* **Note 2**). Database of common contaminants.

3 Methods

The bacterial surfaceome workflow is summarized in Fig. 1.

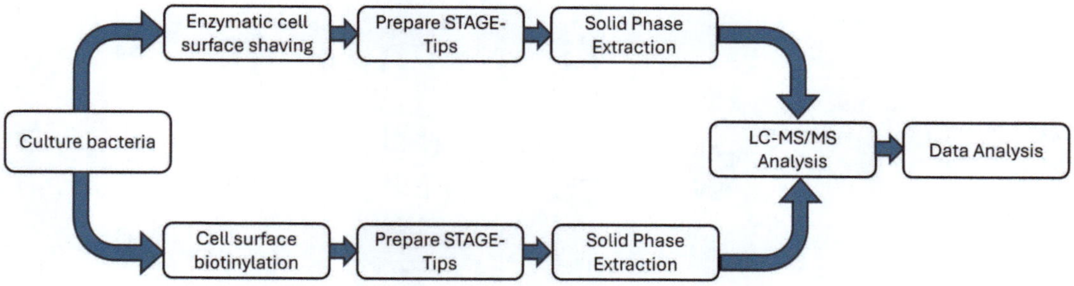

Fig. 1 Surfaceome workflow

3.1 Enzymatic Cell Surface Shaving

1. Culture bacteria until log phase (*see* **Notes 3** and **4**).
2. Centrifuge cells at 4000× *g* for 5 min, pour off media.
3. Wash cells by adding 15 mL warm PBS, rocking gently, then centrifuging at 4000× *g* for 5 min and pouring out waste. Repeat three times.
4. Using a 0.1 g cell pellet (wet weight), resuspend cells in 5 mL warm PBS.
5. Warm trypsin solution 1 for 15 min at 37 °C in water or dry bath.
6. Add 50 μL of trypsin solution 1 (50 μg/mL final concentration) to sample.
7. Incubate at 37 °C for exactly 5 min.
8. Place sample on ice to halt trypsin activity and then centrifuge at 4000× *g* for 10 min at 4 °C.
9. Pour supernatant into new tube and centrifuge supernatant at 10,000× *g* for 10 min at 4 °C (*see* **Note 5**).
10. Pour supernatant into a new tube and reduce and alkylate supernatant with 5 mM TCEP and 20 mM acrylamide monomers for 10 min at 95 °C, then quench with 20 mM DTT, if required.
11. Allow to cool on bench to room temperature (RT) for 1 h.
12. Perform a protein assay (use BCA or NanoDrop), as per manufacturer's instructions, on supernatant to determine protein concentration.
13. Add Trypsin solution 2 to sample in a 1:20–1:100 ratio (trypsin:protein).
14. Incubate overnight at 37 °C.
15. Perform solid phase extraction (SPE) to recover and desalt peptides (*see* Subheading 3.4).
16. Run aliquot on LC-MS/MS (*see* Subheading 3.5).

Each step enzymatic cell surface shaving protocol is illustrated in Fig. 2.

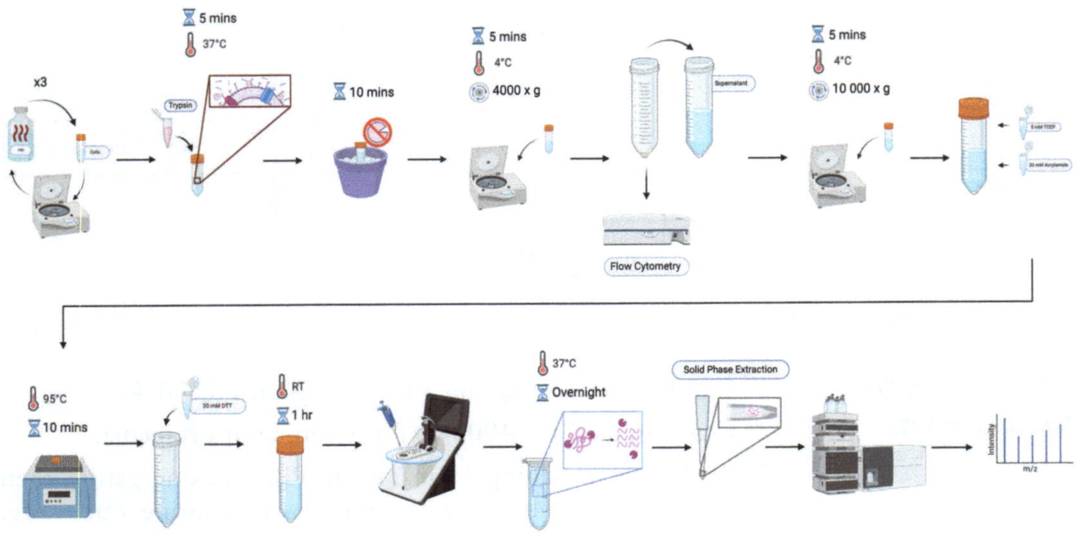

Fig. 2 Diagram depicting step-by-step enzymatic cell surface shaving methodology

3.2 Cell Surface Biotinylation

1. Centrifuge cells from culture.
2. Wash pelleted cells 3× by adding 15 mL ice cold PBS, rocking gently, centrifuging at 4000× *g* for 5 min, pouring out waste and repeating. On the final wash, pipette out as much PBS as possible.
3. Add 5 mL ice-cold EZ-Link Sulfo-NHS-LC-Biotin (0.5 mg/mL) in PBS.
4. Resuspend cells and rock gently on ice for 30 s.
5. Stop reaction with 1 M Tris–HCl (pH 7.2) to a final concentration of [50 mM].
6. Incubate for 10 min at RT with gentle rocking every 2 min.
7. Centrifuge cells and discard as much solution as possible using a pipette.
8. Wash pelleted cells 3× as in **step 2**.
9. Resuspend cells in 2 mL of 1% SDC in HEPES.
10. Boil sample at 95 °C while reducing and alkylating at the same time with 5 mM TCEP and 20 mM acrylamide.
11. Add DTT to a final concentration of 20 mM to quench remaining acrylamide.
12. Leave to cool on the bench for 1 h.
13. Wash and prepare Dynabeads™ MyOne™ Streptavidin T1 magnetic beads by:
    (a) Taking out 100 μL of bead solution for each sample (making sure to resuspend all beads appropriately before aliquoting out).

   (b) Wash beads 3× by adding 1 mL of PBS + 0.1% BSA or PBS + 0.1% Tween, vortex lightly, and place beads on magnet for 1 min. Discard supernatant.

14. Incubate the beads with sample by adding all sample to beads.
15. Incubate samples for 30 min at RT while rotating gently on a rotating mixer.
16. Separate the avidin coated beads with the magnet for 3 min.
17. Wash (*see* **Note 6**) the beads 4× by:
    (a) Adding 1 mL of PBS + 0.1% BSA or PBS + 0.1% Tween.
    (b) Vortex lightly.
    (c) Place beads on magnet for 1 min and discard supernatant.
18. Elute 4× with 30% ACN + 0.4% TFA, collecting each elution.
19. Perform a protein assay using NanoDrop to determine protein concentration.
20. Carry out an overnight trypsin digestion at 1:100 ratio (trypsin:protein).
21. Perform solid phase extraction (*see* Subheading 3.4) for downstream analysis by the LC-MS/MS (*see* Subheading 3.5).

Each step of the cell surface biotinylation protocol is illustrated in Fig. 3.

3.3 Making STAGE-Tips

1. Preparing collection tubes (*see* **Note 7**):
    (a) Use one 2 mL tube per sample as the collection tube.
    (b) Use drill to make hole in tube lid that is the same size as the pipette tip barrel. Alternatively, 3D-printed tube inserts can be used. For large sample numbers, a 3D-printed stage tip holder can be manufactured that uses 96-well plates.
    (c) Remove the excess plastic from the hole and inside of the tube.
    (d) Fit a pipette tip in hole. The hole should allow the pipette tip to be placed inside but not fall through.
    (e) Label collection tube using solvent resistant marker.
2. Preparing STAGE-Tips (*see* **Note 8**):
    (a) Place the 47 mm SDB-RPS disk in a clean plastic petri dish.
    (b) Using a flat ended needle (from a manual chromatography injection valve) fitted with the syringe plunger, take a "core" of the 47 mm disk by pressing the needle through the disk and into the plastic dish. Rotate the needle while pressing down to cut through the disk.

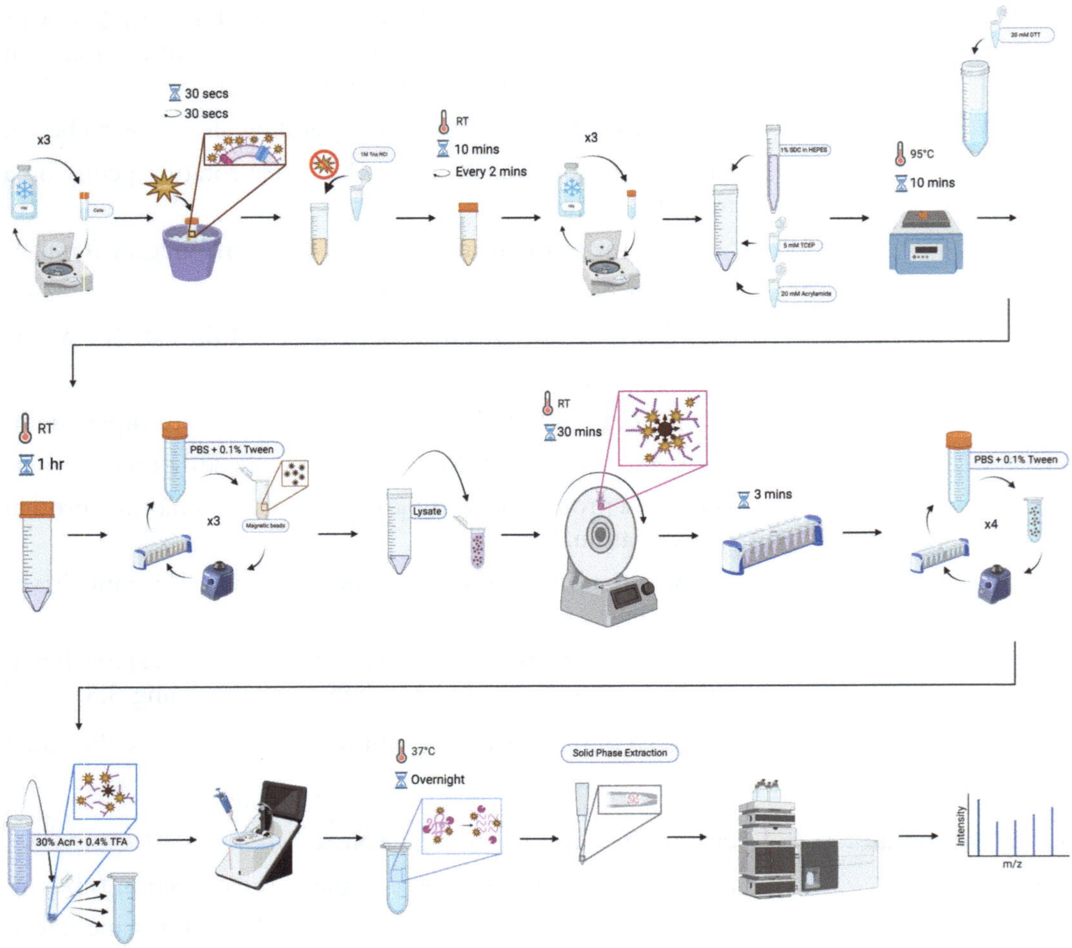

Fig. 3 Diagram depicting step-by-step cell surface biotinylation and purification with Dynabeads™ MyOne™ Streptavidin T1 methodology

(c) Place needle into the top of a yellow pipette tip (typically used for 20–200 μL pipettes) and slide needle down inside tip until it stops. Push the cut disk out into the pipette tip with the plunger. The disk should sit ~2–3 mm from the end of the tip, which will retain ~10 μg peptides.

3. Preparing STAGE-Tips in autovial insert (*see* **Note 9**):

   (a) Obtain one vial insert per sample.

   (b) Remove the spring from the vial insert but retain the spring.

   (c) Put the vial insert without the spring in the collection tube and close the collection tube.

The procedure of making STAGE-Tips is depicted in Fig. 4.

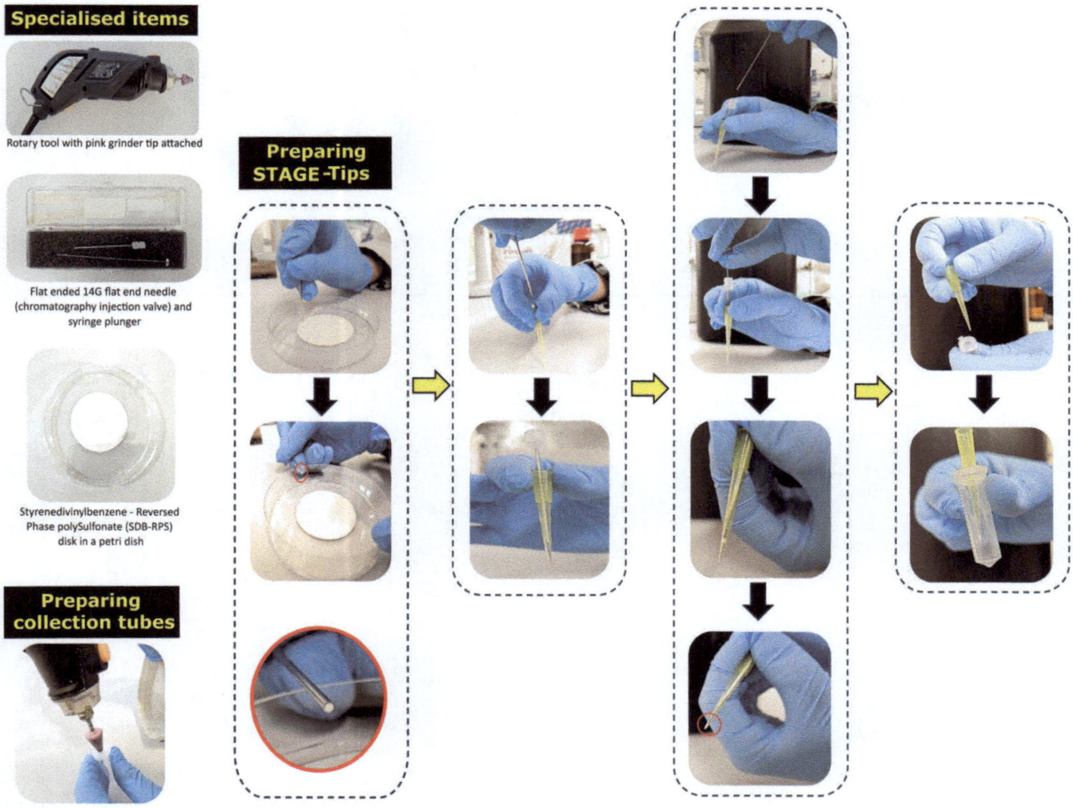

Fig. 4 Step-by-step photos demonstrating how to prepare STAGE-Tips

3.4 Solid Phase Extraction (SPE)

1. Remove the volume equivalent to 15 μg from the digest into a new tube.
2. Add 10× the volume (*see* **Note 10**) of the digest of SPE load buffer to the 15 μg of sample.
3. Centrifuge 10 min at 20,000× *g* to pellet insoluble.
4. Transfer supernatant to STAGE-Tip. Spin the liquid through the tip at an adequate g-force until dry, usually 5000× *g* for 2 min. If all liquid has not moved through, centrifuge again.
5. Wash with 100 μl of SPE load buffer. Spin until dry.
6. Wash with 100 μl of SPE wash buffer. Spin until dry.
7. Elute bound peptides with 50 μl of SPE elution buffer directly into injection vials.
8. Evaporate to dryness in a centrifugal vacuum concentrator such as a SpeedVac.
9. When ready to be injected (*see* **Note 11**), resuspend peptides in Buffer A* and inject the quantity of peptides that is appropriate for your LC-MS/MS system.

Each step of the SPE protocol is illustrated in Fig. 5.

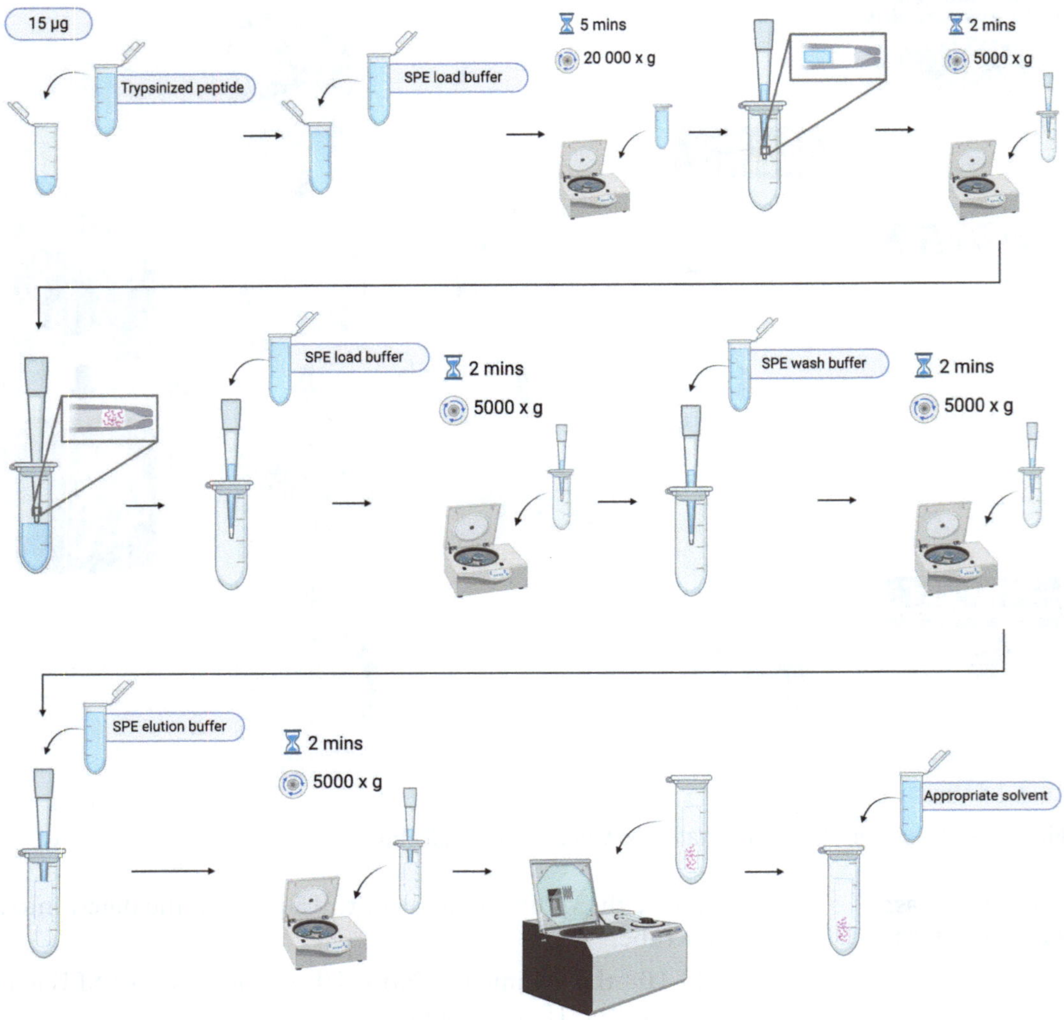

Fig. 5 Diagram depicting step-by-step solid phase extraction methodology

3.5 LC-MS/MS Analysis (See Note 12)

1. Sample preparation:
   (a) Sample loading: Load 5 µL of the sample onto the nanoEase Symmetry C18 trapping column. Flow rate: 15 µL/min. Duration: 3 min.
   (b) Column setup: Trapping column: nanoEase Symmetry C18 (180 µm × 20 mm). Analytical column: PicoFrit column (75 µm ID × 100 mm) packed with SP-120-1.7-ODS-BIO resin. Column temperature: 45 °C.
2. Chromatographic separation.
   (a) Transfer and washing: Wash the sample from the trapping column onto the PicoFrit analytical column.

(b) Elution program:
   (i) Step 1: 5–30% MS buffer B (100% acetonitrile) over 90 min.
   (ii) Step 2: 30–80% MS buffer B over 3 min.
   (iii) Step 3: 80% MS buffer B for 2 min.
   (iv) Step 4: 80–5% MS buffer B over 3 min.

3. Mass spectrometry analysis.
   (a) Peptide ionization: Ionize the eluting peptides at 2400 V.
   (b) Data-dependent MS/MS (dd-MS2):
      (i) Survey scan: Mass range: 350–1500 Da. Resolution: 70,000. Target peptides: charge state 2+ or higher. AGC target: 3e6. Maximum injection time: 50 ms.
   (c) Top 12 peptide selection:
      (i) Fragmentation in HCD cell. Isolation window: 1.4 m/z. AGC target: 1e5. Maximum injection time: 100 ms.
   (d) Fragment scanning:
      (i) Resolution: 17,500. Mass range: 120–2000 Da.
   (e) Exclusion: Mass of the precursor peptide excluded for 30 s.

3.6 Data Analysis of MS/MS Spectra

1. Parameter settings:
   (a) Fixed modifications: none.
   (b) Variable modifications:
      (i) Propionamide.
      (ii) Oxidized methionine.
      (iii) Deamidated asparagine.
      (iv) NHS-LC-biotin (for biotinylated samples).
   (c) Enzyme: semi-trypsin.
   (d) Number of allowed missed cleavages: 3.
   (e) Peptide mass tolerance: 10 ppm.
   (f) MS/MS mass tolerance: 0.05 Da.
2. Data analysis procedure:
   (a) Prepare databases: Ensure the organism database and contaminant database are specific to your experiment and up-to-date and correctly formatted for the search software being used.
   (b) Import data: Load the MS/MS data files into the chosen search software (Peaks Studio 11 or FragPipe v22).
   (c) Set search parameters:

(i) No fixed modifications.
(ii) Variable modifications: propionamide, oxidized methionine, deamidated asparagine, NHS-LC-biotin (if applicable).
(iii) Enzyme: semi-trypsin.
(iv) Number of allowed missed cleavages: 3.
(v) Peptide mass tolerance: 10 ppm.
(vi) MS/MS mass tolerance: 0.05 Da.

3. Run search: Execute the search against the appropriate organism database and the database of common contaminants.
4. Filter results:
   (a) Filter the search results to include only peptides with a –log10P score (Peaks Studio) that corresponds to a false discovery rate (FDR) of <1%.
   (b) Ensure that decoy database search matches are <1% of the total matches. The search software should automatically create decoys.
5. Validate and review results:
   (a) Review the filtered results to confirm the accuracy and relevance of the identified peptides.
   (b) Validate the peptide identifications by checking the –log10P scores and ensuring the FDR is below 1%.
6. Analyze the results:
   (a) Select proteins or peptides that are at least in 75% technical and biological replicates.
   (b) Remove proteins that have been found in the at least 75% technical and biological replicates of the control (which were not trypsin-shaved).
   (c) Compare both dataset from cell surface shaved and biotinylation analysis. Only keep proteins that are common in both datasets.

4 Notes

1. Needs to be made fresh at time of use with NH_4OH that is less than 12 months old as NH_4OH loses its ionic strength over storage.
2. Results will be optimal if the proteome database is strain specific. If required, sequence, assemble, and annotate the genome.

3. Set up technical replicates and use a duplicate as control which will not be shaved.
4. Take an aliquot to either count colony formed units (CFUs) or dye cells by LIVE/DEAD™ BacLight™ Bacterial Viability Kits [Invitrogen].
5. Take a second aliquot to either count CFUs or dye cells by LIVE/DEAD™ BacLight™ Bacterial Viability Kits [Invitrogen]. Compare this aliquot to the one prepared in **Note 4** (*see* **Note 4**). This comparison will determine if cell shaving lysed the cells.
6. Keep all washes.
7. Prepare before commencing SPE (*see* Subheading 3.4).
8. Prepare before commencing SPE..
9. Prepare after SPE.
10. If the sample volume is higher than 50 μl, then add ×0.1 volume 1% TFA in the sample and load it on the Stage-Tip.
11. If samples cannot be injected into the LC-MS/MS, keep them dried at −20 °C or − 80 °C.
12. Ensure all equipment is calibrated and maintained according to manufacturer guidelines, and handle samples and solvents with care to avoid contamination.

Acknowledgments

We would like to acknowledge the following for their contributions to developing this protocol: Dr. Ania Deutscher, Dr. Benjamin B.A. Raymond, Dr. Cheryl Jenkins, Dr. Daniel Bogema, Dr. Jessica Tacchi, Dr. Lisa Seymour, Dr. Michael Widjaja, and Dr. Tracey Kuit (née Burnett).

References

1. Raymond BB, Jenkins C, Seymour LM, Tacchi JL, Widjaja M, Jarocki VM et al (2015) Proteolytic processing of the cilium adhesin MHJ_0194 (P123J) in mycoplasma hyopneumoniae generates a functionally diverse array of cleavage fragments that bind multiple host molecules. Cell Microbiol 17(3):425–444
2. Tacchi JL, Raymond BB, Haynes PA, Berry IJ, Widjaja M, Bogema DR et al (2016) Post-translational processing targets functionally diverse proteins in mycoplasma hyopneumoniae. Open Biol 6(2):150210
3. Berry IJ, Widjaja M, Jarocki VM, Steele JR, Padula MP, Djordjevic SP (2021) Protein cleavage influences surface protein presentation in mycoplasma pneumoniae. Sci Rep 11(1):6743
4. Burnett TA, Dinkla K, Rohde M, Chhatwal GS, Uphoff C, Srivastava M et al (2006) P159 is a proteolytically processed, surface adhesin of mycoplasma hyopneumoniae: defined domains of P159 bind heparin and promote adherence to eukaryote cells. Mol Microbiol 60(3):669–686
5. Wilton J, Jenkins C, Cordwell SJ, Falconer L, Minion FC, Oneal DC et al (2009) Mhp493 (P216) is a proteolytically processed, cilium and heparin binding protein of mycoplasma

hyopneumoniae. Mol Microbiol 71(3): 566–582

6. Artini M, Papa R, Scoarughi GL, Galano E, Barbato G, Pucci P et al (2013) Comparison of the action of different proteases on virulence properties related to the staphylococcal surface. J Appl Microbiol 114(1):266–277
7. Widjaja M, Harvey KL, Hagemann L, Berry IJ, Jarocki VM, Raymond BBA et al (2017) Elongation factor Tu is a multifunctional and processed moonlighting protein. Sci Rep 7(1): 11227
8. Jarocki VM, Raymond BBA, Tacchi JL, Padula MP, Djordjevic SP (2019) Mycoplasma hyopneumoniae surface-associated proteases cleave bradykinin, substance P, neurokinin a and neuropeptide Y. Sci Rep 9(1):14585
9. Jarocki VM, Santos J, Tacchi JL, Raymond BB, Deutscher AT, Jenkins C et al (2015) MHJ_0461 is a multifunctional leucine aminopeptidase on the surface of mycoplasma hyopneumoniae. Open Biol 5(1):140175
10. Robinson MW, Buchtmann KA, Jenkins C, Tacchi JL, Raymond BB, To J et al (2013) MHJ_0125 is an M42 glutamyl aminopeptidase that moonlights as a multifunctional adhesin on the surface of mycoplasma hyopneumoniae. Open Biol 3(4):130017
11. Li Y, Wang J, Liu B, Yu Y, Yuan T, Wei Y et al (2022) DnaK functions as a moonlighting protein on the surface of mycoplasma hyorhinis cells. Front Microbiol 13:842058
12. Yu Y, Wang H, Wang J, Feng Z, Wu M, Liu B et al (2018) Elongation factor Thermo unstable (EF-Tu) moonlights as an Adhesin on the surface of mycoplasma hyopneumoniae by binding to fibronectin. Front Microbiol 9:974
13. Nunomura K, Nagano K, Itagaki C, Taoka M, Okamura N, Yamauchi Y et al (2005) Cell surface labeling and mass spectrometry reveal diversity of cell surface markers and signaling molecules expressed in undifferentiated mouse embryonic stem cells. Mol Cell Proteomics 4(12):1968–1976

Part II

Methods and Protocols Related to Bioelectricity

Chapter 10

Functional Characterization of Ion Channels in Planar Lipid Bilayers

Andras Szollosi

Abstract

Ion channels form transmembrane protein pores allowing ion fluxes across biological membranes. Their function is indispensable for normal homeostasis; therefore, channelopathies lead to a large variety of diseases affecting almost all tissues and organs. In the last decades, the structure of hundreds of ion channels has been solved. In the majority of the solved structures the pore is closed, and for most of the protein preparations used in these studies, currently little data is available to support that the purified protein forms a functional ion channel capable of gating. Planar lipid bilayer technique is a powerful tool to address this issue. Purified ion channel protein is incorporated into a membrane bilayer separating two compartments in an experimental chamber. Channel currents are recorded between the two sides. Such setup allows characterization of gating and permeation properties of ion channels unbiased by auxiliary components present in native membranes. The planar lipid bilayer technique is thus discussed herein together with the robust protein expressing BacMam system.

Key words Ion channels, BacMaM, Planar lipid bilayer, Sf9, HEK293 GnTI

1 Introduction

Ion channels are discrete protein pores in membranes that allow the passive flux of ions across the membrane [1]. Thousands of ion channels are found in the human body and underlie a plethora of physiological functions, such as generation of action potentials, regulation of Ca^{2+} homeostasis, hormone and chemokine secretion, immune response, heat and mechanical sensation, among others [2–6]. Lack or dysfunction of channels correspondingly lead to a large variety of pathological situations.

Opening and closing of channels, a process called gating, affect the open probability of ion channels, which eventually determine the time-dependent ion flux across the membrane. Certain channels activate upon ligand binding to the protein [7–11], while others respond to membrane potential changes [3, 12–14]. Besides gating, the nature and velocity of the permeating ion is an

Csaba Matta et al. (eds.), *The Surfaceome: Methods and Protocols*, Methods in Molecular Biology, vol. 2908,
https://doi.org/10.1007/978-1-0716-4434-8_10,

important determinant of ion channels, a property called permeation. Some channels allow a very high conductance of either cations or anions through their pore and yet strictly select for a single type of ion, while others are nonselective.

Ion channel function is studied by biophysical (electrophysiological measurements, structure determination methods, or titration assays such as isothermal titration calorimetry, surface plasmon resonance, etc.), biochemical (flux assays, immunohistochemistry, etc.), and molecular biology (RNA level determination, etc.) means. Since the "resolution revolution" in the cryo-EM technology [15], according to the Protein Data Bank, out of ~230,000 experimentally determined protein 3D structures, ~24,000 is determined by electron microscopy tools, out of which hundreds are ion channels [8, 9, 11, 14, 16–19]. Functional characterization of most channels is done in parallel studies by electrophysiological means, typically by the patch clamp technique employed either in the whole-cell or in cell-free configurations [20].

Most structural information on channels has been obtained from overexpressed, purified, detergent solubilized proteins. But whether such protein preparations contain functionally active protein has rarely been tested. Moreover, for many ion channels, auxiliary subunits strongly contribute to functional properties in living cells, rendering their responsiveness to different activating stimuli [8, 9, 14, 16, 17], which can contribute to the obtained biophysical properties of channels studied in patches.

Purely intrinsic properties of ion channel proteins can be studied in the planar lipid bilayer system. The major benefit of this technique is that the function of the same ion channel preparation that has been used for structural analysis can be examined. The function of several large-sized ion channels has been studied by the planar bilayer method [10, 13, 21–24]. Channel proteins are incorporated into a small bilayer portion of multilayer lipid films painted on an aperture separating the two sides of a bilayer chamber. Electrodes are dipped into buffers loaded into the two compartments, and channel currents between the two sides are measured.

Functional measurements in planar lipid bilayers require the expression and purification of the target ion channel protein. A robust expression method is the BacMam technique that has been successfully applied for expression of hundreds of large membrane-integrant proteins [25]. In brief, baculovirus is added to protein expressing mammalian HEK cells (hence the name BacMam expression system). Recombinant baculoviruses are generated in insect cells following transfection of cells with a "bacmid," a large viral genomic DNA bearing the gene of the target channel inserted by transposition. HEK cells internalize baculoviruses, and upon induction with a histone-deacetylase blocker, butyrate, they start to express the ion channel protein recognizing the CMV promoter upstream to the gene.

The present chapter reviews the general workflow of the BacMam technique with detailed methods for maintenance of insect and mammalian cell lines, transfection, and heterologous expression. It also discusses an example protein purification procedure and operation of the planar lipid bilayer system.

2 Materials

2.1 General Note

Most laboratories follow the manual of the Bac-to-Bac® Baculovirus Expression System by Life Technologies; therefore, certain materials are herein indicated as the originally recommended products; however, users can adapt the protocol to their own needs and specifications. Also, the DNA construct used for the BacMam expression system can differ among research groups. The target ion channel must be purified, yet the purification process (i.e., applied tags and chromatography) can be customized. We hereby describe an example purification protocol that employs an eGFP-fusion protein and the corresponding affinity chromatography (Subheading 3.6).

2.2 Sf9 Insect Cell Culture

1. Sf9 cell line.
2. Supplemented insect cell medium (serum-free insect cell medium, 5% heat-inactivated fetal bovine serum, 1% antibiotic-antimycotic reagent) (recommended: Sf-900™ II SFM Gibco™).
3. Screw-cap flat bottom Erlenmeyer flasks (100–500–1000–2500 mL).
4. dH_2O.
5. 70% ethanol.
6. Domestic bleach.
7. DMSO.
8. Consumables: 1.5 mL reaction tubes, pipette tips, 15–50 mL conical tubes, cryovials.
9. Devices: autoclave, 27–28 °C full-door cabinet incubator with orbital shaker, laminar flow hood, vacuum aspirator, 37 °C water bath, centrifuge, automated pipette, liquid nitrogen tank, isopropanol-containing freezing container, −80 °C freezer, hemocytometer or Bürker chamber.

2.3 HEK 293S GnTI$^-$ Culture

1. HEK 293S GnTI$^-$ cell line.
2. Supplemented HEK_{adh} cell medium (serum-free DMEM-F12 medium, 10% heat-inactivated fetal bovine serum, 1% antibiotic-antimycotic reagent).

3. Supplemented HEK_{sus} cell medium (serum-free FreeStyle™ 293 Expression Medium Gibco™, 2% heat-inactivated fetal bovine serum, 1% antibiotic-antimycotic reagent).
4. Baffled Erlenmeyer flasks (500–1000–2500 mL).
5. Ca^{2+} /Mg^{2+} free Dulbecco's phosphate-buffered saline (D-PBS).
6. 0.25% (w/v) Trypsin–0.53 mM EDTA solution.
7. dH_2O.
8. 70% ethanol.
9. Domestic bleach.
10. DMSO.
11. Consumables: 75 and 175 cm^2 cell culture flasks, 1.5 mL reaction tubes, pipette tips, 15–50 mL conical tubes, cryovials.
12. Devices: autoclave, 37 °C 5% CO_2 cabinet incubator, CO_2-resistant orbital shaker (placed into incubator), laminar flow hood, vacuum aspirator, 37 °C water bath, centrifuge, automated pipette, liquid nitrogen tank, isopropanol-containing freezing container, −80 °C freezer, hemocytometer or Bürker chamber.

2.4 Transfection of Sf9 Cells with Bacmid and Virus Production

1. Transfer vector with insert (CMV promoter, chimeric intron, suitable tag, antibiotic resistance gene, Tn7 R and L transposition sites).
2. MIDIprep DNA purification kit.
3. MAX Efficiency™ DH10Bac Competent Cells (Gibco™).
4. LB-broth.
5. LB-agar.
6. dH_2O and ddH_2O (RNAse, DNAse free).
7. 1000× antibiotic stocks.
8. IPTG 100× stock: 100 mM in H_2O (0.119 g/5 mL) - 1 mL frozen aliquots.
9. Bluo-GAL 100× stock: 30 mg/mL in 50% DMSO +50% H_2O (0.15 g/5 mL) - 1 mL frozen aliquots.
10. S.O.C. medium.
11. Isopropanol.
12. 70% ethanol.
13. Cellfectin™ II Reagent (Gibco™).
14. Screw-cap flat bottom Erlenmeyer flasks (2500 mL).
15. Consumables: 1.5–2 mL reaction tubes, pipette tips, 15–50 mL conical tubes, 10 cm Petri dishes, 6-well cell culture

plates, 175 cm^2 cell culture flasks, 0.22 μm filtration system (for 500 mL volume).

16. Devices: autoclave, 37 °C bacterial cabinet incubator, table-top heated mixer (for 37 °C and 42 °C), 27–28 °C full-door cabinet incubator with orbital shaker, laminar flow hood, vacuum aspirator, centrifuge, automated pipette, hemocytometer or Bürker chamber, analytical balance, microvolume UV–Vis spectrophotometer.

2.5 Transduction of HEK 293S GnTI– Cells with P3 Virus

1. Supplemented HEK_{sus} cell medium (serum-free FreeStyle™ 293 Expression Medium Gibco™, 2% heat-inactivated fetal bovine serum, 1% antibiotic-antimycotic reagent).
2. Baffled Erlenmeyer flasks (2.5–3 liter).
3. dH_2O.
4. Na-butyrate (1 M stock: 1.1 g/10 mL dH_2O; filter it).
5. 500 mL screw-cap centrifuge tubes.
6. Consumables: 0.22 μm syringe filter, 10 mL syringe, pipette tips.
7. Devices: autoclave, 37 °C 5% CO_2 cabinet incubator, 30 °C 5% CO_2 cabinet incubator, CO_2-resistant orbital shaker (placed into incubator), laminar flow hood, centrifuge, automated pipette, analytical balance.

2.6 Protein Purification

1. Lauryl Maltose Neopentyl Glycol (LMNG) powder.
2. Cholesteryl Hemisuccinate (CHS) Tris Salt powder.
3. Digitonin powder.
4. dH_2O.
5. Isopropanol.
6. 96% ethanol.
7. Dithiothreitol (1 M stock in dH_2O).
8. Phenylmethylsulfonyl fluoride (PMSF), 100× stock: 100 mM in isopropanol.
9. Leupeptine, 1000× stock: 1 mg/mL in dH_2O.
10. Pepstatine A, 1000× stock: 1 mg/mL in 96% ethanol.
11. Lysis buffer: 50 mM Tris–HCl pH 7.0–8.0, 200 mM NaCl, 2 mM $MgCl_2$, 20% glycerol, 1 mM DTT, 1× PMSF, 1× leupeptin, 1× pepstatin A.
12. SEC buffer: 20 mM Tris–HCl pH 7.0–8.0, 150 NaCl, 1 mM DTT, ~0.06% digitonin, set pH to 7.0–8.0; see preparation at Subheading 3.6, **steps 1** and **2**.
13. NHS-sepharose-immobilized GFP-nanobody resin.
14. PreScission protease.

15. Glass chromatography column.
16. SEC column.
17. Ultrafiltration device.
18. Gel coloring reagent (e.g., Coomassie brilliant blue).
19. Consumables: 1.5 mL reaction tubes, pipette tips, 15–50 mL conical tubes, 0.22 μm syringe filter, 10 mL syringe.
20. Devices: FPLC purification system in low-temperature cabinet, analytical balance, pH meter, SDS-PAGE system, water bath sonicator, nutating platform mixer, centrifuge (for 48.000 rcf), 4 °C cabinet for nutation, microvolume UV–Vis spectrophotometer.

2.7 Ion Channel Reconstitution in Planar Lipid Bilayers

1. *Cis* buffer (dependent on the given ion channel; e.g., 20 mM Tris–HCl pH 7.0–8.0; 140 mM NaCl, stimulating ligand).
2. *Trans* buffer (dependent on the given ion channel; e.g., 20 mM Tris–HCl pH 7.0–8.0; 140 mM NaCl).
3. Phosphatidylethanolamine (16:0–18:1 POPE) chloroform solution.
4. Phosphatidylserine (16:0–18:1 POPS) chloroform solution.
5. Phosphatidylcholine (16:0–18:1 POPC) chloroform solution.
6. Glass vial.
7. Glass pipette tips or Hamilton syringe.
8. n-Decane.
9. Fire-polished close-tip glass capillary.
10. Slim copper wire (to puncture lipid film on cuvette).
11. dH_2O.
12. 70% ethanol.
13. Detergent solution for cuvette cleaning.
14. Chloroform.
15. Devices: argon gas bottle; *bilayer setup*: bilayer chamber, magnetic stirrer bars, cuvette, metal insulating shell (box or Faraday cage), magnetic stirrer, antivibration table, amplifier with headstage electrodes, A/D converter, PC with analysis software.

3 Methods

The general guideline for cell culture work is to handle all materials under sterile conditions. Do dilutions and passage of cells under a laminar flow (i.e., cell culture) hood. Prior to placing any consumables, pipette tips, vials into the hood, wipe them with a 70% ethanol-soaked paper wiper. Expose the chamber to UV light for

~15 min before and after cell culture manipulation. Let medium stand on the bench before using it to bring it close to room temperature.

3.1 The Workflow of the BacMam System and Ion Channel Reconstitution

The workflow of the BacMam system is illustrated in Fig. 1. The gene of the target ion channel protein first must be cloned into a suitable transfer vector together with an appropriate fusion tag for later purification. The transfer vector is transformed into $DH10_{Bac}$ cells, which generate recombinant bacmid via transposition. Sf9 insect cells are then transfected with the purified bacmid DNA and will produce recombinant baculoviruses. Virus aliquots are added to HEK 293S $GnTI^{-}$ mammalian cells in suspension. Mammalian expression is induced, and cell pellets are used for subsequent protein purification. In the present protocol, the example ion channel is purified by eGFP-affinity and subsequent size exclusion chromatography (Subheading 3.6). Purified protein is added to pre-established planar bilayer membranes and reconstituted channel currents are measured.

3.2 Sf9 Insect Cell Culture

Use supplemented media for all steps indicated. Serum-free medium is only used for cell washes during the transfection protocol (*see* Subheading 3.4, **steps 31–35**).

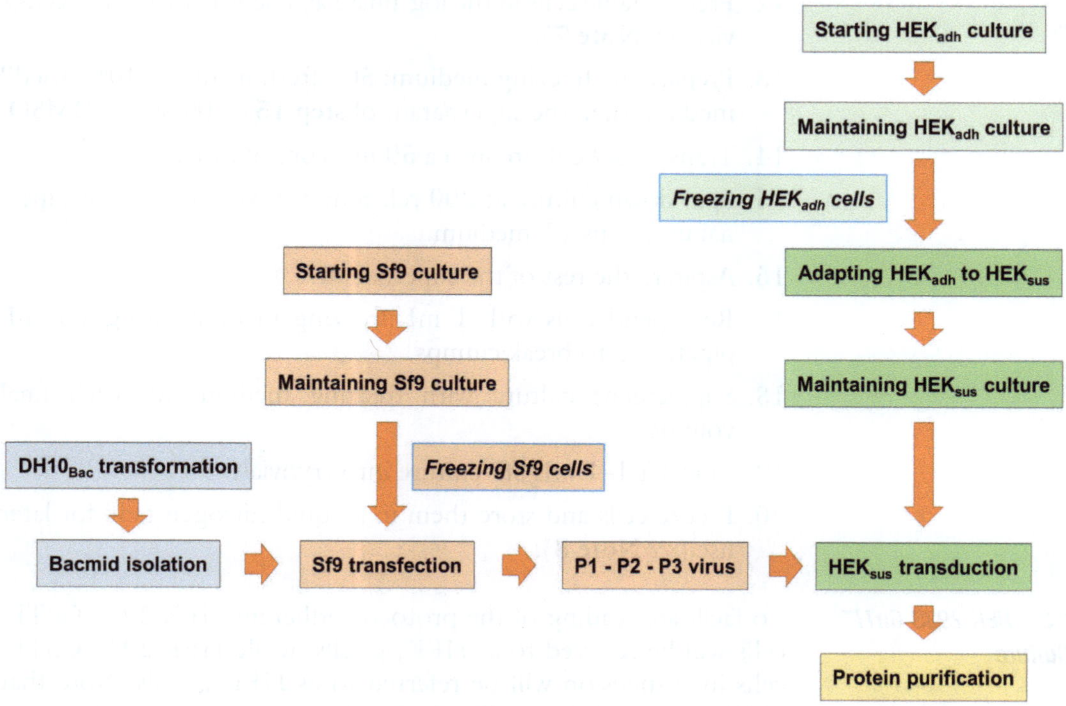

Fig. 1 Workflow of the BacMam system. Following bacmid transfection into Sf9 insect cells, recombinant baculoviruses are added to HEK cells in solution. Ion channel protein expression is induced by sodium butyrate

1. Use frozen aliquots of Sf9 cells for initiating a new culture at a cell density of 1×10^7 cells/mL.
2. Add 30 mL medium to a 50 mL conical tube.
3. Quickly thaw 2 vials of cells (removed from liquid nitrogen) in a 37 °C water bath for ~1–1.5 min but avoid complete thawing of cells.
4. Transfer cells from both vials into the medium.
5. Spin down the tube at 200 rcf, 5 min at room temperature. Cells will sediment to the bottom (*see* **Note 1**).
6. Aspirate the supernatant.
7. Resuspend cells with a 1 mL pipette tip to break clumps. Add 19 mL more medium and transfer the suspension into a 100 mL Erlenmeyer flask.
8. Shake cell culture in a heated cabinet (27–28 °C) at 90–130 rpm (*see* **Notes 2**, **3**, and **4**).
9. Count cell density in a hemocytometer or Bürker chamber (*see* **Note 5**).
10. Passage cells at a density of $2–3 \times 10^6$ cells/mL. Dilute culture to reach a seeding density of $5–10 \text{ x } 10^5$ cells/mL (*see* **Note 6**).
11. Add bleach to discarded culture. Rinse vials also with bleach and water before autoclaving them.
12. Freeze viable cells in the log-phase at a density of 1×10^7 cells/vial (*see* **Note** 7).
13. Prepare the freezing medium: 50% fresh medium, 40% "used" medium (i.e., the supernatant of **step 15**), 10% sterile DMSO.
14. Transfer Sf9 culture into a 50 mL conical tube.
15. Spin down culture at 200 rcf, 5 min. Save some of the supernatant as "used" medium.
16. Aspirate the rest of the supernatant.
17. Resuspend cells with 1 mL freezing medium using a 1 mL pipette tip to break clumps.
18. Supplement culture with freezing medium to reach final volume.
19. Transfer 1–1 mL Sf9 culture into cryovials.
20. Freeze cells and store them in a liquid nitrogen tank for later use (*see* **Note 8**).

3.3 HEK 293S GnTI⁻ Culture

To facilitate reading of the protocol, adherent HEK 293S GnTI$^-$ cells will be referred to as HEK_{adh} cells, while HEK 293S GnTI$^-$ cells in suspension will be referred to as HEK_{sus} cells. Note that HEK_{adh} and HEK_{sus} cells require different culture medium (*see*

Subheading 2; Subheading 3.3, **steps 2** and **37**). ***All media are complete and supplemented as indicated in Materials.***

1. Use frozen aliquots of HEK cells for initiating a new culture at a cell density of 3–5 × 10^6 cells/mL.
2. Add 30 mL ***DMEM F-12*** medium to a 50 mL conical tube.
3. Quickly thaw a vial of cells (removed from liquid nitrogen) in a 37 °C water bath for ~1–1.5 min but avoid complete thawing of cells.
4. Transfer cells into the medium.
5. Spin down the tube at 200 rcf, 5 min at room temperature. Cells will sediment to the bottom (*see* **Note 1**).
6. In parallel, add 14 mL DMEM F-12 to a 75 cm^2 cell culture flask.
7. When centrifugation is done, aspirate the supernatant.
8. Resuspend cells in 1 mL of DMEM F-12 medium with a 1 mL pipette tip to break clumps.
9. Pipette resuspended cells into the flask (volume adds up to 15 mL) and gently agitate it to disperse cells.
10. Incubate cells in a 5% CO_2 incubator at 37 °C.
11. When cells reach 80–90% confluency, passage them (*see* **Note 9**).
12. Aspirate medium from the culture flask.
13. Add 5 mL D-PBS (*see* **Note 10**).
14. Aspirate D-PBS.
15. Add 2 mL Trypsin-EDTA to the cells. Let it stand for 5 min in the incubator (*see* **Note 11**).
16. Add 8 mL medium to the flask to dilute out trypsin. Pipette it up-and-down gently to break clumps.
17. Dilute cells with media in a 75 cm^2 cell culture flask to reach a seeding density of 3–5 × 10^6 cells/mL.
18. Incubate cells in a 5% CO_2 incubator at 37 °C.
19. To expand adherent culture, make dilution in a 175 cm^2 flask (follow **steps 12–18**). Final volume is 25 mL.
20. Freeze viable cells at the log-phase (*see* **Note 12**).
21. Grow cells in flask (either 75 or 175 cm^2) until 80–90% confluency prior to freezing (*see* **Note 13**).
22. Aspirate medium from the culture flask.
23. Add 5 mL D-PBS (*see* **Note 10**).
24. Aspirate D-PBS.

25. Add 2 mL Trypsin-EDTA to the cells. Let it stand for 5 min in the incubator (*see* **Note 11**).
26. Add 8 mL medium to the flask to dilute out trypsin. Pipette it up-and-down gently to break clumps.
27. Transfer cells into a 15 mL conical tube and count the total cell number in a hemocytometer or Bürker chamber (*see* **Note 5**).
28. Calculate the required amount of medium to yield a final cell count of 3–5 × 10^6 cells/mL for freezing.
29. Prepare the calculated amount of freezing medium: 90% DMEM F-12, 10% DMSO.
30. Spin down the cell suspension from **step 27** (200 rcf, 5 min) aspirate supernatant and resuspend cells in the prepared freezing medium.
31. Transfer 1–1 mL HEK_{adh} cultures into cryovials.
32. Freeze cells and store them in a liquid nitrogen tank for later use (*see* **Note 8**).
33. To adapt HEK_{adh} to HEK_{sus} cells, expand HEK_{adh} culture to 4–5× 175 cm^2 flasks and grow cells to 70–80% confluency.
34. Aspirate media from all flasks.
35. Add 10–10 mL D-PBS (*see* **Note 10**).
36. Aspirate D-PBS.
37. Add 25 mL ***Freestyle*** medium to all flasks (***without trypsin***). Resuspend cells with a 10 mL pipette tip to break clumps.
38. Transfer all resuspended cells into a ***baffled*** flask (*see* **Note 14**).
39. Incubate cells in a 5% CO_2 incubator at 37 °C with ***continuous shaking at 135 rpm***.
40. Passage cells by diluting the culture at a 2–3 × 10^6 cells/mL density to a seeding density of 5 × 10^5 cells/mL.
41. Add bleach to discarded culture. Rinse vials also with bleach and water before autoclaving them.

3.4 Transfection of Sf9 Cells with Bacmid and Virus Production

1. Prior to transformation, prepare LB-agar plates: 20–25 mL LB-agar/10 cm Petri-dish. Autoclave LB-agar and add antibiotics, IPTG and BLUO-Gal to cooled medium before solidifying. Final concentrations: kanamycin 50 μg/mL, gentamicin 7 μg/mL, tetracycline 10 μg/mL, BLUO-Gal 100 μg/mL, IPTG 40 μg/mL.
2. Thaw one vial of $DH10_{Bac}$ competent cells on ice.
3. Add 2–300 ng transfer vector DNA (*see* **Note 15**).
4. Incubate cells for 30 min on ice.
5. Expose the mixture to 42 °C (heat-shock) for 45 s.
6. Cool mixture for 2 min on ice.

7. Add 900 μL S.O.C. medium and shake cells at 37 °C at 220 rpm for 4 h.
8. Split 100–250–650 μL aliquots of culture into 3× 1.5 mL tubes. Spin tubes at 2200 rcf for 5 min.
9. Drain supernatant and resuspend cells in 200–200 μL S.O.C. medium.
10. Plate cells on LB-agar (Kan^r-Gent^r-Tet^r, BLUO-Gal, IPTG) plates and incubate them in a 37 °C bacterial cabinet incubator for 2–3 days.
11. Recombinant colonies on plates should look solid white in contrast to blue/purple nontransformed colonies.
12. With a 200 μL pipette tip, collect a large white colony and dip it into 10 mL LB (Kan^r-Gent^r-Tet^r) (*see* **Note 16**).
13. Grow culture at 37 °C at 220 rpm overnight (*see* **Note 17**).
14. For bacmid preparation, use the solutions (P1–3; or solubilization, lysis and neutralization buffers) of any MIDIprep DNA purification kit. Pre-chill all solutions, keep tubes on ice and do centrifugation steps at 4 °C.
15. Spin down the culture (**step 13**) at 5000 rcf for 10 min. Keep pellet and discard supernatant.
16. Prepare two 2 mL tubes with 800–800 μL isopropanol and place them on ice for **step 22** later.
17. Resuspend cell pellet (**step 15**) in 500 μL P1 and transfer sample into a 2 mL plastic tube.
18. Add 500 μL P2 and mix. Incubate for 5 min on ice.
19. Add 500 μL P3 and mix. Incubate for 5 min on ice.
20. Spin for 10 min at 14,000 rcf. A large pellet should be visible. Keep the supernatant.
21. Spin down the supernatant for a second time (same settings), no major pellet should be visible.
22. Split the supernatant into 750–750 μL aliquots and add them directly into the 800–800 μL isopropanol (prepared in **step 16**).
23. Mix tubes by flipping 5–6 times and incubate them for 15 min on ice.
24. Spin samples for 15 min at maximum speed (e.g., 30,000 rcf). Simply drain supernatant by flipping tube.
25. Wash pellet with 500 μL 70% ethanol and spin for 2 min at maximum speed.
26. Drain ethanol and spin for a second time (same settings). Aspirate all ethanol and air-dry tube for 10–15 min at room temperature.

27. Dissolve both pellets in 30–30 μL ddH_2O by gentle pipetting. Do not vortex.
28. After 20–30 min, merge the 2 tubes (yielding 60 μL bacmid sample) and measure DNA concentration in a microvolume UV–Vis spectrophotometer (*see* **Note 18**).
29. To transfect Sf9 insect cells, prepare a 6-well plate for each bacmid transfection.
30. Add 1.5×10^6 Sf9 insect cells/well. Let them adhere for at least 30 min (*see* **Note 19**).
31. Prepare mixes for transfection in sterile 1.5 mL plastic tubes. ***Use serum-free medium!***
    (a) **4×** [Per well: 10 ug bacmid - e.g., 2 μL of 5 μg/μL - into 100 μL media]: total 400 μL
    (b) **4×** [Per well: 8 μL well resuspended Cellfectin in 100 μL media]: total 400 μL
32. Add bacmid mix (**step 31a**) into Cellfectin mix (**step 31b**) and mix thoroughly by inverting tube 5–6 times.
33. Let the reaction mixture stand at room temperature for 30 min.
34. In parallel, wash plated Sf9 cells with ***serum-free medium (SFM)***: aspirate medium and carefully add 2 mL SFM to the wells. Aspirate medium again and load 2 mL SFM for the second time.
35. Add 200 μL transfection mixture (**step 32**) to each of the 4 wells (*see* **Note 20**).
36. Incubate plate, protected from light, at 27 °C at least for 4 h, then aspirate media and change it to 2 mL ***supplemented medium***.
37. Incubate plate at 27 °C for 3 days.
38. Collect the supernatant from all 4 wells of the 6-well plate and merge them (*see* **Note 21**).
39. Plate 3.5×10^7 cells in a 175 cm^2 plate. Let it settle for at least 30 min (*see* **Note 22**).
40. Aspirate medium completely and add 3 mL P1 virus (from **step 38**) to the cells without addition of extra medium.
40. Incubate cells with P1 virus at 27 °C for 1 h, while gently agitating it every 20 min.
41. Add 22 mL medium and incubate plate, protected from light, at 27 °C for 3 days (*see* **Note 23**).
42. To obtain the P3 virus, simply pour the 25 mL media from the plate (**step 42**) into 500 mL of Sf9 suspension culture (in a 2.5-liter flat bottom Erlenmeyer) at density of 1.5×10^6 cells/mL.

43. Shake cells at 27 °C for 3 days at 110–120 rpm.
44. Following the 3 days incubation spin down the culture at 2500 rcf, 20 min in a 500 mL centrifuge tube.
45. Keep supernatant as it is the P3 virus fraction (*see* **Note 24**).

3.5 Transduction of HEK 293S GnTI– Cells with P3 Virus

1. Grow 800–1000 mL HEK_{sus} cells in a 2–3 l baffled flask to a density of 2.5–3.5 × 10^6 cells/mL in Freestyle medium (*see* **Note 14**).
2. Add 10% P3 virus (*see* **Note 25**).
3. Shake cells for 12 h at 135 rpm in a 5% CO_2 incubator at 37 °C.
4. Add 9 mL freshly prepared Na-butyrate from sterile 1 M stock (*see* **Note 26**).
5. Shake cells for 48 h at 135 rpm in a 5% CO_2 incubator at ***30 °C***.
6. Harvest cells after 48 h of expression at 2500 rcf, 20 min.
7. Discard supernatant and keep pellet.
8. Gently wash pellet with D-PBS and use immediately for protein purification or store pellet for several months in a – 80 °C freezer.

3.6 Protein Purification

1. Prepare detergent stocks (5% LMNG +0.5% CHS and 0.06% digitonin) in dH_2O on the day before purification and nutate them at 4 °C overnight.
2. On the day of purification, filter digitonin, then water bath sonicate it for 10 min. Add SEC buffer components, bring to volume and set pH to 7.0–8.0.
3. Use fresh or thawed 800 mL HEK_{sus} cell pellet (*see* Subheading 3.5, **step 8**). Resuspend it in 10 mL Lysis buffer, then bring volume to 30 mL.
4. Add the 10 mL of 5% LMNG-0.5% CHS stock (**step 1**).
5. Solubilize membrane proteins by nutation for 2 h at 4 °C.
6. Spin down the lysate at 48.000 rcf, 1 h at 4 °C (*see* **Note 27**).
7. Mix the supernatant with 2 mL bed volume of GFP-nanobody-resin (preequilibrated with 10 column volume H_2O and 10 column volume Lysis buffer without detergent).
8. Bind protein by nutation in a 50 mL conical tube for 2–3 h at 4 °C.
9. Pour the protein-resin mixture to a glass chromatography column and wash the beads with 20 column volume SEC buffer (*see* **Note 28**).
10. Resuspend beads 1:1 with SEC buffer and add PreScission protease (*see* **Note 29**).
11. Cleave eGFP-tag by nutation overnight at 4 °C.

12. Collect flow-through with 2 column volume excess SEC buffer.
13. Concentrate protein to 0.5–1 mL with a proper molecular weight cut off ultrafiltration device.
14. Perform size-exclusion chromatography with SEC buffer at 4 ° C using a column with appropriate molecular weight resolution for the target protein.
15. Collect elution peaks at the expected retention volume and concentrate protein using a proper molecular weight cut off ultrafiltration device.
16. Check the purity of the ion channel protein by SDS-PAGE and determine final concentration by UV or colorimetric spectroscopy.
17. Adjust the concentration of the protein to 2–3 mg/mL with SEC buffer.

3.7 Ion Channel Reconstitution in Planar Lipid Bilayers

Horizontal and vertical bilayer setups are both used in laboratories. The present protocol introduces a horizontal lipid bilayer setup consisting of a bilayer chamber with a cup (or cuvette) (Fig. 2). An aperture on the front wall of the cuvette with a variable diameter (typically 150 or 200 μm) separates the *cis* and *trans* compartments. Lipid films are painted on the aperture, and ion channel proteins are embedded into bilayer zones of this multilayer lipid film. A sample and a reference electrode are dipped into buffers loaded into the *cis* and *trans* sides, and channel currents between the two sides are measured. The chamber is placed into an electrically insulating shell (metal box or Faraday cage), which is placed on an antivibration table to avoid mechanical noise. Proper grounding of the setup is inevitable for low-noise channel current recordings.

1. Use high purity chloroform stocks of membrane lipids (*see* **Note 30**). Mix phosphatidylethanolamine (POPE), phosphatidylserine (POPS), and phosphatidylcholine (POPC) at a ratio of 5:4:1 (*see* **Note 31**).
2. Dry the lipid mixture with a blow of argon gas (*see* **Note 32**).
3. Resuspend the dried lipid mixture in n-decane at a final concentration of 20 mg/mL. Vortex for 1 min to completely dissolve lipids. Keep lipid/decane mixture on ice during measurement (*see* **Note 33**).
4. Paint lipid on the aperture of a 1 mL cuvette. Pipette 0.5 μL lipid/decane mixture with a small tip and just touch it to the aperture without releasing any of it. It will add sufficient amount of lipids to the aperture.
5. Puncture the lipid paint with a narrow filament (e.g., a slim copper wire).

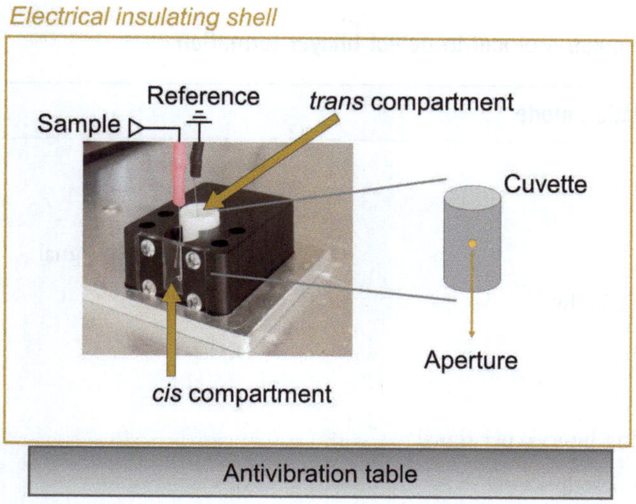

Fig. 2 The planar bilayer setup. The front wall of the cuvette, bearing the aperture, separates the *cis* and *trans* compartments of the bilayer chamber. During measurement, the sample electrode is dipped into the *cis* while the reference electrode into the *trans* compartment filled with buffer. Bilayer is formed in areas of the multilayer lipid film painted on the aperture

6. Let the cuvette stand for 10 min in a laid down position with aperture pointing up.
7. Puncture the lipid paint again prior to chamber assembly.
8. Fix the lipid-coated cuvette between the *cis* and *trans* compartments of the chamber.
9. Load appropriate buffers to both sides, add magnetic stirrers and stir solutions for 1 min (*see* **Note 34**).
10. Dip the sample (to *cis* side) and reference (to *trans* side) electrodes attached to the headstage of the amplifier into the buffers.
11. Load an extra layer of lipids (from the n-decane solution) onto the aperture, by dipping a closed-tip glass capillary into the lipid mix and then touch it to the aperture (immersed into buffer). Do not stir solution at this point.
12. Wait for bilayer formation. To detect it, currents are measured between the sample (*cis*) and reference (*trans*) electrodes using an episodic 5 mV voltage pulse protocol (Table 1) (*see* **Notes 35**, **36**, and **37**).
13. Pipette 1–5 μL of purified channel protein (from Subheading 3.6, **step 17**) slightly beyond the aperture. At this point keep membrane potential at 0 mV.
14. Wait a few minutes without agitation then stir solution for 1 min.

Table 1
Voltage pulse protocol to detect bilayer formation

Acquisition mode		Episodic stimulation
Trials	Runs/trial	10.000
	Sweeps/run	1
	Samples/sweep/signal	360
Trial hierarchy	First holding	5 samples (0.5 ms)
	Epochs	350 samples (35 ms)
	Last holding	5 samples (0.5 ms)
Sampling interval per signal		100 μs
Epoch description	Type	Ramp
	First level (mV)	± 5
	Delta level (mV)	0
	First duration (samples)	50
	Delta duration (samples)	0

15. Apply −80 mV to the *cis* side and then apply alternating ±200 mV potential pulses with external command (*see* **Note 38**).
16. Upon channel opening, switch off the external command, apply the desired membrane voltage, close the insulating box (or Faraday cage), and start recording.
17. Currents are digitized at 3–10 kHz and filtered at 1 kHz by an eight-pole low-pass Bessel filter. Use a gain of 0.5–10 on the amplifier.
18. Analyze recordings with an appropriate display software.
19. Between measurements, disassemble bilayer chamber, wipe the aperture of the cuvette with an ethanol-soaked wiper then rinse it with water. Dry it thoroughly prior to subsequent lipid painting. Paint new lipid on aperture following the instructions in **steps 4–7**. Start new reconstitution.
20. At the end of a measurement day, wash cuvette thoroughly with a detergent solution and a big excess of water and dry it. Also rinse bilayer chamber and parts with ethanol and water. Rinse the lipid containing glass vial with chloroform to get rid of residual lipids.

4 Notes

1. **Steps 1–5** are important for a rapid wash-out of DMSO.
2. Shaking velocity depends on the shape of the vial; the aim is to have a regular circular mixing of the suspension instead of local trembling in the vial.
3. Use a maximum of 1:5 v:v ratio for culture volume vs total Erlenmeyer capacity for good aeration. Erlenmeyer is a flat-bottom flask (i.e., nonbaffled).
4. Viable Sf9 cells normally duplicate overnight. After initiation of new culture from frozen stock, repeat passages until such division rate is accomplished.
5. Remove 10–30 μL of the culture under the hood for cell counting. Agitate culture before removal not to let cells to sediment.
6. Calculate dilution factor to reach seeding density and mix Sf9 culture with fresh medium. Transfer mixture into a new autoclaved Erlenmeyer flask. A good routine is to use the 5×10^5 cells as seeding density and grow them for 2–3 days without changing medium.
7. Sf9 cells can be frozen to maintain an in-house cell bank. Freeze vials depending on needs and storage capacity. The protocol is described in **steps 12–20**, which are therefore optional. Here is an example calculation: cell density of the culture is 2×10^6 cells/mL. Spin down 50 mL culture and resuspend cells in 10 mL freezing medium (*see* **step 13**). It will provide for 10× frozen aliquots at the desired cell density.
8. Freeze cells in an isopropanol-containing container decreasing temperature by 1 °C/min. Transfer vials into the container at room temperature then place it to a − 80 °C freezer. Keep cells in freezer at least for 4 h (ideally overnight), then transfer vials into a liquid nitrogen tank.
9. After the first 1–2 passages, cells will grow with a regular division rate. A good routine is to dilute cells *5×* for 3 days grow between Tuesdays and Fridays and a *10×* dilution for 4 days grow between Fridays and Tuesdays.
10. Add D-PBS to the side of the flask in order not to disturb the cells, then roll it over the cells and agitate mildly.
11. Note that tapping the flask might promote clumping so instead wait a bit longer for the cells to detach.
12. HEK_{adh} cells can be frozen to maintain an in-house cell bank. The protocol is described in **steps 20–32**, which are therefore optional. One can skip these steps and directly move to **step 33**.

13. The number of cells on a confluent plate is typically 8–9 × 10^6 (T75) or 2–3 × 10^7 (T175) cell.
14. Grow HEK_{sus} in baffled Erlenmeyer flasks and use a maximum of 1:5 v:v ratio for culture volume vs total Erlenmeyer capacity for good aeration.
15. Prepare a clean DNA sample for your transfer vector. MID-Iprep DNA concentration is typically 500–1000 ng/ul, yet using lower concentrations is also appropriate.
16. *Optional Step*: Instead of picking a full recombinant colony for the bacterial culture, use only half of it. Pick the other half of the same colony with another tip and simply re-streak it on an LB-agar (Kan^r-$Gent^r$-Tet^r, BLUO-Gal, IPTG) plate. It should produce white colonies again in 2–3 days. This is a confirmation step.
17. Start culture already in the morning since growing might be long. You can even prolong it for an extra day.
18. Wait at least 20 min before measuring the bacmid concentration since it takes time to dissolve the large chromosomal DNA. Too early measurement can lead to underestimation of the DNA concentration.
19. An example: dilute Sf9 culture to 5 × 10^6 cells/mL and pipette 3 mL to 4 wells of the plate. Leave 2 wells empty.
20. Add the mixture to cells dropwise by a circular movement of the pipette tip dipped into the medium. Agitate plate gently.
21. It is the P1 virus fraction. No need to filter it. Ideally use P1 immediately for P2 virus generation, otherwise, P1 fraction can be stored in a 4 °C fridge for a few weeks.
22. An example: dilute Sf9 culture to 5 × 10^6 cells/mL and pipette 7 mL to the plate already containing 18 mL medium.
23. Note that the dilution corresponds to ~12% P1. The 25 mL supernatant of this step is the P2 virus fraction.
24. Filter the supernatant (using a 0.22 μm sterile filter) and keep the P3 fraction in a 4 °C fridge. It can be kept and used for 1–3 months.
25. An example: in a 2.5-liter baffled flask: add 80 mL P3 virus to 800 mL HEK_{sus} culture.
26. A good routine is: if virus was added to culture in the evening, then add butyrate on the next morning; if virus was added to the culture in the morning, then add butyrate in the evening.
27. Keep small volume aliquots at each purification step to monitor the progress of purification by SDS-PAGE later.

28. It is crucial to use a large excess of SEC buffer to facilitate detergent exchange in micelles.
29. The volume of protease depends on its specific activity. Check the datasheet of your product for specific activity and unit definition, then calculate an amount that can cleave several mg of protein.
30. Handle chloroform-based solutions with glass or metal equipment (glass pipette tips or Hamilton syringe) as chloroform can react with plastic and might bring contamination to the samples.
31. Example: mix 50 μL POPE, 35 μL POPS, and 20 μL POPC from 25 mg/mL chloroform-based stocks in a glass vial. Rigidity of the bilayer can be increased by adding slightly more, e.g., 25 μL POPC.
32. This step takes a few minutes. Use a low gas stream not to disperse the lipids on the wall of the vial.
33. In our example (also *see* **Note 31**), it is 50 μL.
34. Select buffers (ionic strength, pH, etc.) that accommodate well to your target protein. Also, ion channel insertion into the membrane is a stochastic process and can happen in two orientations. To achieve selective detection of channels with identical orientations in all experiments, add stimulatory ligands only to one side of the chamber.
35. The typical current amplitude of the multilayer lipid membrane is 50–60 pA (for a 150 μm aperture) or 100–150 pA (for a 200 μm aperture) upon the 5 mV voltage pulse.
36. Bilayer formation is monitored by a stretching of the lipid and a corresponding enlargement of the current amplitude to 80–100 pA (for a 150 μm aperture) or 200–220 pA (for a 200 μm aperture) upon voltage pulse. It can happen spontaneously.
37. If bilayer is not formed spontaneously, facilitate its formation by applying alternating ±300 mV voltage pulses with external command. One should observe the increase in current amplitude. 400 mV applied potential should cease the bilayer, otherwise it is too rigid and unable for reconstitution. If it is resistant to such high potential, tap the chamber to break the bilayer mechanically (infinitely large current will be measured due to a short circuit) and paint more lipids on it with the glass capillary as in Subheading 3.7, **step 11**.
38. Channel protein insertion into the bilayer typically occurs upon a switch from positive to negative voltages.

Acknowledgments

This work was supported by EU Horizon 2020 Research and Innovation Program grant 739593 and National Research, Development and Innovation Fund grants KKP 144199.

References

1. Hille B (1978) Ionic channels in excitable membranes. Current problems and biophysical approaches. Biophys J 22:283–294
2. Bartok A, Csanady L (2022) Dual amplification strategy turns TRPM2 channels into supersensitive central heat detectors. Proc Natl Acad Sci USA 119:e2212378119
3. Catterall WA, Lenaeus MJ, Gamal El-Din TM (2020) Structure and pharmacology of voltage-gated sodium and calcium channels. Annu Rev Pharmacol Toxicol 60:133–154
4. Csanady L, Vergani P, Gadsby DC (2019) Structure, gating, and regulation of the CFTR Anion Channel. Physiol Rev 99:707–738
5. Miller C (2000) An overview of the potassium channel family. Genome Biol 1: REVIEWS0004
6. Szollosi A (2021) Two decades of evolution of our understanding of the transient receptor potential Melastatin 2 (TRPM2) Cation Channel. Life 11(5):397
7. Clayton GM, Silverman WR, Heginbotham L, Morais-Cabral JH (2004) Structural basis of ligand activation in a cyclic nucleotide regulated potassium channel. Cell 119:615–627
8. Lee CH, MacKinnon R (2018) Activation mechanism of a human SK-calmodulin channel complex elucidated by cryo-EM structures. Science 360:508–513
9. Lee KPK, Chen J, MacKinnon R (2017) Molecular structure of human KATP in complex with ATP and ADP. elife 6:e32481
10. Wang W, Whorton MR, MacKinnon R (2014) Quantitative analysis of mammalian GIRK2 channel regulation by G proteins, the signaling lipid PIP2 and Na+ in a reconstituted system. elife 3:e03671
11. Zhang Z, Toth B, Szollosi A, Chen J, Csanady L (2018) Structure of a TRPM2 channel in complex with Ca(2+) explains unique gating regulation. elife 7:e36409
12. Sait LG, Sula A, Ghovanloo MR, Hollingworth D, Ruben PC, Wallace BA (2020) Cannabidiol interactions with voltage-gated sodium channels. elife 9:e58593
13. Tao X, MacKinnon R (2008) Functional analysis of Kv1.2 and paddle chimera Kv channels in planar lipid bilayers. J Mol Biol 382:24–33
14. Wu J, Yan Z, Li Z, Qian X, Lu S, Dong M, Zhou Q, Yan N (2016) Structure of the voltage-gated calcium channel Ca(v)1.1 at 3.6 a resolution. Nature 537:191–196
15. Henderson R (2018) From electron crystallography to single particle CryoEM (Nobel lecture). Angew Chem 57:10804–10825
16. Sun J, MacKinnon R (2017) Cryo-EM structure of a KCNQ1/CaM complex reveals insights into congenital long QT syndrome. Cell 169(1042–1050):e1049
17. Tao X, Hite RK, MacKinnon R (2017) Cryo-EM structure of the open high-conductance Ca(2+)-activated K(+) channel. Nature 541: 46–51
18. Zhang Z, Chen J (2016) Atomic structure of the cystic fibrosis transmembrane conductance regulator. Cell 167:1586–1597. e1589
19. Zhao C, MacKinnon R (2021) Molecular structure of an open human K(ATP) channel. Proc Natl Acad Sci USA 118(48): e2112267118
20. Sakmann B, Neher E (1984) Patch clamp techniques for studying ionic channels in excitable membranes. Annu Rev Physiol 46:455–472
21. Eckford PD, Li C, Ramjeesingh M, Bear CE (2012) Cystic fibrosis transmembrane conductance regulator (CFTR) potentiator VX-770 (ivacaftor) opens the defective channel gate of mutant CFTR in a phosphorylation-dependent but ATP-independent manner. J Biol Chem 287:36639–36649
22. Lukacs V, Rives JM, Sun X, Zakharian E, Rohacs T (2013) Promiscuous activation of transient receptor potential vanilloid 1 (TRPV1) channels by negatively charged intracellular lipids: the key role of endogenous phosphoinositides in maintaining channel activity. J Biol Chem 288:35003–35013
23. Magyar ZE, Hevesi J, Groom L, Dirksen RT, Almassy J (2023) Function of a mutant ryanodine receptor (T4709M) linked to congenital myopathy. Sci Rep 13:14659

24. Szollosi A, Almássy J (2023) Functional characterization of the transient receptor potential melastatin 2 (TRPM2) cation channel from *Nematostella vectensis* reconstituted into lipid bilayer. Sci Rep 13:11471

25. Goehring A, Lee CH, Wang KH, Michel JC, Claxton DP, Baconguis I, Althoff T, Fischer S, Garcia KC, Gouaux E (2014) Screening and large-scale expression of membrane proteins in mammalian cells for structural studies. Nat Protoc 9:2574–2585

Chapter 11

Direct Assessment of Temperature Sensitivity of the TRPM2 Channel with the Patch Clamp Technique

Adam V. Toth and Adam Bartok

Abstract

The transient receptor potential (TRP) family of ion channels contains a large array of receptors of pain, taste, or temperature. Thus far, several TRP channels have been confirmed as warm or cold sensors; however, the exact molecular mechanism of heat detection remains obscure. The methods of investigation of temperature sensitivity involve several different techniques, e.g., in vivo experiments, fluorescence-based Ca^{2+} measurements, and electrophysiology. Here we describe an electrophysiological configuration that allows the direct assessment of the temperature sensitivity of the TRPM2 channel by measuring the temperature dependence of microscopic and macroscopic currents, apparent ligand binding affinities, and the gating parameters of the channel under the precise control of temperature and activating ligand concentrations. The technique described here may be adapted to explore temperature sensitive behavior of various other ion channels.

Key words Temperature sensitive ion channels, Thermoreceptors, Patch clamp, TRP channel, TRPM2

1 Introduction

TRPM2 (Transient receptor potential melastatin 2) is a strictly ligand gated, Ca^{2+} permeable, nonselective cation channel with confirmed intrinsic temperature sensitivity [1]. For opening, the channel requires intracellular Ca^{2+} ions and ADP-ribose (ADPR) and phosphatidylinositol-4,5-bisphosphate (PIP_2) in the inner leaflet of the plasma membrane [2]. The opening of the channel is intrinsically endothermic, and its temperature threshold is set by the concentrations of its activating ligands. Functional TRPM2 channels are homotetramers with structural resemblance to voltage-gated cation channels. TRPM2 orthologues are present in different eukaryotic species ranging from taxa as early as choanoflagellates to man [3]. TRPM2 is referred to as a chanzyme, as at the C-terminus of the protein, a domain homologous to the mitochondrial ADP-ribose pyrophosphatase enzyme NUDT9, called

Csaba Matta et al. (eds.), *The Surfaceome: Methods and Protocols*, Methods in Molecular Biology, vol. 2908, https://doi.org/10.1007/978-1-0716-4434-8_11,

NUDT9H, is located. TRPM2 NUDT9H is catalytically active in invertebrates; however, in vertebrates it is catalytically inactive although it binds ADPR [4].

Human TRPM2 was first cloned from brain tissue [5]. Besides its high expression level in the brain, channel expression has been demonstrated with high relative levels in macrophages and bone marrow, and lower but significant levels in several other human tissues such as pituitary, heart, lung, liver, kidney, skeletal muscle, stomach, intestine, spleen, PBMC, adipose tissue, placenta, pancreas, and prostate cells [6]. TRPM2 is involved in immune responses [7], insulin secretion [8], and pathological cell death [9]. Genetic mutations of the channel have been associated with mental disorders and neurodegenerative diseases [10].

Temperature-sensitive behavior of TRPM2 was first demonstrated in HEK293 cells transiently expressing the channel, and, in the same study, TRPM2 was shown to mediate heat-induced insulin release in pancreatic β-cells [11]. Further research in primary cell models found that TRPM2 activation by heating initiated membrane potential oscillations in pancreatic β-cells [12], and, in macrophages, the temperature threshold of the channel was affected by redox signals [13]. It was recently found that, in neurons, TRPM2 plays important physiological roles in the central regulation of body temperature [14, 15] and in the peripheral detection of non-noxious heat [16].

Despite the fact that heat-sensitive activation of the channel was shown in a number of in vitro experiments, there were contradictions and gaps in the description of the channel's heat-sensitive behavior [17]. A serious limitation of studies in intact cellular systems is that the intracellular concentrations of the agonists of TRPM2 (Ca^{2+} and ADPR) cannot be directly regulated. Therefore, to overcome this limitation and to allow precise regulation of ligand concentrations as well as temperature, we conducted electrophysiological measurements on isolated, inside-out membrane patches, where the intracellular domain of the channel is accessible for superfusion by a temperature-regulated gravity-driven perfusion system (Fig. 1).

Using the method described here, we found that TRPM2 gating is intrinsically temperature sensitive, with a standard enthalpy difference between the closed and open states of ~180 kJ/mol. We could demonstrate that increasing temperature alone, up to 40 °C, could not activate the channel in the absence of its agonists, ADPR, and Ca^{2+}. A crucial factor that regulates the TRPM2 channel's temperature sensitivity is the Ca^{2+} influx through the channel, which can form a high-[Ca^{2+}] microdomain in the direct vicinity of the intracellular Ca^{2+} binding site. This high-[Ca^{2+}] microdomain exerts a strong positive feedback on channel activity that provides extremely high temperature

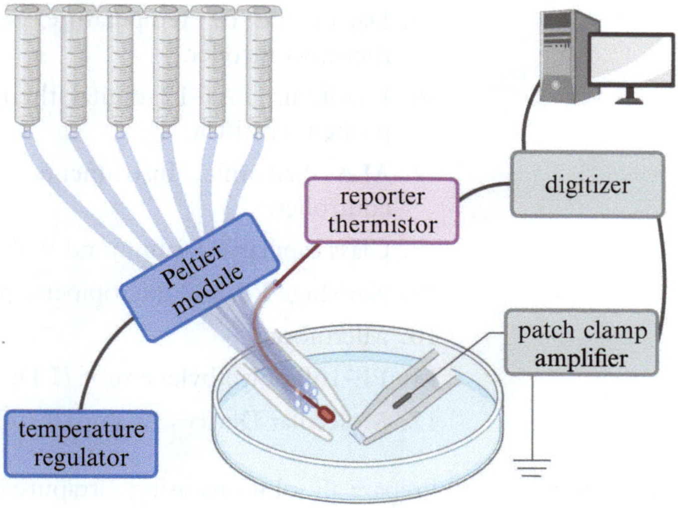

Fig. 1 Scheme of the array of the patch clamp setup. A Peltier module regulates the temperature of the solutions in the perfusion system. The temperature of the perfused solution is monitored by a reporter thermistor positioned in the flow stream within the perfusion manifold. The isolated membrane patch is positioned in the outlet orifice of the perfusion manifold. Ionic currents and temperature are digitalized and recorded simultaneously. Created with BioRender.com

sensitivity (Q_{10} ~ 1000) to TRPM2 in the range of 37–40 °C, thus we could confirm that the channel is a suitable receptor for the regulation of core body temperature [1].

2 Materials

2.1 Cell Culture and Transfection

1. HEK-293 t cell line.
2. DMEM medium.
3. OptiMEM medium.
4. Fetal bovine serum (FBS).
5. L-glutamine (200 mM solution).
6. Penicillin/streptomycin solution (10,000 U/ml).
7. FuGENE transfection reagent.
8. 5 mm round glass coverslip.
9. 35 mm cell culture Petri dish.

2.2 Electrophysiology

1. Vibration isolation table with Faraday cage.
2. Molecular Devices Axopatch 200B amplifier.
3. Molecular Devices Digidata 1550 digitizer.
4. Sutter mp-285 micromanipulator.

5. Dagan TC-10 temperature regulator with Peltier-driven thermocontroller.
6. Physitemp BAT-12 digital thermometer with IT-18 microprobe thermistor.
7. ALA Scientific Instruments ALA-VM8 electronic valve controller.
8. Glass capillary (Drummond 9–000-2321).
9. Narishige PP-830 micropipette puller.
10. Microforge.
11. PE-160 polyethylene tube (I.D: 1.14 mm, O.D: 1.57 mm).
12. Molecular Devices Pclamp10 software.

2.2.1 Solutions for Electrophysiology

Prepare all solutions using ultrapure deionized water.

1. Pipette (tip-fill, extracellular) solution: 140 mM Na-gluconate, 2 mM Mg-gluconate$_2$, 10 mM PIPES, 1 mM EGTA, pH 7.4 with NaOH (*see* **Notes 1**, **2**, and **3**).
2. Pipette (back-fill, electrode) solution: 140 mM NaCl, 2 mM $MgCl_2$, 10 mM PIPES, pH 7.4 with NaOH.
3. Bath (intracellular) solution: 140 mM Na-gluconate, 2 mM Mg-gluconate$_2$, 10 mM PIPES, pH 7.1 with NaOH (*see* **Notes 2** and **3**). All bath solutions also contain 200 μM AMP and 10 μM dioctanoyl-PIP$_2$ (*see* **Note 4**).

   To obtain ~1.7–1400 μM free [Ca^{2+}], 0–10 mM Ca-gluconate$_2$ is added to the bath solution. To obtain "zero" (~4 nM) free [Ca^{2+}]), 1 mM EGTA is added to the bath solution. To obtain 1 μM, 300 nM, 100 nM, or 30 nM free [Ca^{2+}], 1 mM EGTA with 760 μM, 500 μM, 240 μM, or 70 μM Ca-gluconate$_2$ is added to the bath solution, respectively (*see* **Note 5**).
4. 3 M KCl solution for the reference electrode.

3 Methods

3.1 Cells and Ion Channel Gene Expression

3.1.1 Cell Cultures

Culture cells at 37 °C in 5% CO_2 in DMEM medium with 4.5 g/L glucose, supplemented with 10% FBS, 2 mM glutamine, 100 units/mL penicillin/streptomycin.

3.1.2 Ion Channel Expression in HEK-293 T Cells

Expression vectors for ion channels were stably or transiently transfected in HEK-293 t cells with FuGENE reagent (*see* **Note 6**).

1. Grow cells on a 35 mm cell culture dish until ~80% confluency.
2. Prepare 129 μL of 0.020 μg/μL plasmid DNA solution in OptiMEM medium by adding 2.8 μg DNA in 129 μL medium.

Mix carefully. To identify transfected cells for electrophysiological measurements, there are several options: (a) use fluorescently tagged ion channel coding vector, (b) use bicistronic expression vector, and (c) mix the vector of untagged ion channel with a fluorophore expression vector in a 10:1 ratio.

3. Add 8.3 μl FuGENE HD reagent to obtain 3:1 FuGENE:DNA ratio (μl/μg). Vortex briefly.
4. Incubate mixture for 5–10 min at room temperature.
5. Add mixture to the cells and mix carefully.
6. Incubate cells for 4–8 h at 37 °C in 5% CO_2.
7. Suspend cells with pipetting (*see* **Note 7**).
8. Place 4–5 pieces of 5 mm poly-L-lysine treated glass coverslips in a new 35 mm cell culture dish with 2 mL DMEM medium. Plate approx. 1/3 of cell suspension onto 5 mm poly-L-lysine treated glass coverslips in 35 mm dish.
9. Incubate cells for 12–48 h at 37 °C in 5% CO_2 prior to electrophysiological experiments.

3.2 Electrophysiology

1. Fabricate patch pipettes on the day of the experiments. Pull and heat polish pipettes to obtain a tip resistance of 3–5 MΩ.
2. Fill pipettes partly with extracellular solution then layer the electrode solution carefully on top.
3. Place the reference electrode in 3 M KCl and position an agar bridge between the 3 M KCl and the recording chamber (*see* **Note 8**).
4. Wash cells on 5 mm coverslips in extracellular solution then transfer to the recording chamber on the stage of the microscope. Select cells for measurements based on morphology and fluorescent reporter protein (GFP) expression.
5. After obtaining a gigaohm seal, isolate an inside-out patch from the cell membrane. Navigate the pipette tip with the membrane patch directly to the outlet of the perfusion system. During the electrophysiological measurements, ion channel activity can be regulated by perfusing the membrane patch with agonists at desired temperatures. Temperature of the bath solution is set with a Peltier-driven temperature controller unit (Dagan TC-10) (*see* **Note 9**). Perfusion lines (PE-160 polyethylene tubes) are directed through the heater/cooler unit positioned as close as possible to the perfusion manifold to minimize temperature deviations from the set value. A micro thermistor (Physitemp BAT-12) is guided in through one of the inlets to the central cavity of the perfusion manifold where the solutions converge. Such arrangement of the perfusion system and the

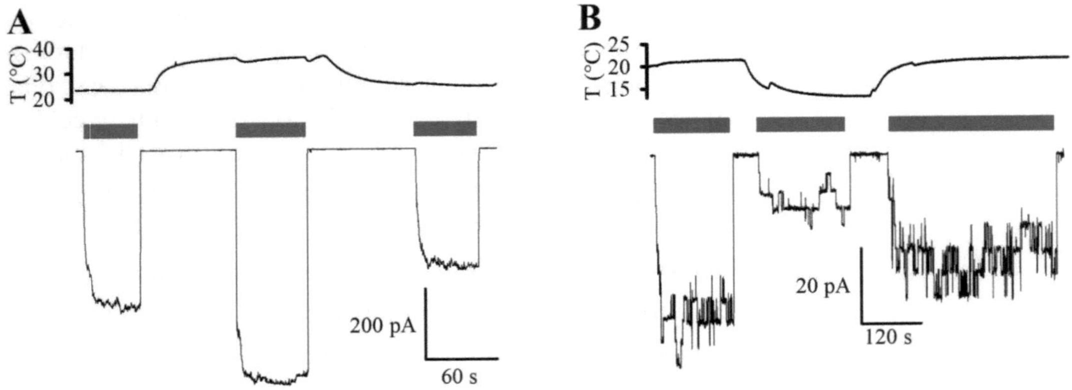

Fig. 2 Simultaneous recordings of TRPM2 current and temperature. Steady-state ionic currents recorded on the noninactivating pore mutant of the human TRPM2 channel (TRPM2-T5L, [2]). TRPM2 channels are activated with agonists in saturating concentrations (400 μM Ca^{2+} and 32 μM ADPR, labeled by gray bars). (**a**) Effect of warming from room temperature to 37 °C on macroscopic TRPM2 currents. (**b**) Effect of cooling from room temperature to 15 °C on microscopic TRPM2 currents

independent reporter thermistor allows the real time monitoring of the temperature of the bath solution (Fig. 1).

6. Record microscopic or macroscopic currents at a bandwidth of 2 kHz (Axopatch 200B), digitize at 10 kHz (Digidata 1550), and save to disk (Pclamp10) (Fig. 2).
7. Analyze data in Clampfit [1].

4 Notes

1. Regulating [Ca^{2+}] in the extracellular solution is important. A well-known feature of the channel is that it conducts Ca^{2+}. This property may lead to difficulties in data interpretation in different experimental models. In intact cells, channel activity strongly depends on extracellular Ca^{2+} concentration as high extracellular [Ca^{2+}] was shown to increase, while the absence of extracellular Ca^{2+} diminished channel activity [13, 18–20]. This phenomenon is the result of the architecture of the channel which allows direct access for the Ca^{2+} ions entering through the pore to the intracellular Ca^{2+} binding sites while creating a high-concentration Ca^{2+} microdomain in the intracellular cavity of the channel [21, 22], thus generating a positive feedback mechanism on channel activation [1, 18].
2. Temperature-dependent changes in pH of the pipette solution were kept minimal by the use of PIPES buffer ($\Delta pK_a/10\ °C = -0.085$).
3. In our electrophysiological measurements, we used the low-affinity buffer gluconate to buffer Ca^{2+} in our solutions.

The measured free [Ca^{2+}] of the solution was ~1–2 μM [1, 23]. Gluconate containing solutions (extracellular and intracellular) can be stored at 4 °C after filtering with 0.22 μm pore size sterile filter.

4. HEK cells endogenously express a Ca^{2+} activated low-conductance cation channel that can be inhibited by 200 μM AMP. PIP_2 is an essential activator of TRPM2, therefore we applied it at a saturating concentration (10 μM di-octanoyl-PIP_2).
5. Kd of Ca-gluconate between 14–40 °C is slightly temperature dependent. At room temperature, the final free [Ca^{2+}] of the bath solution is approximately 1/eighth of the added [Ca-gluconate$_2$]. For higher accuracy, we calculated the actual free [Ca^{2+}] of the bath solutions at different temperatures. For further details, refer to [1, 21].
6. Other transfection methods may be applied (e.g., Lipofectamine, etc.).
7. HEK cells adhere to the dish lightly, there is no need to add trypsin to detach the cells.
8. To fabricate agar bridges, bend glass capillaries over flame to desired shape. Fill the capillaries with hot 4% agar in 3 M KCl. Avoid forming air bubbles within the capillary. After the agar solidifies, agar bridges can be stored in 3 M KCl for several months.
9. Dagan TC-10 temperature regulator is powered by a 12 V DC battery to minimize electrical noise during the measurements.

Acknowledgments

Supported by the EU Horizon 2020 Research and Innovation Program grant 739593 and National Research, Development and Innovation Fund NKFIH KKP_22 grant 144199 and ADVANCED grant 149640. A.B. was supported by the János Bolyai Research Scholarship of the Hungarian Academy of Sciences (BO/00103/20) and the New National Excellence Program (ÚNKP) Bolyai+ scholarship of the Ministry of Human Capacities of Hungary (ÚNKP-20-5-SE-6, ÚNKP-21-5-SE-10, and ÚNKP-22-5-SE-12).

References

1. Bartok A, Csanady L (2022) Dual amplification strategy turns TRPM2 channels into supersensitive central heat detectors. Proc Natl Acad Sci USA 119(48):e2212378119
2. Toth B, Csanady L (2012) Pore collapse underlies irreversible inactivation of TRPM2 cation channel currents. Proc Natl Acad Sci USA 109(33):13440–13445

3. Szollosi A (2021) Two decades of evolution of our understanding of the transient receptor potential Melastatin 2 (TRPM2) Cation Channel. Life (Basel) 11(5):397
4. Iordanov I et al (2019) Enzyme activity and selectivity filter stability of ancient TRPM2 channels were simultaneously lost in early vertebrates. elife 8:8
5. Nagamine K et al (1998) Molecular cloning of a novel putative Ca2+ channel protein (TRPC7) highly expressed in brain. Genomics 54(1):124–131
6. Fonfria E et al (2006) Tissue distribution profiles of the human TRPM cation channel family. J Recept Signal Transduct Res 26(3):159–178
7. Syed Mortadza SA et al (2015) TRPM2 channel-mediated ROS-sensitive ca(2+) signaling mechanisms in immune cells. Front Immunol 6:407
8. Uchida K, Tominaga M (2014) The role of TRPM2 in pancreatic beta-cells and the development of diabetes. Cell Calcium 56(5): 332–339
9. Belrose JC, Jackson MF (2018) TRPM2: a candidate therapeutic target for treating neurological diseases. Acta Pharmacol Sin 39(5): 722–732
10. Toth AV, Bartok A (2024) Reviewing critical TRPM2 variants through a structure-function lens. J Biotechnol 385:49–57
11. Togashi K et al (2006) TRPM2 activation by cyclic ADP-ribose at body temperature is involved in insulin secretion. EMBO J 25(9): 1804–1815
12. Uchida K et al (2011) Lack of TRPM2 impaired insulin secretion and glucose metabolisms in mice. Diabetes 60(1):119–126
13. Kashio M et al (2012) Redox signal-mediated sensitization of transient receptor potential melastatin 2 (TRPM2) to temperature affects macrophage functions. Proc Natl Acad Sci USA 109(17):6745–6750
14. Song K et al (2016) The TRPM2 channel is a hypothalamic heat sensor that limits fever and can drive hypothermia. Science 353(6306): 1393–1398
15. Kamm GB et al (2021) A synaptic temperature sensor for body cooling. Neuron 109(20): 3283–3297 e11
16. Tan CH, McNaughton PA (2016) The TRPM2 ion channel is required for sensitivity to warmth. Nature 536(7617):460–463
17. Bartok A, Csanady L (2024) TRPM2 – an adjustable thermostat. Cell Calcium 118: 102850
18. McHugh D et al (2003) Critical intracellular Ca2+ dependence of transient receptor potential melastatin 2 (TRPM2) cation channel activation. J Biol Chem 278(13):11002–11006
19. Starkus J et al (2007) Regulation of TRPM2 by extra- and intracellular calcium. J Gen Physiol 130(4):427–440
20. Kashio M, Masubuchi S, Tominaga M (2022) Protein kinase C-mediated phosphorylation of transient receptor potential melastatin type 2 Thr738 counteracts the effect of cytosolic ca (2+) and elevates the temperature threshold. J Physiol 600(19):4287–4302
21. Csanady L, Torocsik B (2009) Four Ca2+ ions activate TRPM2 channels by binding in deep crevices near the pore but intracellularly of the gate. J Gen Physiol 133(2):189–203
22. Huang Y et al (2019) Ligand recognition and gating mechanism through three ligand-binding sites of human TRPM2 channel. elife 8:8
23. Tóth B, Iordanov I, Csanády L (2020) Selective profiling of N- and C-terminal nucleotide-binding sites in a TRPM2 channel. J Gen Physiol 152:e201912533

Chapter 12

Measuring Molecular Interactions with Subcellular Resolution: Single-Cell FRET Using the Quantitative Three-Filterset (Intensity-Based) Approach

István Rebenku, Cameron B. Lloyd, and György Vereb

Abstract

The actual interaction between signaling species in cellular processes is often more important than their expression levels. Förster resonance energy transfer (FRET) is a popular tool for studying molecular interactions, since it is highly sensitive to proximity in the range of 2–10 nm. Spectral spillover-corrected quantitative (three-filterset) FRET is a cost-effective and versatile approach, which can be applied in flow cytometry and various modalities of fluorescence microscopy, where it provides pixel-by-pixel quantitative FRET efficiency values as well as FRET-corrected fluorescent intensities corresponding to the expression level of the two labeled interacting molecular species. Here, we provide a protocol for implementing such measurements, starting from the preparation of appropriate samples including controls, through a step-by step guide for image processing and derivation of microscopic FRET maps, to various aspects of interpreting the results. The image processing described is aided by a freely available ImageJ/Fiji plugin RiFRET which has been developed with the biology-focused users in mind. The interface guides the user through the evaluation process in a friendly manner and provides the option for automatic processing of large image libraries once the various calibration factors have been correctly set.

Key words Förster resonance energy transfer (FRET), Fluorescence resonance energy transfer, FRET microscopy, Single-cell pixel-by-pixel FRET analysis, Molecular interactions, Molecular proximity, EGFR, Integrin A5

1 Introduction

Molecular interactions are among the emerging potential targets of today's medicine both in diagnosis and therapy. Although several biochemical methods are available to show these interactions, these are not always applicable, for example, when the interaction is not stable enough, i.e., it is not preserved after applying biochemical techniques. A further drawback of biochemical approaches is that

István Rebenku and Cameron B. Lloyd have equally contributed to this chapter.

Csaba Matta et al. (eds.), *The Surfaceome: Methods and Protocols*, Methods in Molecular Biology, vol. 2908, https://doi.org/10.1007/978-1-0716-4434-8_12,

the cells are destroyed during the process, and no information is provided on the subcellular localization of events induced by the drug. Förster resonance energy transfer (FRET, also known as fluorescence resonance energy transfer) [1] is an excellent and widely used tool for determining interactions between two molecules, not only in solution but also in situ in cells in their natural state. Furthermore, FRET methods have been adapted for microscopy, allowing the analysis of subcellular localization at the single-cell level.

FRET is a nonradiative transfer of energy from an excited donor fluorophore to an acceptor which needs to be at a distance of ~1–10 nm for conventional fluorescent tags [2–5]. Energy transfer is characterized by the FRET efficiency E, which is the ratio of excited-state donor molecules relaxing by FRET to the total number of excited donors.

The rate of energy transfer is dependent on the negative sixth power of the distance R between the donor and the acceptor, and R_0 is the distance where $E = 0.5$, that is, where there is a 50% chance that the energy of the excited donor will be transferred to the acceptor:

$$E = \frac{R_0^6}{R_0^6 + R^6} \tag{1}$$

Since the efficiency of FRET depends on the inverse sixth power of the distance, it is not just a tool for determining molecular interactions, but it can also be used as a spectroscopic ruler [6] in homogeneous systems; however, in cells, the heterogeneity of molecular assemblies precludes the direct translation of FRET E values into distances. Nonetheless, the relatively short range over which FRET occurs makes it an excellent tool for the visualization of molecular interactions or even conformational changes at the ensemble level. The occurrence of FRET has profound consequences on the fluorescence properties of both the donor and the acceptor. An additional process to dissipate the energy captured by the excited fluorophore is introduced in the donor, which decreases the fluorescence lifetime and the quantum efficiency of the donor, rendering it less fluorescent. The decrease in donor fluorescence (often termed donor quenching) can be one of the most easily measured spectroscopic characteristics that indicates the occurrence of FRET. Also, the simplest way to determine E is to measure donor fluorescence in the absence (F_D) and in the presence ($F_{D(+A)}$) of acceptors:

$$E = 1 - \frac{F_{D(+A)}}{F_D} \tag{2}$$

Additionally, since the acceptor is excited as a result of FRET, those acceptors that are fluorescent will emit photons (proportional to their quantum efficiency) also when FRET occurs. This is called sensitized acceptor emission and can also be a good measure of FRET. To quantitate FRET efficiency, numerous approaches have evolved so far [7]. In flow cytometric FRET [8, 9], we can obtain cell-averaged statistics for large cell populations, while the subcellular details can be investigated with various microscopic approaches. Most of these are based on donor quenching and/or acceptor sensitization, and a few on measuring emission anisotropy of either the donor or the acceptor [10]. Some of these methods can be combined to extend the information content of the measurement, for example, two-sided FRET [11] involves both acceptor depletion [16] and donor photobleaching rate measurements [12] on the same sample consecutively to assess the association state of two molecular pairs out of three distinct molecules in a correlated manner. Other combination techniques, like the three color FRET [13] and variations on the direct measurements of fluorescence lifetimes in microscopic imaging (FLIM) [13, 14], have also been developed and applied in the assessment of protein interactions.

In this chapter, we demonstrate the implementation of the quantitative three-filterset (intensity-based) ratiometric method [4, 15] which has the advantage over the acceptor photobleaching FRET method [16] that it is amenable to time series measurements as well as 3D FRET imaging. It involves measuring both the donor and acceptor fluorescence upon exciting with a donor specific wavelength, as well as the acceptor fluorescent upon exciting with an acceptor specific wavelength on samples labeled with both the donor and the acceptor, and calculating the FRET efficiency based on spectral correction factors derived from appropriate single-labeled samples. The result is the actual (absolute) FRET efficiency defined as

$$E = \frac{\text{the number of quanta transfered from D to A}}{\text{the number of quanta absorbed by D}} \tag{3}$$

which is not mathematically distorted by donor and acceptor concentration and stoichiometry unlike various other semiquantitative FRET indices (for a comprehensive overview of these latter, *see* [17]). Of note, the observed FRET efficiency itself can be influenced by, and inform about, stoichiometry.

In order to obtain reliable results, it is of basic importance to have a sound knowledge of this technique to avoid the various pitfalls. Here, we describe in detail the application of this method for confocal microscopy and provide all the necessary information to make its application precise and fruitful.

Our model system for demonstrating this application is a potentially important interaction of the ErbB family of receptor tyrosine kinases—that of the epidermal growth factor receptor (EGFR, or ErbB1) and Integrin α5 [15].

2 Materials

2.1 Cells

Any kind of cells, adherent and also nonadherent, can be grown on or attached to (e.g., using poly-l-lysine coated) coverslips or chamber wells. In the example experiment, we used the A172 cell line (ATCC catalog number CRL-1620), grown on cover glass.

2.2 Fluorophores

1. The donor and acceptor fluorophores, similarly to flow-cytometric FRET, must be appropriate for the lasers available [18], and it is advisable to use fluorophores that emit in the red and far red spectral range, where cellular autofluorescence becomes progressively weaker [19].
2. The most straightforward way of labeling cells is using fluorophore-conjugated antibodies that are usually available with the most commonly used dyes (*see* Table 1 and **Notes 1** and **2**).
3. Labeling the proteins of interest with fluorescent proteins is another feasible option. However, this requires special strategies for determining the spectral correction factors which is beyond the scope of this protocol [4, 20]. The most used fluorescent protein pairs are also listed in Table 1.
4. In the following sections, we describe the use of fluorescently labeled antibodies and demonstrate the measurement using indirect immunofluorescent labeling, which can be practical when target proteins are expressed at a low level, or direct-labeled antibodies are not available. The rationale to this approach is that in the case of indirect labeling, simply the second cycles of labeling described here need to be omitted.

Table 1
The most frequently used donor-acceptor fluorophore pairs

Donor (excitation, emission nm)	Acceptor (excitation, emission nm)
AlexaFluor488 (495, 519), Cy2 (492, 510), Atto488 (501, 523)	AlexaFluor546 (556, 573), AlexaFluor555 (555, 565), Cy3 (554, 658), Atto550 (554, 576)
Cy3 (554, 658), AlexaFluor546 (556, 573), AlexaFluor555 (555, 565), Atto550 (554, 576)	Cy5 (650, 665), AlexaFluor633 (621, 639), AlexaFluor647 (650, 668), Atto647N (644, 669)
CFP and variants, such as cerulean (433, 475)	YFP and variants, such as Venus (515, 528)
eGFP (488, 509)	tagRFP (555, 584), mStrawberry (574, 596)

2.3 Materials for the Labeling of Proteins Using Fluorescent Antibodies

1. HEPES buffer: 20 mM HEPES, 123 mM NaCl, 5 mM KCl, 1.5 mM $MgCl_2$, 1 mM $CaCl_2$, pH 7.2 (*see* **Note 3**).
2. HEPES + 5 mM glucose (for extracellular labeling).
3. HEPES + 0.1% BSA + 0.1% (v/v) Triton X-100 (for intracellular labeling).
4. HEPES + 1% formaldehyde (for fixation).
5. Fluorescently labeled antibodies, which are conjugated with donor and acceptor dyes, respectively.
6. In the case of indirect immunofluorescence labeling, unconjugated primary and fluorophore-conjugated secondary antibodies. The two primary antibodies need to come from different species, and the secondary antibodies need to be highly cross absorbed to avoid nonspecific binding.

2.4 Evaluation Software

Spectrally corrected intensity-based quantitative (three-filter) FRET measurements can be efficiently analyzed using the RiFRET ImageJ plugin [4]. To accompany this protocol, the original plugin has been updated and modified to be compatible with the latest versions of ImageJ 2.0/Fiji and support new image formats, thanks to the Bio-Formats plugin. The new version 2.x also includes pixel-by-pixel autofluorescence correction and offers fully automatic processing of larger image sets [15]. The source code is available on GitHub (https://github.com/CellMoTher/RiFRET) and the plugin is available via the FRET Imaging update site in Fiji and is licensed under the GNU General Public License v3.0.

3 Methods

3.1 Sample Preparation

1. In general, the following samples are necessary:
   (a) Sample 1. Unlabeled cells.
   (b) Sample 2. Cells with protein A labeled with the donor-conjugated antibody.
   (c) Sample 3. Cells with protein B labeled with the acceptor-conjugated antibody.
   (d) Sample 4. Cells with protein A labeled with the donor-conjugated and protein B labeled with the acceptor-conjugated antibody. Distinct samples of such labeling should be prepared for various treatments and for control (*see* **Notes 4** and **5**).
2. It is advisable to have a negative and a positive control with known molecular distances or FRET efficiencies obtained earlier using the same method. For example, in the case of a directly labeled antibody labeled with the donor dye, a straightforward positive control would be a secondary antibody against

the primary labeled with the acceptor dye. Such a sample will be used in the example demonstrated below. In the case that only indirect labeling is used, a mixture of donor- and acceptor-conjugated secondary antibodies can be attached to the unlabeled primary antibody. Both approaches provide a good positive control.

3.2 Labeling Extracellular Epitopes

1. Wash cells twice with ice-cold HEPES + 5 mM glucose. (Optional: fix and wash again if antibodies bind well after fixation.)
2. Label cells with (usually) 5–50 μg/mL final concentration of antibodies (should be above saturating concentration that has to be determined previously in a dose-response titration) in HEPES (in the case of prefixed samples) or HEPES + 5 mM glucose (in the case of live cells) for 10–30 min. Labeling is usually done on ice, but for prefixed samples, room temperature also works well (*see* **Note 6**).
3. Wash cells twice with ice-cold HEPES or HEPES + 5 nm glucose.
4. Postfix cells with HEPES + 1% formaldehyde for 5 min on ice (live cells only) and then 10 min at room temperature (all samples).
5. Store samples at 4 °C until measurement. Samples can be stored at least for a week unless a nuclear counterstain was used which could leak over time depending on the type of stain.

3.3 Labeling Intracellular Epitopes

1. Wash cells with ice-cold HEPES + 5 mM glucose.
2. Fix cells with HEPES + 1% formaldehyde for 5 min on ice and then 10 min at room temperature.
3. Permeabilize cells with PBS + 0.1% BSA + 0.1% (v/v) Triton X-100.
4. Label cells with (usually) 5–50 μg/mL final concentration of antibodies (should be above saturating concentration that has to be determined previously) in HEPES + 0.1% BSA + 0.1% (v/v) Triton X-100 for 30–60 min at room temperature.
5. Wash cells twice with HEPES + 0.1% BSA + 0.1% (v/v) Triton X-100 (RT).
6. Postfix cells with PBS + 1% formaldehyde for 20 min (RT).
7. Store samples at 4 °C until measurement. Samples can be stored at least for a week.

3.4 Measurement Using a Confocal Microscope

1. Before the measurement, set up the confocal microscope including the objective, lasers, and detection channels according to the fluorophores used. The donors AlexaFluor546, AlexaFluor555, and Cy3 can be excited at 543 nm and

detected through a 560–615 nm emission filter. Acceptors Cy5, AlexaFluor633, and AlexaFluor647 can be excited at 633 nm and detected through a 650 nm longpass filter.

2. The following images need to be taken:
   - Donor excitation/donor emission (donor).
   - Donor excitation/acceptor emission (transfer).
   - Acceptor excitation/acceptor emission (acceptor).
3. Measure (1) unlabeled, (2) donor only, (3) acceptor only, (4) either the donor or the acceptor epitope labeled with the other complementary dye (the latter three are needed for calculating correction factors), and (5) donor + acceptor samples of untreated and treated cells. Take at least ten images of each sample.

3.5 Evaluation of FRET Images

Analysis of confocal images and calculation of energy transfer efficiencies can be done using the RiFRET ImageJ plugin which uses the following calculations:

For a spectrally corrected intensity-based quantitative (three-filter) FRET measurement, three fluorescent channels should be measured: I_1 is the donor channel, which should be spectrally appropriate for the donor dye. I_2 is the transfer channel with donor excitation and acceptor detection range. I_3 is the acceptor channel with excitation and detection suitable for the acceptor dye. Equations below describe the contribution of various components to each fluorescence channel. I_D and I_A are the fluorescence intensity of the unquenched donor and of the acceptor in their own detection channels, respectively. The other components are calibration factors described below.

$$I_1 = I_D(1 - E) + I_A \cdot S_4 + I_D \cdot E \cdot \alpha \cdot \frac{S_4}{S_2} \tag{4}$$

$$I_2 = I_D(1 - E) \cdot S_1 + I_A \cdot S_2 + I_D \cdot E \cdot \alpha \tag{5}$$

$$I_3 = AF \cdot B_3 + I_D(1 - E) \cdot S_3 + I_A + I_D \cdot E \cdot \alpha \cdot \frac{1}{S_2} .epsR \tag{6}$$

The *epsR* is the product of the ratios of the donor and acceptor dye's molar absorption coefficients at each other's excitation wavelengths to their absorption coefficients at their own excitation wavelengths. For most dye pairs, this numeric is negligibly small.

$$epsR = \frac{\varepsilon^D_{A.exc} \cdot \varepsilon^A_{D.exc}}{\varepsilon^D_{D.exc} \cdot \varepsilon^A_{A.exc}} \tag{7}$$

S_1 and S_3 are determined using a sample containing only the donor dye:

$$S_1 = \frac{I_2^D}{I_1^D} \quad (8)$$

$$S_3 = \frac{I_3^D}{I_1^D} \quad (9)$$

S_2 and S_4 are determined on an acceptor-only sample:

$$S_2 = \frac{I_2^A}{I_3^A} \quad (10)$$

$$S_4 = \frac{I_1^D}{I_3^A} \quad (11)$$

The alpha parameter corrects for the difference in the quantum efficiency Q_A and Q_D and the detection efficiency η_A and η_D of the donor and the acceptor, respectively, and can be determined on donor- and acceptor-labeled samples. When using labeled cells for this purpose, the ratio of expression levels of donor and acceptor molecules should be constant across the measured cells. Ideally, the same epitope is used, targeted once with donor-labeled antibodies and once with acceptor-labeled antibodies. I_2^D and I_3^A are to be measured on these samples and averaged for at least 5–10 images. L_D and L_A are the mean number of dye molecules attached to the donor- and the acceptor-conjugated antibody, BD and BA are the mean number of receptors per cell labeled by the donor- and the acceptor-conjugated antibody, and ε_D and ε_A are the molar absorption coefficients of the donor and acceptor dyes at the wavelength of donor excitation.

$$\alpha = \frac{Q_A}{Q_D} \cdot \frac{\eta_A}{\eta_D} = \frac{I_3^A}{I_2^D} \cdot \frac{B_D}{B_A} \cdot \frac{L_D}{L_A} \cdot \frac{\varepsilon_D}{\varepsilon_A} \quad (12)$$

FRET efficiency is calculated as

$$\mathrm{E} = \frac{S_2(I_1S_1 + I_3S_2 - S_2I_1S_3 - I_3S_1S_4 + I_2S_3S_4 - I_2)}{\alpha\cdot(epsR - 1)\cdot(I_1S_2 - I_2S_4) + S_2(I_1S_1 + I_3S_2 - S_2I_1S_3 - I_3S_1S_4 + I_2S_3S_4 - I_2)} \quad (13)$$

Prospective users should download the latest version of the Fiji distribution of the free, open source image processing package, ImageJ. To install the RiFRET plugin, users should launch Fiji and choose *Help > Update...* from the Fiji menu bar. From here, click the "Manage update sites" button and enable the "FRET Imaging" update site by clicking the neighboring checkbox. Click the "Close" button and, finally, click the "Apply changes" button. Once downloaded and installed successfully, you will be prompted to restart ImageJ. Once restarted, run the RiFRET plugin by choosing *Plugins > FRET Imaging > RiFRET*. This method uses the ImageJ Updater to check and ensure that you always have the latest up-to-date version of the RiFRET plugin whenever ImageJ is launched. The plugin leads the user through the FRET analysis process, which includes opening and registering the images, optional Gaussian blur, background subtraction, thresholding, determining the spectral correction factors, creating the FRET image, and making measurements on it.

After running the RiFRET plugin, a new window will open, presenting the main RiFRET user interface (UI). The UI divides the workflow for FRET image analysis into logical steps as shown in Fig. 1. The top panel provides the option to calculate the necessary spectral correction (spill over) factors; S_1, S_2, S_3, and S_4, respectively, as well as the α factor, by clicking the relevant buttons. Alternatively, these factors can be entered into their respective text fields, if already determined. As each step is performed, the relevant button color turns green to reflect this. These will reset when "Close images" is clicked, in preparation for the next image. Alternatively, choose *File* > *Reset All...* in the plugin menu bar.

In the *S* factor calculation windows, images must be set as donor, transfer, and acceptor channels, respectively, by selecting an image slice and clicking the respective "Set image" button. Once set, the channel images are automatically converted to 32-bit float grayscale, image titles are renamed and timestamped. Optionally, a region of interest (ROI) on the donor channel image can then be drawn in a cell-free area using ImageJ's rectangle, polygonal or freehand tool, and this can be copied to the transfer and acceptor channels if desired by clicking the "Copy background ROI" button—ensuring that the same ROI is subtracted for each image. If a ROI is drawn, the mean intensity will be subtracted from every pixel within the ROI. Additionally, if an autofluorescence (AF) value is entered in the respective text field, this will be subtracted also. Alternatively, AF values can also be subtracted without subtracting a ROI containing average instrument noise. The results of this step will be output to the log. Images can be thresholded by clicking the "Set threshold" button. Note that the threshold will only be applied after clicking "Apply," then clicking "Set to NaN." To undo the applied threshold, click "Reset" next to the "Set threshold" button in the S factor calculation windows (Fig. 2). Thresholding can only be reset if background subtraction has been performed in the previous step.

When calculating S_1/S_3 factors, the donor channel should be thresholded as a minimum. Conversely, when calculating S_2/S_4 factors, the same applies to the acceptor channel. Provided is also a checkbox (disabled by default) with the option to show S_1 and S_3 images for manual calculation (these are opened upon clicking "Calculate"). Once set, click "Calculate", followed by "Set S_1 and S_3"—the *S* factor values will be set in the RiFRET main window (displayed to three decimal places) and the "Calculate S_1, S_3" button will change to green indicating that this step is complete.

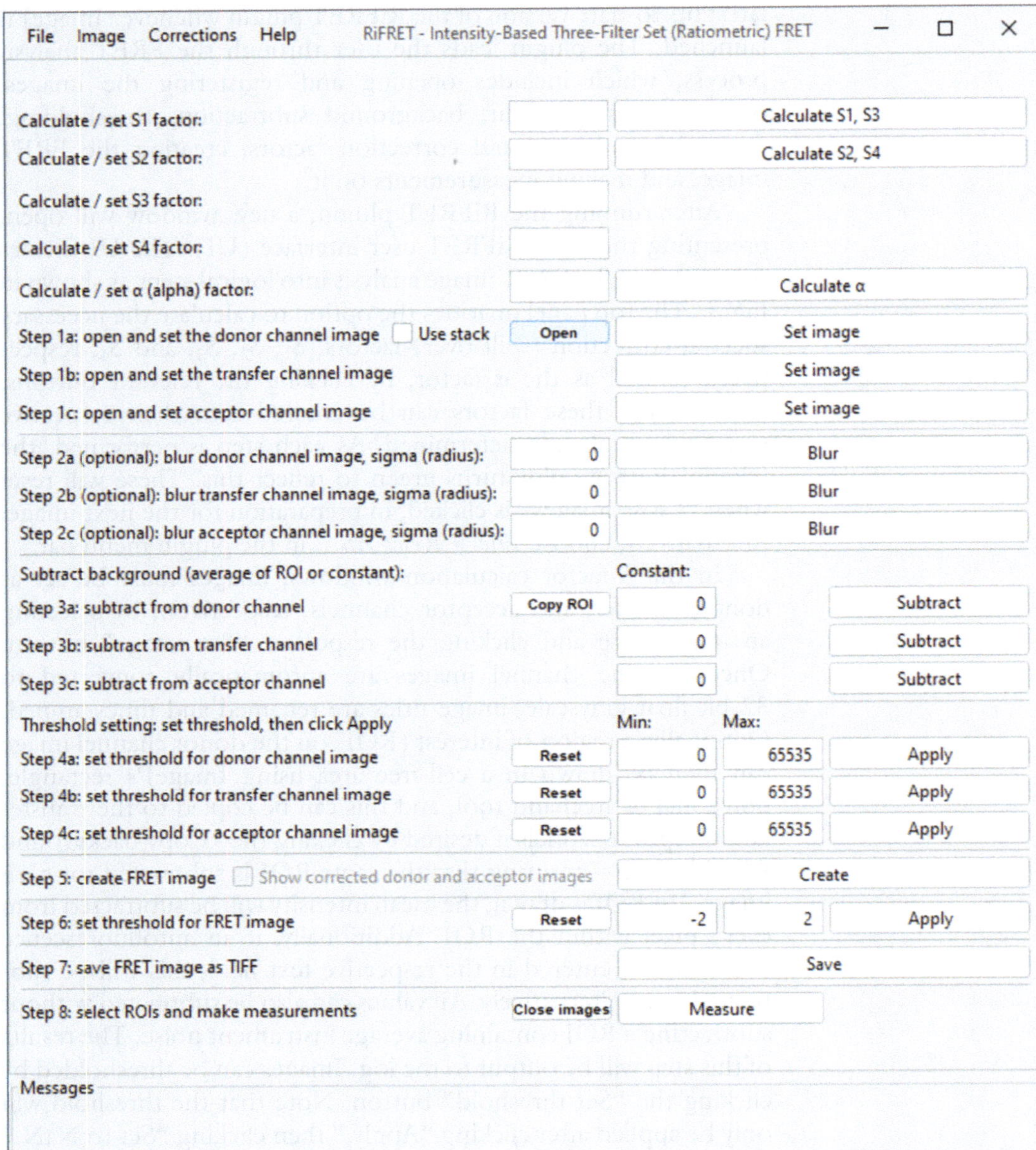

Fig. 1 Main window of the RiFRET plugin used for three-filterset spectrally corrected quantitative FRET. The main window is divided into image analysis steps including opening and setting channel images, optional blurring, background subtraction, and thresholding. Spectral correction (spillover) factors can also be calculated or entered if already known

S factor calculation can be reset at any time by clicking the "Reset" button in the bottom-right-hand corner (however, autofluorescence values entered will be preserved).

Step 1 in the main window is further divided into three parts for each respective channel used for the FRET measurement. Images for analysis can be opened by dragging and dropping onto the

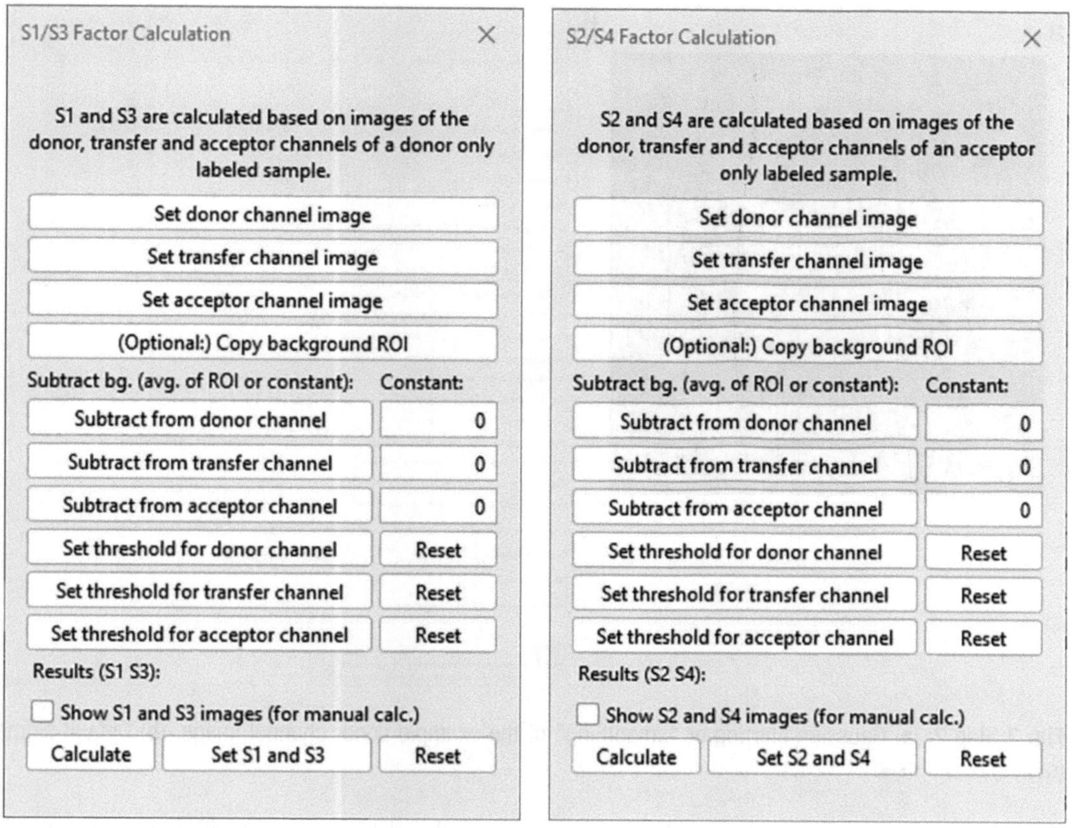

Fig. 2 S_1/S_3 (left) and S_2/S_4 (right) factor calculation windows. Here images can be set, background subtracted via a user-defined region of interest (also additional autofluorescence correction can be done if desired), as well as thresholded. The workflow in both calculation windows is essentially the same, with the exception of channel thresholding (depending on whether donor- or acceptor-only- labeled sample is used). Spectral correction factor images can also be shown for manual calculation

ImageJ toolbar, by choosing *File* > *Open…* in the RiFRET menu or alternatively by clicking the "Open" button in the RiFRET main window. Once open, the image stack or hyperstack must be split into its constituent images, or "slices" by choosing *Image* > *Stack to Images* in the plugin menu. Select each slice and click the respective "Set image" button. RiFRET also supports non-native formats, Zeiss CZI and LSM images, allowing for stacks of these formats to be opened and set automatically (in the order of Donor, Transfer, and Acceptor channels). This feature has been tested using images acquired with a Zeiss LSM 880/ZEN 2.1 SP1 (black) version 12.0.0.0.

Step 2 allows for optional blurring or "smoothing" of the images by applying a Gaussian filter (Fig. 3). Smoothing images in this way will reduce unwanted noise, generally improving the signal-to-noise ratio up to a point but may also come at the expense of loss of detail and disappearance of more extreme low or high

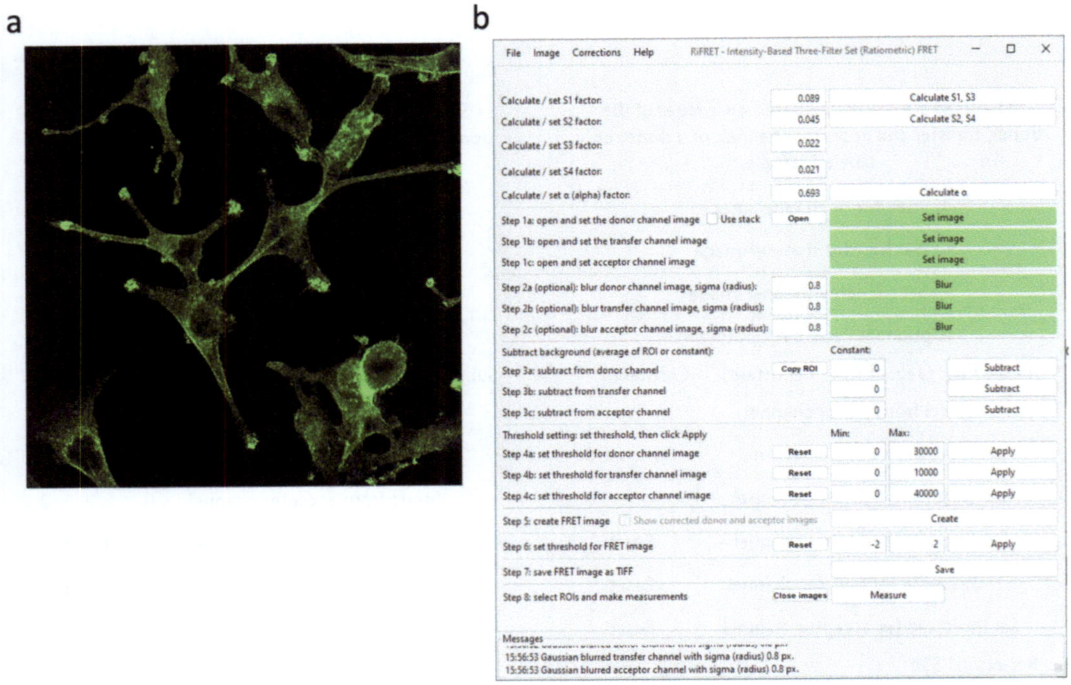

Fig. 3 Step 2: (**a**) Gaussian blurring or "smoothing" of the example donor channel image. (**b**) Default sigma (radius) of 0.8 used

values. Ideally, new images should be acquired if they contain too much noise. The default value used here for smoothing is a sigma (radius) of 0.8 pixels. If you wish to undo the blurring step, it is recommended to reload the images.

Step 3 allows for subtraction of average background signal using a ROI in the analyzed image (Fig. 4). After marking an appropriate ROI (cell-free area) in the Donor image, clicking the "Copy ROI" button will copy the same region of interest (ROI) from the Donor channel to the Transfer and Acceptor channels to ensure the same ROI is subtracted in each channel. Then click the Subtract button: once background subtraction has been performed on each channel, the ROIs are automatically cleared to prevent any further unintended behavior.

As an alternative approach, average cellular autofluorescence in each channel can be manually determined from unlabeled samples exposed in the same manner in the three channels, and these values can be entered into the text boxes and will be subtracted when the "Subtract" button is clicked. Please note that when values are entered into the text fields and also ROIs are active, their content is additive so both will be subtracted. If you wish to undo this step, simply repeat the background subtraction using the inverse/opposite value.

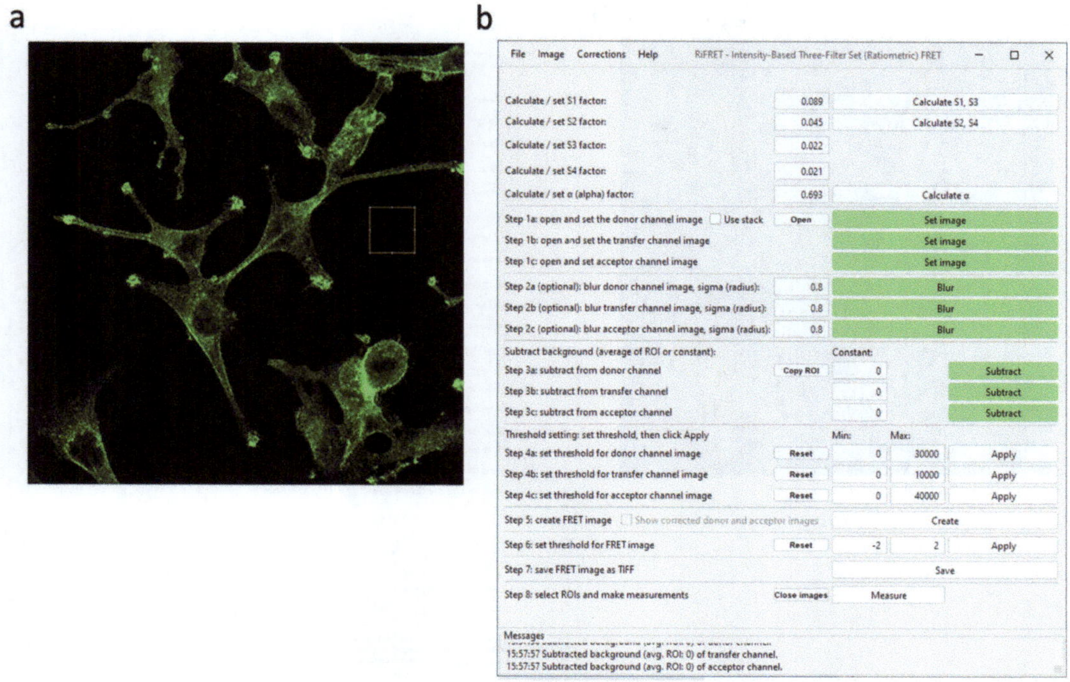

Fig. 4 Step 3: (**a**) Background subtraction of the donor channel image with region of interest drawn in cell-free region to be considered as background. (**b**) Green highlight in the GUI shows that the calculation has been performed

Step 4 allows for thresholding of each channel (Fig. 5). The min and max values can be defined for all the fluorescence channels. With the "Apply" button, the pixels outside the given maximum and minimum range will be set to NaN. FRET *E* values are only calculated on numeric values, so if pixel is NaN in any of channels, it is set to NaN in the FRET *E* image as well. With the "Reset" button, all pixels are reset to their original value before any threshold operation.

Step 5 allows for creation of transfer (FRET) images. On clicking the "Create" button, a new results window will also appear with measurements of the transfer (FRET) image as a whole (including pixels, not NaN pixels, mean, median, standard deviation, min, and max—Fig. 6). Creation of FRET images can only be performed if *S* factors have been given and channel images have been set in Step 1.

In Step 6, an option is available to threshold with given min and max values to exclude extreme FRET efficiencies (enabled by default), while Step 7 saves these images as TIFF files to the directory specified by the user.

Finally, Step 8 allows for any additional measurements to be made on specific regions of interest (such as leading edges) once a transfer (FRET) image has been created.

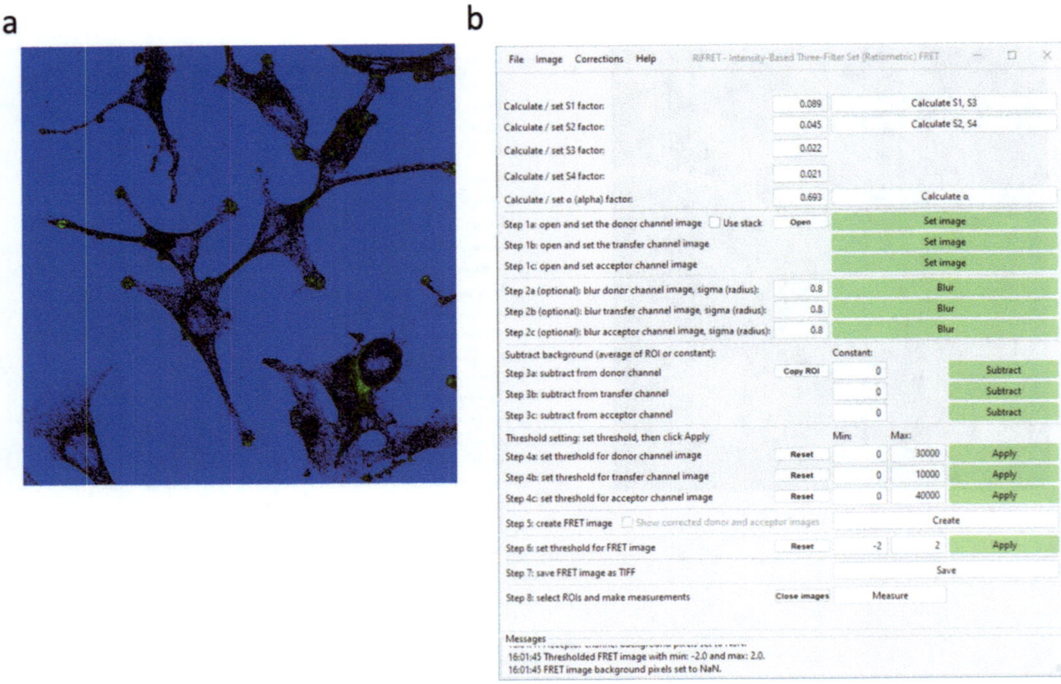

Fig. 5 (**a**) Preview of thresholding the donor channel image. Example shown uses ImageJ's built-in "Default" thresholding method (based on a modified IsoData/iterative intermeans algorithm) and the Over/Under display mode—pixels below a given lower threshold are shown in blue; thresholded pixels are shown in grayscale with "Green" lookup table applied. (**b**) The GUI shows that thresholding is completed, as well as threshold values for each channel. The 0 value for the lower threshold is coherent with the previously subtracted average background for each channel

3.6 Interpretation of FRET Efficiencies

By using this method, we can obtain reliable results without the need for expensive instrumentation and software. However, it is always important to use positive and negative controls to prove the reliability of the results, and negative FRET data need to be interpreted with caution. Since FRET has very sharp distance dependence, it is difficult to measure relatively long distances because FRET efficiency becomes very low. Therefore, the absence of FRET is not a direct proof of the absence of molecular proximity. Also, absence of FRET can be the result of constraints imposed by the location or orientation of fluorophores. On the other hand, presence of FRET to any appreciable extent above the experimental error of measurement (about 0.05) is strong evidence of molecular interactions.

Figure 7 shows the results of the analysis of the subcellular localization dependent interaction between the active conformation of integrin α5 and the epidermal growth factor receptor (EGFR). The two molecules were labeled indirectly on A172 glioblastoma multiforme cell line. The EGFR was labeled with Ab-11

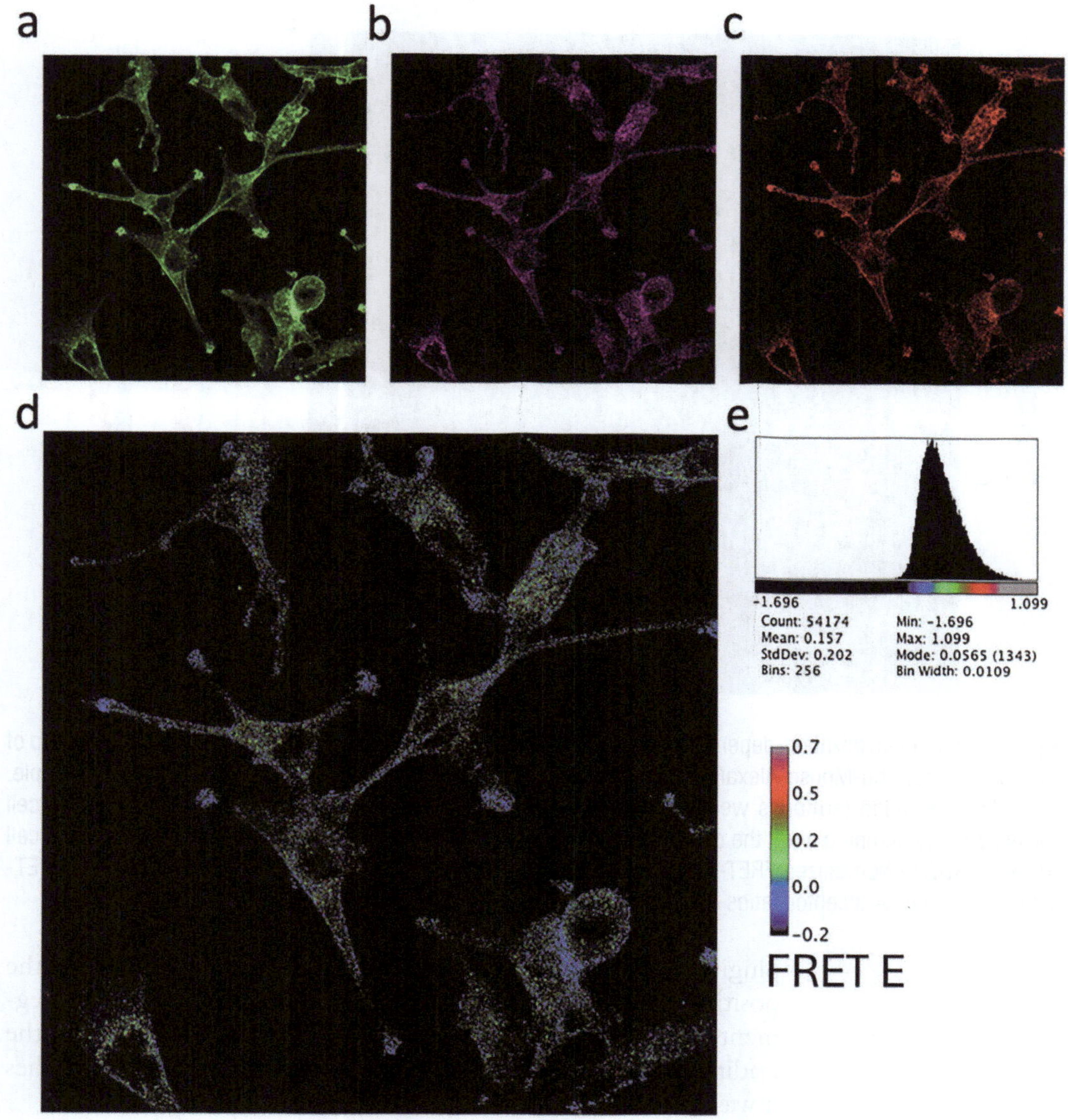

Fig. 6 (**a–c**) Example images showing donor, transfer, and acceptor fluorescence intensity of A172 glioblastoma cells at 40× magnification indirectly labeled with SNAKA-51 and Ab-11 and Goat anti-Mouse AlexaFluor546 and Goat anti-Rabbit AlexaFluor647, respectively—evaluated using the ratiometric intensity-based approach using the RiFRET ImageJ plugin. Acquisition parameters were kept constant throughout. (**d**) Generated transfer (FRET) image with rainbow RGB lookup table applied. (**e**) Histogram of transfer (FRET) image with "Rainbow RGB" lookup table applied

and Goat anti-Mouse AlexaFluor546, the integrin was labeled with SNAKA51 and Goat anti-Rabbit AlexaFluor647. As a positive control, indirect labeling was used where the primary antibody was TS2/16 directly conjugated with Alexafluor546 and the secondary antibody was Goat anti-Mouse AlexaFluor647. The FRET efficiency map was constructed in the presented way using the RiFRET

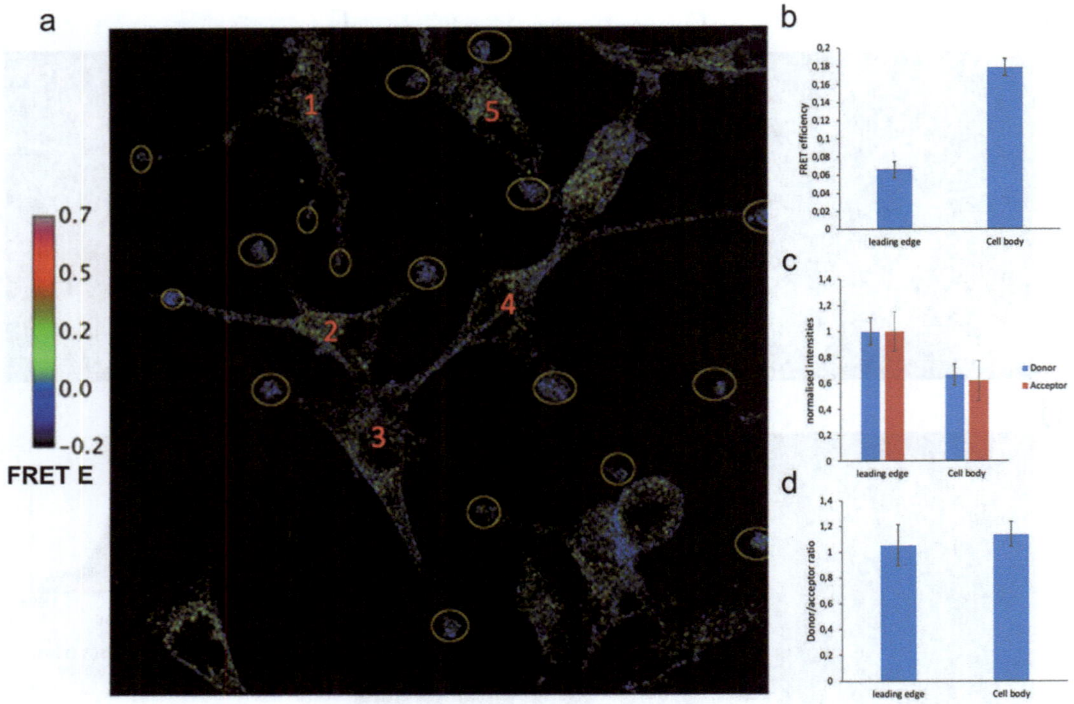

Fig. 7 Analysis of localization-dependent interaction of active integrin $\alpha5$ and EGFR. (**a**) FRET efficiency map of the Ab-11–Goat anti-Mouse AlexaFluor546 and SNAKA51–Goat anti-Rabbit AlexaFluor647 labeled sample. The yellow encircled territories were considered leading edges, all the other pixels were evaluated as cell bodies. The red numbers mark the cells evaluated. (**b**) Average FRET efficiencies in the leading edges and cell bodies ± SD. (**c**) Normalized FRET-corrected intensities of the donor and acceptor channel ± SD. (**d**) FRET-corrected donor-to-acceptor ratios ± SD

plugin. The whole image FRET efficiency was 0.1573, while the positive control FRET efficiency was 0.5698. The sample was segmented into two territories: cell bodies and leading edges—in the leading edges, the FRET efficiency was 0.068, and in the cell bodies it was 0.1671.

Analysis was performed on a cell-by-cell basis. Five cells were chosen - the FRET efficiencies; FRET-corrected donor and acceptor intensities; and the donor-to-acceptor ratio were determined in their cell bodies and their leading edges. The FRET efficiency in the leading edges was 0.066 ± 0.0089; in the cell bodies 0.179 ± 0.0095. FRET efficiency correlates with the average distance between the studied molecules. Theoretically, the smaller the distance, the higher FRET that could be measured. Two molecules can be close to each other as a merely statistical phenomenon, without any interaction between them. In those cases, the higher the relative molecule density, the higher the FRET that efficiency can be measured. We excluded this possibility by measuring the average donor and acceptor intensities. Both intensities were significantly lower in the cell bodies compared to the leading edges, which suggests that the higher FRET efficiency in the cell bodies is not a statistical phenomenon in this case. Another effect to consider when interpreting the FRET result is the changes in the donor-to-acceptor ratio. If the number of acceptor dyes is low compared to the donor dyes, the FRET efficiency will be low (*see* **Note 4**). In the presented experiment there is only a slight change in the donor-to-acceptor ratios between the two studied cellular regions, so we can conclude that the significantly higher FRET efficiencies in the cell bodies are the result of stronger interactions between the two molecules.

4 Notes

1. The selection of dyes is based not only on spectral characteristics, but it is also important to choose dyes that are reasonably photostable, especially if a time sequence or a 3D FRET image is to be generated, which requires multiple exposure of the same areas of the sample. Excitation and emission spectra of molecules for donor-acceptor pairs can be found at one of the following web sites: Becton-Dickinson Fluorescence Spectrum Viewer (*BD® Spectrum Viewer | Spectral Analyzer (bdbiosciences.com)*), Thermo Fisher Fluorescence Spectra Viewer (Fluorescence SpectraViewer (thermofisher.com)).
2. Direct labeling with conjugated primary antibodies is usually preferable. Indirect labeling can either diminish FRET owed to the greater separation of fluorophores, but it is also possible that FRET occurs between otherwise more distant targets if a large molecular labeling complex is built up [16]. However, changes in FRET upon various treatments are still indicative of a change of the degree of molecular interactions.
3. PBS buffer: 137 mM NaCl, 3 mM KCl, 10 mM Na_2HPO_4, 1.8 mM KH_2PO_4, pH 7.4 can also be used instead of HEPES, both for extra- and intracellular labeling.
4. It is advisable to measure FRET in both directions (from epitope A to B, and from epitope B to A). If one epitope is expressed in great molar excess of the other, labeling that one with the acceptor usually yields a more efficient FRET, since all donors can have an acceptor pair. However, this can also overestimate the biological significance of the measured FRET efficiency, since out of many acceptors some could randomly colocalize at the molecular level with the donor.
5. Although low expression levels limit the application of FRET methods, the spectrally corrected intensity-based ratiometric FRET can be successfully used with proteins expressed at as low as 100–150 thousand per cell, and when the measurement is

corrected for autofluorescence, a few tens of thousands [15]. In the latter case, the same RiFRET plugin can be used for evaluation; all one needs to do is enable Pixelwise Autofluorescence Correction by choosing *Corrections > Pixel-wise Autofluorescence Correction* in the plugin menu bar.

6. Time of labeling with fluorescent antibodies should be optimized previously. In the case of extracellular epitopes, incubation beyond 10 min using saturating concentration usually does not increase labeling intensity by more than 10%.

Acknowledgments

We acknowledge the financial support from GINOP-2.2.1-15-2017-00072 (co-financed by the Hungarian State and the European Regional Development Fund), NKFIH OTKA grants K135938 and K143771 (the National Research, Development and Innovation Office, Hungary), the HUN-REN Hungarian Research Network, and the University of Debrecen Program for Scientific Publication.

References

1. Forster T (1946) Energiewanderung und Fluoreszenz. Naturwissenschaften 33(6):166–175
2. Clegg RM (2002) FRET tells us about proximities, distances, orientations and dynamic properties. J Biotechnol 82(3):177–179
3. Koushik SV et al (2006) Cerulean, Venus, and VenusY67C FRET reference standards. Biophys J 91(12):L99–L101
4. Roszik J et al (2009) Evaluation of intensity-based ratiometric FRET in image cytometry–approaches and a software solution. Cytometry A 75(9):761–767
5. Shrestha D et al (2015) Understanding FRET as a research tool for cellular studies. Int J Mol Sci 16(4):6718–6756
6. Stryer L, Haugland RP (1967) Energy transfer: a spectroscopic ruler. Proc Natl Acad Sci USA 58(2):719–726
7. Nagy P et al (2006) Chapter 12: *Measuring FRET in flow cytometry and microscopy.* Curr Protoc Cytom 12:Unit12.8
8. Vereb G, Nagy P, Szollosi J (2011) Flow cytometric FRET analysis of protein interaction. Methods Mol Biol 699:371–392
9. Ujlaky-Nagy L et al (2018) Flow Cytometric FRET analysis of protein interactions. Methods Mol Biol 1678:393–419
10. Jares-Erijman EA, Jovin TM (2006) Imaging molecular interactions in living cells by FRET microscopy. Curr Opin Chem Biol 10(5):409–416
11. Fazekas Z et al (2008) Two-sided fluorescence resonance energy transfer for assessing molecular interactions of up to three distinct species in confocal microscopy. Cytometry A 73(3):209–219
12. Szentesi G et al (2005) Computer program for analyzing donor photobleaching FRET image series. Cytometry A 67(2):119–128
13. Sun Y et al (2011) FRET microscopy in 2010: the legacy of Theodor Forster on the 100th anniversary of his birth. ChemPhysChem 12(3):462–474
14. Buranachai C et al (2008) Rapid frequency-domain FLIM spinning disk confocal microscope: lifetime resolution, image improvement and wavelet analysis. J Fluoresc 18(5):929–942
15. Rebenku I et al (2023) Pixel-by-pixel autofluorescence corrected FRET in fluorescence microscopy improves accuracy for samples with spatially varied autofluorescence to signal ratio. Sci Rep 13(1):2934
16. Roszik J, Szollosi J, Vereb G (2008) AccPbFRET: an ImageJ plugin for semi-automatic, fully corrected analysis of acceptor

photobleaching FRET images. BMC Bioinformatics 9:346
17. Berney C, Danuser G (2003) FRET or no FRET: a quantitative comparison. Biophys J 84(6):3992–4010
18. Horvath G et al (2005) Selecting the right fluorophores and flow cytometer for fluorescence resonance energy transfer measurements. Cytometry A 65(2):148–157
19. Sebestyen Z et al (2002) Long wavelength fluorophores and cell-by-cell correction for autofluorescence significantly improves the accuracy of flow cytometric energy transfer measurements on a dual-laser benchtop flow cytometer. Cytometry 48(3):124–135
20. Szabo A et al (2018) The effect of fluorophore conjugation on antibody affinity and the Photophysical properties of dyes. Biophys J 114(3):688–700

Chapter 13

Quantifying the Electrophysiological Properties of Cells Using Dielectrophoresis

Krista S. P. Clarke, Oreoluwa V. Griffiths, Kai Hoettges, Rebecca Lewis, and Fatima H. Labeed

Abstract

The 3DEP Reader (DEPtech) is a commercial noninvasive, high throughput, easily adoptable benchtop microfluidic platform that utilizes the principles of dielectrophoresis to analyze cells in suspension and determine their electrical properties. This protocol describes the operation, value, and common setup of the platform to provide users with the ability to characterize three important electrophysiological parameters of approximately 20,000 cells in under 2 min. Dielectrophoresis (DEP) is an electrokinetic phenomenon which describes the movement of suspended polarizable cells, in a nonuniform electric field. The 3DEP plots the relative DEP force experienced by cells across a range of AC frequencies generating a spectrum through which membrane conductance, membrane capacitance, and cytoplasmic conductivity are derived.

Key words Dielectrophoresis (DEP), 3DEP, Effective membrane conductance (G_{eff}), Effective membrane capacitance (C_{eff}), Cytoplasmic conductivity (σ_{cyto}), Electric field (EF), Potassium ion (K^+), Sodium ion (Na^+), Weight per volume (w/v)

1 Introduction

Changes in cellular electrophysiology can provide insight into cell health and function, offering information about transitions from healthy to diseased states, cell viability, functional changes, effect of pharmacological treatments, and tracking cell differentiation [1–4].

Currently, patch clamp is considered the gold-standard technique for cellular electrophysiological quantification; it is versatile and provides high-resolution measurements of ionic currents at a single-cell level. However, it has significant limitations: it is highly invasive, often rupturing the cell; it is time-consuming, requires substantial equipment and financial investment, is low-throughput, and necessitates expert operators [5]. As a consequence, there is a

Krista S. P. Clarke and Oreoluwa V. Griffiths have equally contributed to this chapter.

Csaba Matta et al. (eds.), *The Surfaceome: Methods and Protocols*, Methods in Molecular Biology, vol. 2908, https://doi.org/10.1007/978-1-0716-4434-8_13,

demand for more accessible methods to measure the electrical properties of cells. One such method arises through the application of the electrostatic phenomenon known as *dielectrophoresis* (DEP). When applied to cell characterization, DEP enables the quantification of the electrical properties of cells in a high-throughput, rapid, and low-cost format.

DEP is defined as the motion of a dielectric particle polarized in a nonuniform electric field (EF). It was first named by Herbert Pohl in a study of the behavior of inorganic uncharged suspensoids [6–8]; the first application of DEP to biological cells in suspension was the separation of live and dead yeast cells [8]. When compared to classical electrostatic techniques in the biological sciences, such as electrophoresis, DEP has an advantage that it can be used with both alternating current (AC) and and direct current (DC) fields. In the AC case, the behavior of suspensoids such as cells can be observed at different frequencies; since the overall DEP response at a given frequency depends on the electrical properties, as well as surface charge, shape, and size [9], determination of the variation in DEP response over a range of frequencies allows estimation of many of these properties.

In a nonuniform EF with field gradient ∇E, a neutral polarizable particle will acquire a dipole moment p (due to the displacement of net positive and negative charges). Interaction of the charged poles with local EFs of different magnitudes (due to the presence of a gradient) results in a force, $\boldsymbol{F}_{\mathbf{DEP}}$, [10] producing translational movement [10], shown in Eq. 1:

$$\boldsymbol{F}_{\mathbf{DEP}} = (p \cdot \nabla) E \tag{1}$$

A dipole involves the redistribution of surface and cytoplasmic charges and depends on the relative polarizability of the suspensoid [11] and the medium [10]. When the polarizability of the particle exceeds that of the medium, it experiences positive DEP (pDEP) and moves toward increased EF magnitude; when the opposite is the case, the net dipole will orient counter to the field and the particle will experience negative DEP (nDEP) and move toward lower EF density [12].

For AC fields, the permittivities of both suspensoid and medium will change as a function of angular frequency ω (given by $2\pi f$, where f is the frequency of the applied electric field). The relationship giving the net polarizability of a spherical suspensoid is defined by the Clausius–Mossotti factor (CM or $K(\omega)$) shown in Eq. 2 [10]:

$$K(\omega) = \frac{\varepsilon_p^* - \varepsilon_m^*}{\varepsilon_p^* + 2\varepsilon_m^*}, \text{ where } \varepsilon_p^* = \varepsilon_p - \frac{j \cdot \sigma_p}{\omega} \text{ and } \varepsilon_m^* = \varepsilon_m - \frac{j \cdot \sigma_m}{\omega} \tag{2}$$

where * denotes a complex variable [10], j is the square root of -1, and σ is the conductivity [12]. It is then clear that the sign of CM factor sign is the determining factor for the presence of either pDEP or nDEP. The net time-averaged force $\boldsymbol{F}_{\mathbf{DEP}}$ on a particle suspended in a medium is given by Eq. 3:

$$\boldsymbol{F}_{\mathbf{DEP}} = 2 \cdot \pi \cdot r^3 \cdot \varepsilon_m \cdot \mathrm{Re}[K(\omega)] \cdot \nabla E^2 \quad (3)$$

where Re denotes the real part. When the suspensoid is a cell, the relationship in Eq. 2 is inadequate, as it treats the cell as a single homogeneous sphere. However, it can be adapted to consider the properties of both membrane and cytoplasm separately by replacing the homogeneous parameter ε_p^* in Eq. 2 with a "shell model" of a cell with radius r, membrane thickness t, and complex permittivities for cytoplasm $\varepsilon^*_{\mathrm{cyto}}$ and membrane $\varepsilon^*_{\mathrm{mem}}$, thus [13]:

$$\varepsilon_p^* = \varepsilon_m^* \frac{\left(\frac{r+t}{r}\right)^3 + 2\frac{\varepsilon^*_{\mathrm{cyto}} - \varepsilon^*_{\mathrm{mem}}}{\varepsilon^*_{\mathrm{cyto}} + 2\varepsilon^*_{\mathrm{mem}}}}{\left(\frac{r+t}{r}\right)^3 - \frac{\varepsilon^*_{\mathrm{cyto}} - \varepsilon^*_{\mathrm{mem}}}{\varepsilon^*_{\mathrm{cyto}} + 2\varepsilon^*_{\mathrm{mem}}}} \quad (4)$$

Using the above principles, the electrical properties of biological cells can be derived empirically [14]. These properties can be extracted from frequency spectra showing the change in DEP force experienced across a range of frequencies, to which a single-shelled spherical model is fitted [15], by finding optimally fitting values in Eq. 4. This can be achieved using a measurand which correlates with the DEP force, such as velocity or collection rate. The frequency at which $\boldsymbol{F}_{DEP}$ is equal to zero, where $\varepsilon_p = \varepsilon_m$ [16] is termed the "cross-over frequency." The first cross-over frequency, at lower frequencies in the DEP spectrum, relates to membrane properties σ_{mem} and $\varepsilon_{\mathrm{mem}}$ [10], the second cross-over frequency occurs at higher frequencies and concerns σ_{cyto} [10]. The values of σ_{mem} and $\varepsilon_{\mathrm{mem}}$ are commonly translated to more physiologically relatable values for specific (per unit area) effective membrane conductance G_{eff} and capacitance C_{eff}, as described in **Note 7**.

The velocity and direction of movement of the particle in the nonuniform field depends on the shape, size, relative polarizability of the particle to the suspension media, and the applied AC EF frequency [17–19]. Plotting the relative $\boldsymbol{F}_{DEP}$ experienced by cells within suspension across a range of frequencies produces a dielectrophoretic spectrum from which the electrophysiological properties of the cells can be measured. These properties are membrane conductance, membrane capacitance, and cytoplasmic conductivity (σ_{cyto}) [19, 20].

1.1 Cytoplasm Conductivity

Cytoplasm conductivity (σ_{cyto}) refers to the concentration of free positive and negative ions within the cell cytoplasm, most notably potassium (K^+), sodium (Na^+), calcium (Ca^{2+}), chloride (Cl^-), and magnesium (Mg^{2+}) [21]. It is a measure of the conductivity of cellular cytoplasm [22]. σ_{cyto} may be used to indicate ion channel and pump activity, describing the metabolic activity of a cell [23], or the permeability of a cell membrane; high permeability means greater ease for both ion and water flow across the membrane and can lead to a rapid efflux of either, measurable as a drop in σ_{cyto} [24].

1.2 Membrane Conductance

Membrane conductance refers to the ability of a cell membrane to conduct charge and is a composite of the movement of ions through the membrane (transmembrane conductance) and the conduction of ions across the membrane surface (surface conductance) [25]. Changes in membrane conductance can indicate changes in ion channel activity, charges on the membrane surface, and membrane morphology [26]. Ion channels and permeability are a significant contributor to G_{eff}: more open ion channels increases the permeability of the membrane [27] facilitating mass transport of ions through the membrane, as such G_{eff} increases [28]. It has also been demonstrated that a spike in membrane conductance that occurs in late-stage apoptosis coincides with a drastic increase in membrane permeability as the cell deteriorates, allowing passage of ions across the membrane [29].

1.3 Membrane Capacitance

The impermeability of the lipid bilayer in cell membrane to ions means it functions as a very thin dielectric (~7 nM), separating electrolytes within the conductive cytoplasm from the extracellular fluid. The cell membrane thus forms a basic capacitor, with electric charge accumulating on either side of the membrane and electrostatically interacting across the microscopic distance they are separated [30]. This phenomenon is known as membrane capacitance.

Membrane capacitance signifies the "membrane time constant which dictates how fast the cell membrane potential responds to the flow of ion channel currents" [31]. Capacitance, especially in non-excitable cells, is highly dependent on changes in membrane composition [31], cell surface modifications, and changes in cell radius and morphology with features such as membrane invaginations, ruffles, and microvilli [32]. An increase in membrane capacitance may also indicate an increase in the voltage/charge required to depolarize the excitable cell membrane [31]. Values of whole cell capacitance derived from the 3DEP are compared to values acquired by patch clamp [25].

Effective membrane conductance (G_{eff}) and effective membrane capacitance (C_{eff}) are calculated by dividing specific membrane conductance and membrane capacitance respectively by membrane thickness, removing the influence of membrane thickness on the CM model. Values of whole-cell capacitance derived from the 3DEP are compared to values acquired by patch clamp [25, 26].

1.4 The 3DEP (DEPtech)

The 3DEP (DEPtech) is a commercial benchtop microfluidic platform that utilizes the principles of dielectrophoresis to analyze cells in suspension and determine their electrical properties, namely G_{eff}, C_{eff}, and σ_{cyto}.

The analysis technique in this paper, the DEPtech 3DEP Reader (Heathfield, UK), uses the principle of DEP and light adsorption to calculate F_{DEP} at a range of frequencies. The aggregate speed and direction of cells is tracked by monitoring the change in light absorption across multiple concentric rings (bands) within the wells. pDEP attracts cells to the wall and reduces cell concentration close to the wall, while nDEP is pushing cells into a dark ring moving away from the wall; monitoring the rate of change in this region enables the measurement of DEP force [25, 33–35].

The 3DEP system consists of a disposable 3DEP chip (Fig. 1a) and the benchtop reader (Fig. 1b). The chip houses 20 hydraulically connected cylindrical electrode-wells, each 1.2 mm in diameter and gold-plated for biocompatibility. Hydraulic well connection via a gasket allows injection into one well to fill all the wells simultaneously [35]. 80 μL of cells in suspension, at between 10^4 and 10^8 cells/mL, are introduced to the chip (via blunt dispensing needles), a glass coverslip seals the suspension, the chip is loaded into the 3DEP's reading platform, and the motorized clamp locks the chip in position for analysis. The assessment of thousands of cells at one time in this manner reduces the chance of type 1 errors [25].

The circular orientation of the electrode-wells creates a nonuniform electric field. When energized, nDEP will induce cell movement toward the well center and pDEP toward well edge. The direction of cell movement is tracked via visible light spectrometry, where the rate of change of light intensity is measured over time, this is a parameter proportional to the speed and direction of the cells. Since cells reach their terminal velocity in <10 ms, this is also proportional to $\boldsymbol{F_{DEP}}$ [35].

Each well is served by a different electrode connected to separate DDS-based signal generators so the user may energize each well with independent AC frequency of voltage $1–20_{Vp\text{-}p}$ covering a frequency range between 1 kHz and 45 MHz. The 20-point logarithmically spaced spectrum produced has sufficient resolution to observe all the main features of a DEP spectrum, including the low, middle, and high frequency plateaus, and two dispersions that describe the transitions between positive and negative DEP [25, 33–35]. A single-shelled spherical model is fitted to this high-resolution spectrum of DEP force against frequency; from this, electrophysiological data on G_{eff}, C_{eff}, and σ_{cyto} can be extracted [14].

The 3DEP device has been used to characterize cellular dielectric changes across a plethora of applications, including oral cancer [36–39], bladder cancer [40], characterization of neural and

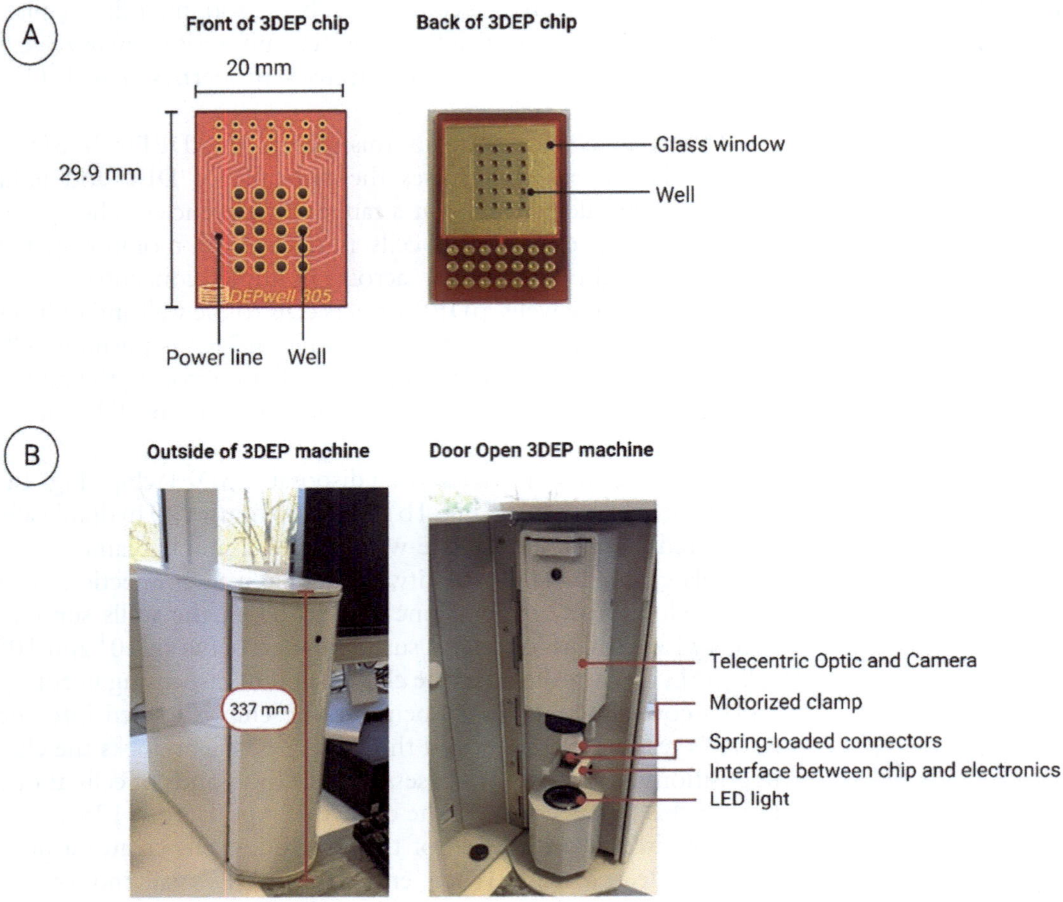

Fig. 1 Diagrams of (**a**) the 3DEP chip and (**b**) the 3DEP machine. (Figure designed by Krista Clarke, adapted from [62])

mesenchymal stem cells (NSPCs) [41–43], characterization of cardiomyocytes [44], dielectric changes in chondrocytes during osteoarthritis [45], characterizing biological rhythms in blood cells [25, 46–48], electrical changes in stored red blood cells [49], mitochondrial dynamics [50], platelet activation state [51], embryogenesis [52], and bacteria [53].

2 Materials

2.1 Equipment

1. Class II biological safety cabinet.
2. Benchtop centrifuge.
3. 3DEP (DEPtech, UK).
4. DEPwell lab-on-a-chip (DEPtech, UK).
5. Microscope.

6. Pipette controller.
7. Gilson pipettes P100, P200, P1000.
8. Beaker.
9. Tube rack.
10. Spatula.

2.2 Consumables

1. Hemocytometer.
2. Gilson pipette tips.
3. 15 mL falcon tubes.
4. Serological pipettes 10 mL.
5. Serological pipettes 25 mL.
6. Paper towel.
7. 1.5 mL Eppendorf Safe-Lock microcentrifuge tubes.
8. Ethanol (Absolute) 2.5 L ((GLC) 0.7897 g/mL, absolute, duty-free for analysis, CertiFied AR).
9. Blunt needles.

2.3 Cell Culture

1. Cell type, appropriate medium.
2. Dry ice.
3. Trypan Blue Solution, 0.4%.

2.4 DEP Medium

1. DI H_2O.
2. Sucrose (8.5% w/v).
3. Dextrose (0.3% w/v).
4. 100 μM $CaCl_2$.
5. 250 μM $MgCl_2$.
6. DPBS.

2.5 Software

1. ImageJ.
2. 3DEP Version 1.5.1.68.

3 Methods

An overview of the 3DEP protocol is shown in Fig. 2 and described below. Note that in parallel with 3DEP data collection, the radius of the cell is a critical input into the 3DEP model, therefore, cell radius measurements should be taken according to your standard laboratory method (*see* **Note 2**).

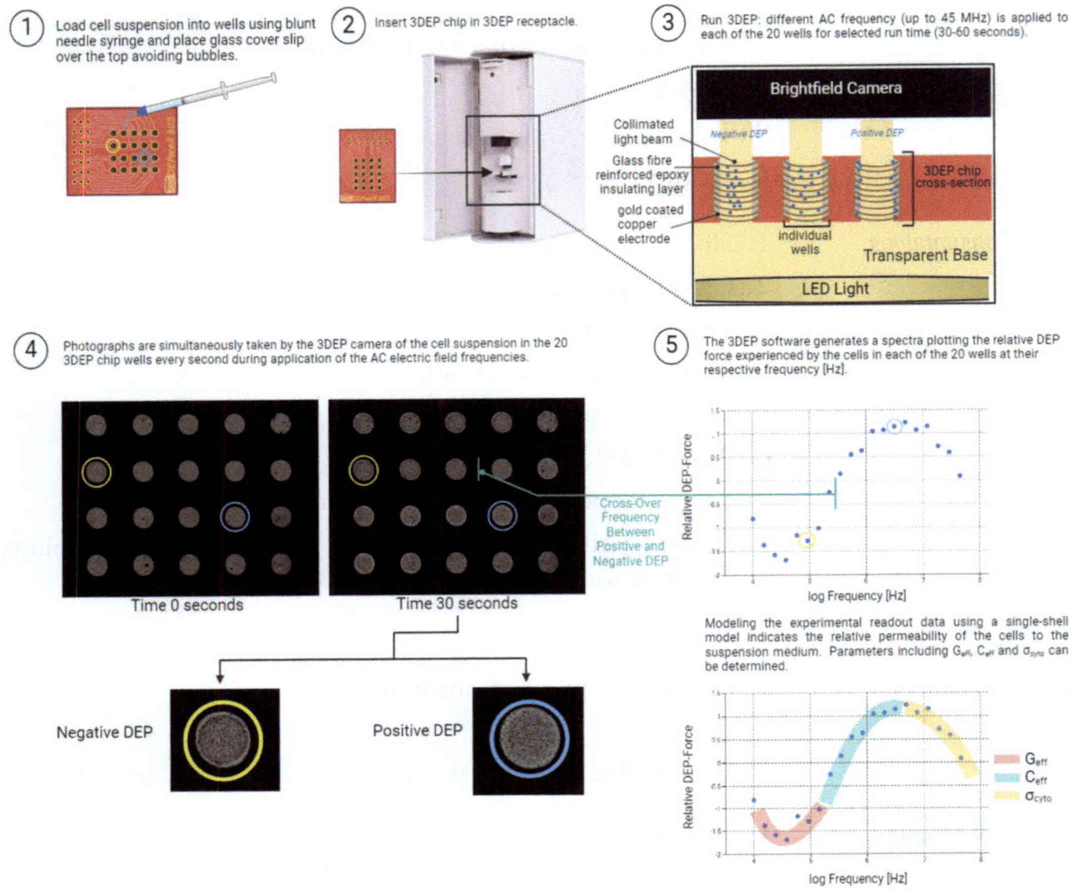

Fig. 2 Standard operation of the 3DEP—an overview of how the 3DEP works. Figure designed by Krista Clarke, adapted from [62]. Created using Biorender.com

3.1 DEP Medium

DEP medium is a simple iso-osmotic sugar solution with added salts to aid in ion channel stability. DEP medium: 8.5% (w/v) sucrose, 0.3% (w/v) dextrose, 250 μM $MgCl_2$, and 100 μM $CaCl_2$ in deionized water. Adjust conductivity to desired value by titrating Dulbecco's phosphate buffered saline (without $CaCl_2$ and $MgCl_2$) and measuring changes in conductivity using a calibrated conductivity meter at room temperature (18–22 °C) (*see* **Note 1**).

3.2 Cell Suspension

Cells to be analyzed are required to be in suspension with minimal clumping; therefore, when adherent cells are used, these need to be disassociated using the desired method, such as trypsinization [54]. Cells should first be pelleted in their media, then the supernatant should be removed, and the pellet resuspended in DEP medium; cells should be pelleted again, and supernatant removed before resuspension at a concentration of 1×10^6 cells/mL in DEP medium for analysis. Cells should be washed in DEP medium at least once to ensure minimal carry over of conductive culture medium that may alter the conductivity of the analysis DEP medium. As a beginner, a suspension volume of 1 mL is commonly necessary to ensure proper chip loading and removal of bubbles, however, as the user becomes proficient, suspension volumes of 300–400 μL should become sufficient to reach four technical repeats (*see* **Note 2**).

3.3 Site Requirements

An adequate location must be found for the 3DEP. The requirements are as follows: a stable and flat surface, sufficient ventilation ensuring all air slots are uncovered (this includes underneath the device), a 10 cm isolation distance from surroundings, and temperature and humidity ranges of 10–30 °C and 0–95%, respectively [55].

3.4 Preparing the 3DEP

Before starting the experiment.

1. Turn on the 3DEP.
2. Clean in between the pins with ethanol, using the spatula covered in paper towel. This additionally prevents crust accumulation from the sugary medium.
3. Enter the following:
   (i) Experiment Name—include cell name and date.
   (ii) Lowest Frequency/Highest Frequency—1 kHz/45 MHz (*see* **Note 3**).
   (iii) Runtime—duration of between 10 s and 60 s (*see* **Note 4**).
   (iv) Signal Voltage—1–20 $V_{p\text{-}p}$ (*see* **Note 5**).
   (v) Chip Type—refers to the type of chip (805 (1 mM wells), 806 (1.2 mM wells)).

The infographic below describes the location of entry for input values of i, ii, iii, iv, and v (Fig. 3).

3.5 3DEP Data Collection

1. Fill a 1 mL syringe with a suspension of cells at a concentration of 1×10^6/mL of treated cells.
2. Dab the surface of a 3DEP chip with a paper towel.
3. Load the 3DEP chip with the cell suspension (×4).
4. Remove any bubbles that form in the wells using the needle (*see* **Note 6**).
5. Position a 18 × 18 mm glass cover slide over the chip. Angle the cover slip over the fluid carefully and drop gently to prevent the formation of bubbles.
6. Transfer the four loaded technical repeats to the 3DEP reader bench.

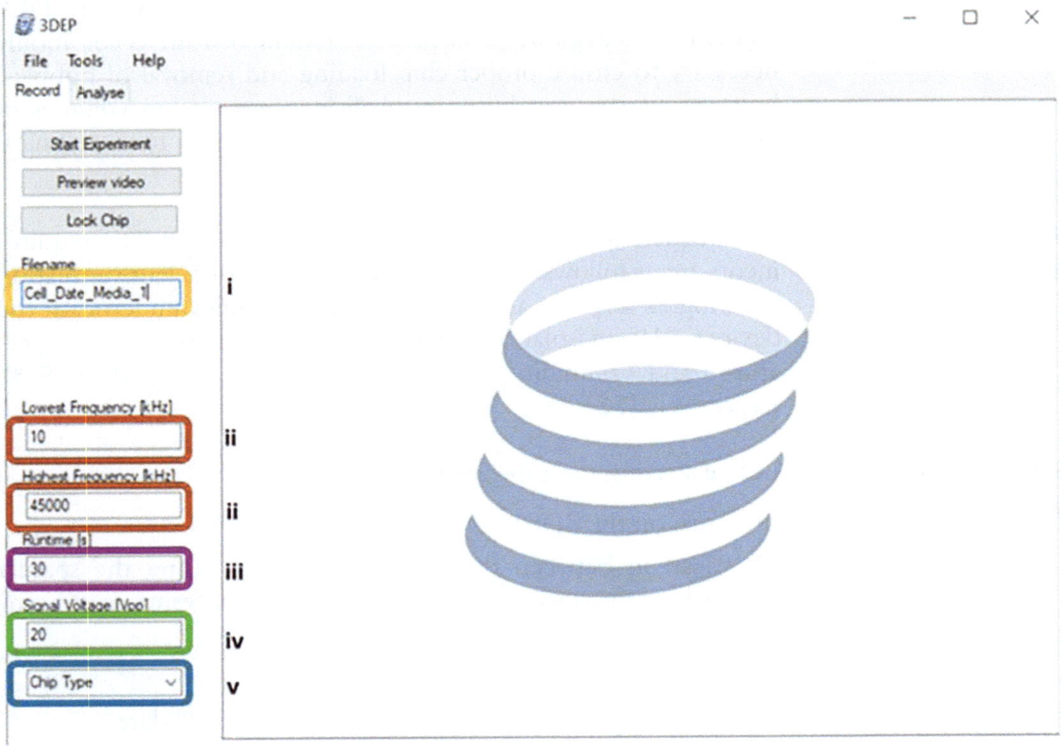

Fig. 3 Infographic demonstrating location of input parameters i, ii, iii, iv, and v required for sample analysis

7. Insert the loaded 3DEP chip into the 3DEP reader such that the contact pads are toward the machine, and the wells are toward the user.
8. Open "Record" tab of software "preview window."
9. View the 20 wells using the "Preview" window of the software to ensure the wells of the chip do not contain any bubbles. If multiple bubbles are detected, remove the chip, flush with DEP medium, and reload with cell suspension. (*see* **Note 6**).
10. Press the "Lock Chip" button of the software while manually holding the chip in place in the machine. Once locked in place, check that the pins have inserted into the holes of the chip and wiggle the chip to ensure there is good contact.
11. Name the file "celltype_date_treatment_medium conductivity_repeat."
12. Close the door of the 3DEP and press "Start Experiment." The power light at the front will turn orange during the experiment; do not open the machine while the light is orange, wait until the power light is green again.

13. Once data appear in the analysis tab, the door can be opened, and the results are automatically saved to the path selected in the "preferences" menu.
14. Dispose of the glass cover slide in a sharp's disposal bin.
15. Repeat for each technical repeat.
16. On experiment completion, turn off the 3DEP and clean pins with 70% ethanol.

3.6 3DEP Data Analysis

3.6.1 3DEP Software

The 3DEP software both records and analyzes data from a cell suspension; controlling the 3DEP during data collection and analyzing the resulting data. After data collection, the Analyse tab on the software may be chosen and the relevant data loaded. During postprocessing, the software fits a single-shelled model to the data taking into account the assumptions described in **Note 7**, from which the electrophysiological parameters are derived (C_{eff}, G_{eff}, and σ_{cyto}).

3.6.2 Parameter Input

For each sample's data, there are four outputs C_{eff}, G_{eff}, σ_{cyto}, and R^2. The R^2 value is the correlation coefficient between the data and a putative best-fit model (*see* **Note 8**), which is displayed below the spectrum after analysis. There are additional four inputs that must be included in the model prior to modeling the data, as follows:

1. The average radius of the cells in suspension.
2. The medium conductivity of the buffer (typically DEP medium).
3. Medium permittivity—this value typically remains at 78 ε, however, if the permittivity of the medium used is known to be different, this value should be input.
4. Cytoplasm permittivity—as standard $60*\varepsilon_0$ (where ε_0 is the permittivity of free space, 8.85×10^{-12} F/m) if known to be different alter as required.
5. Bands set the region for image analysis. The band for concentric rings starting from the center of the well (band 1) to the edge (band10). For most mammalian cells, bands in ranges between 4–9 or 7–9 are appropriate; smaller cells with a more limited range of movement can work better with a range of 8–9, while very large cells may require 6–9. Band 10 is generally excluded since, close to the wall, the cell concentration behaves differently, belcoming empty by the action of either positive or negative DEP.

3.6.3 Modeling

For each run (technical repeat), the 3DEP software generates a DEP spectrum, plotting the relative DEP force experienced by the cells in each of the 20 wells against the respective frequency [Hz]. The 3DEP software is adept at identifying outliers within

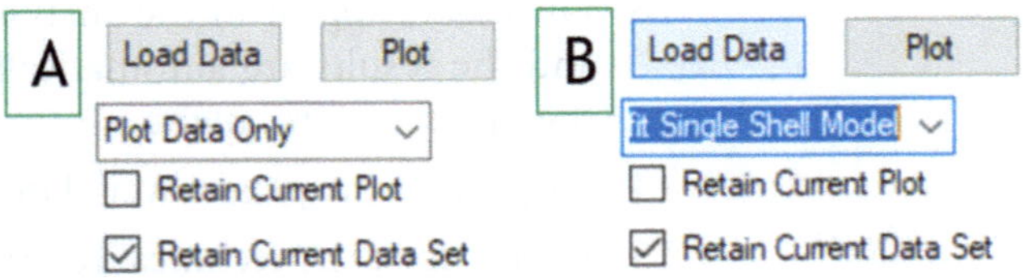

Fig. 4 (**a**) Image showing location of 'Retain Current Data Set' selection box. (**b**) Image showing location of 'Auto-fit Single Shell Model' drop-down selection

spectra, following the methods described in **Note 8**. Upon identification of outliers by the software, these should be removed to minimize erroneous results. These outliers can be removed in the software by loading the relevant data > Tools > Datapoint Editor > Remove all outliers + Remove all bubbles/low conc. This process must be followed for each technical repeat individually prior to averaging, with the edited data either saved separately or overwriting the unedited data.

Aiming for a total of four technical repeats allows for the identification of poor and noisy spectra: noisy spectra should be removed from the final average as these can skew the data. As a general rule, spectra with more than four outliers may be counted as noisy and tipped for removal. If the data appears particularly noisy, excluding the inner bands can improve the spectra quality (*see* **Note 9**). Noisy spectra can sometimes be obtained at higher conductivities, low cell concentration, and with the presence of bubbles in the 3DEP chip.

Once the spectrum of each technical repeat has been edited to remove outliers and the required parameters have been inputted, the repeats may be averaged together, and a single shell model fitted. This may be done by loading the relevant data and selecting retain current data set, the next data sets may now be loaded and will appear on the graph (Fig. 4a). To fit the single-shelled model, the drop-down menu displaying 'Plot Data Only' should be selected and the 'Auto-fit Single Shelled Model' option selected (Fig. 4b).

The resulting spectrum, with four technical repeats included, should look similar to Fig. 5. Once this stage is reached, the resulting spectra can be saved as a new result and electrophysiological parameter extracted from the corresponding excel sheet.

4 Notes

1. **Medium Conductivity Used.**

   Many DEP systems require low σ_{med}, as the current densities caused by high field strengths when using high ionic strength media can result in electrolysis, electrode screening,

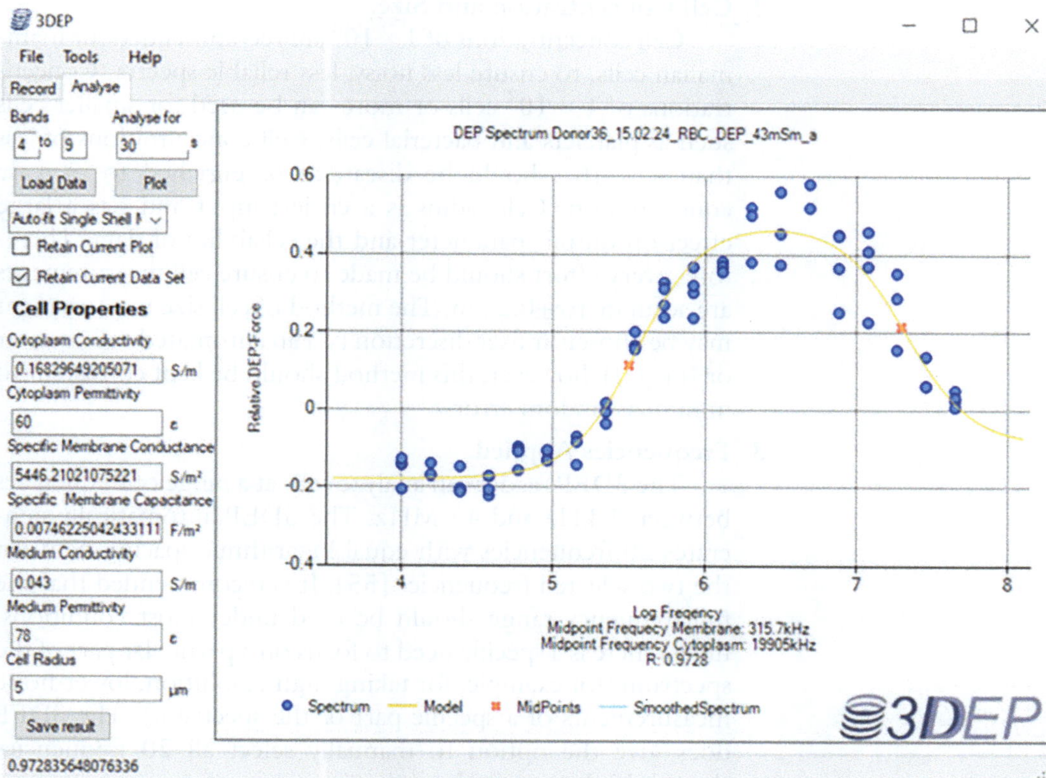

Fig. 5 Demonstration of the expected resultant data after analysis with the 3DEP software

and excess Joule heating at the water–electrode interface [56, 57]. Additionally, increases in σ_{med} flatten the DEP curve under the assumption that σ_{cyto} does not change with σ_{med} [58]. Altogether, these limitations have hindered 3DEP analysis of cells in physiological media, where conductivities are around 1400 mS/m.

However, Hoettges et al. [25] found that the 3DEP was capable of obtaining robust data of platelets suspended in near physiological strength media. Due to the 3DEP chip design that incorporates thick copper layers, high charge concentrations are avoided. This can allow for DEP of cell suspensions in higher-conductivity media. However, generally, DEP media of conductivities lower than 250 mS/m produce reliably low noise spectra. It has been shown that dependent on cell properties some 3DEP spectra can be obtained at conductivities up to 1600 mS/m as is the case for red blood cells [59].

Conductivity is temperature dependent, therefore, conductivity of the media should be measured at the same temperature that the cells will be analyzed.

2. **Cell Concentration and Size.**

   Cell concentrations of 1×10^6 are recommended for mammalian cells, to ensure less noisy, less reliable spectra. Concentrations of 1×10^7 cells or more can be used for smaller cells such as platelets and bacterial cells. Cell concentrations of less than 4×10^5 should be discarded or enriched to increase concentration. Cell radius is a critical input and has a large effect on output parameter and the reliability of data. Therefore, every effort should be made to ensure cell measurements are accurate to ±0.2 μm. The method of cell size measurement may be chosen at user discretion (via an automated cell counter or ImageJ); however, this method should be kept consistent to minimize random error.

3. **Frequencies Applied.**

   The 3DEP reader can analyze cells at a range of frequencies between 1 kHz and 45 MHz. The 3DEP automatically generates 20 frequencies with equal logarithmic spacing between the two selected frequencies [55]. It is recommended that the full frequency range should be used under most conditions, unless there is a specific need to focus on a particular part of the spectrum (for example, for taking high-resolution, lower-noise measurements of a specific part of the spectrum). The 3DEP does give the option to manually select all 20 frequencies should this be required.

4. **Selecting the Duration of Experiment.**

   The runtime of the experiment can be selected between 10 s and 60 s. The system takes an image of the cells every second within the wells. In general, larger cells produce larger movement which can be detected more quickly; a typical duration of data acquisition for mammalian cells is 10–30 s, though times up to 2 min are useful for smaller sized cells such as bacteria [53]. Shorter durations may introduce noise, while durations longer than 2 min can lead to saturation [35]. It should be noted that technical repeats within a sample should be produced with a consistent experimental duration.

5. **Voltage Applied.**

   The 3DEP reader can analyze cells at a range of voltages from $1V_{p\text{-}p}$ to 20 $V_{p\text{-}p}$. The appropriate voltage will depend on the purposes of the experiment and the cell type, however, typically 20 $V_{p\text{-}p}$ is used.

6. **Tips for Reducing Bubble Formation When Loading a 3DEP Chip.**

   Before loading with sample:

1. To improve wettability, it is advisable to precondition chips by pre-wetting with clean DEP buffer.

2. Where it is not possible to store chips in solution, at the start of experiments chips should first be flushed with ethanol to wet their internal surfaces and disinfect, then flushed with the prepared DEP buffer. In between technical repeats, chips should be kept wet with DEP buffer in the wells.
3. Check there are no bubbles in the 3DEP chip before loading the sample by looking through the wells of the chip out through the transparent back. Where bubbles persist after flushing submerge the chip in buffer media, this will displace the bubbles that will then float out of the chip.
4. Make sure there are no bubbles in the syringe containing the sample before trying to load the chip.

Loading the chip with sample:

1. Load the chip slowly and start with placing the needle in one of the four outer wells, and then in a central well.
2. The chip should be filled until a small meniscus forms over all the wells, a blunt needle tip can be used to spread the meniscus to cover the wells.
3. Without a meniscus, there is a greater risk of introducing bubbles when placing the glass cover slip; however, too large a meniscus results in rapidly dwindling sample that may prevent attainment of four technical repeats. Generally, for an experienced user, a suspension of 400 μL yields at least four repeats.
4. Where after loading sample into the chip bubbles appear, inject slightly more fluid to form a larger meniscus on the chip surface and use the blunt needle tip to dislodge the bubble into this meniscus. Then drag the bubble to the edge of the chip and off the side. Continue with coverslip placement and dab excess fluid by pressing the corner of a chip to a paper towel.
5. To place the coverslip, the 3DEP chip should be held at a 45° angle such that the meniscus bulges at the bottom four wells, the coverslip should be brought in contact with the meniscus and then released, allowing the coverslip to move up the chip via capillary action. Attempts should not be made to try to push the slip onto the chip as this may produce excess bubbles.
6. Once the chip is inserted into the 3DEP reader, lightly dab the top and bottom of the chip with a paper towel if the chip has been overfilled to remove excess liquid.
7. If there is more than one well occlusion due to bubbles, the coverslip should be removed and the chip refilled.

7. **Assumptions of the CM-Model.**

Assumptions are made about the dielectric particle and medium in which it is suspended, these include: each particle is of a uniform and spherical shape, the electric field is sinusoidal and polarized linearly, both particle and medium are lossless dielectrics at any given frequency, permittivity remains constant for the duration of analysis [14]. It is also necessary to assume a membrane thickness, though this is very difficult to measure; a value of 7 nM is often assumed.

The effects of these assumptions on the results can be nullified post-analysis. The derived value of σ_{cyto} is unaffected by these assumptions. For membrane properties, if we convert the calculated values of ε_{mem} and σ_{mem} by the assumed membrane thickness in the model, we obtain the specific (per unit area), effective values of membrane capacitance and conductance C_{eff} and G_{eff} neither of which depend on the values assumed in the model. The dependence on the assumed shape can further be removed by multiplying G_{eff} or C_{eff} by the surface area of the presumed sphere (using $4\pi r^2$, where r is the presumed cell radius) to determine whole-cell capacitance and conductance. These values have been shown to closely follow the results obtained by the gold-standard electrophysiology technique of patch clamp, though the determined values of G_{eff} are sometimes higher when determined by DEP due to the contribution of tangential surface conductance [25].

8. **Curve Fitting and Outlier Detection.**

   The obtained DEP spectrum is fitted to the collected data using a built-in Levenberg-Marquardt curve-fitting algorithm [25]. Outliers are detected through a two-step algorithmic process. The first stage uses a locally weighted scatterplot smoothing (LOWESS) algorithm [60] that passes the signal through a low-pass filter that filters the signal to one dispersion per decade. This results in a curve that represents the mathematically fastest observable changes within the single shell model. The residual between the LOWESS curve and the original data is then processed using ASTM E178 Standard Practice for Dealing with Outlying Observations [61], outlier detection algorithm to identify points that have a statistically high outlying residuals. Boundaries for fitting parameters can be set in the "preferences" menu. The default setting is for mammalian cells, so the ranges may need adapting in particular for nonbiological particles.

9. **Bands Analyzed.**

   To determine the net movement of cells in each well, the 3DEP software divides each well into ten concentric bands: light absorbance is measured to determine cell movement within each band [58]. Since the system is axisymmetric, cells only move between bands. In order to reduce noise, it is advisable to exclude bands which do not change during the experiment except for stochastic (Brownian motion) reasons. For example, band 10 (the outermost band) can be excluded; both pDEP and nDEP act on cells in this band very quickly, emptying it of cells either by attraction to the wall (where they are no longer visible) or repelling them to band 9. Similarly, we can exclude band 1—the innermost band (in fact the central circle) since the field is axisymmetric, so no movement will take place here. In fact, at shorter timescales, the DEP force only extends a relatively short distance through the well; only when experiments are run for several minutes can movement be observed on the inner bands [33–35]. This is exacerbated for small cells such as bacteria, where the DEP force is weaker [53]. As a consequence, for typical experiments of 30–60 s duration, it is not uncommon to limit analysis to bands 7–9 for most cells [62] and 8–9 for bacteria. Longer time periods can give less noisy data but risk saturating the instrument [35].

Acknowledgments

The authors acknowledge the University of Surrey for scholarships to KSPC and OVG, and DEParator Ltd. for a partial scholarship to KSPC.

Contributions KSPC and OVG contributed equally to writing the paper. KFH designed the curve fitting algorithm. RL co-supervised the work. FHL led the work and edited the manuscript.

References

1. Levin M (2021) Bioelectric signaling: reprogrammable circuits underlying embryogenesis, regeneration, and cancer. Cell 184(8): 1971–1989
2. Lobikin M, Chernet B, Lobo D, Levin M (2012) Resting potential, oncogene-induced tumorigenesis, and metastasis: the bioelectric basis of cancer in vivo. Phys Biol 9(6):065002
3. Chernet BT, Adams DS, Lobikin M, Levin M (2016) Use of genetically encoded, light-gated ion translocators to control tumorigenesis. Oncotarget 7(15):19575–19588
4. Chong P, Erable B, Bergel A (2021) How bacteria use electric fields to reach surfaces. Biofilms:3
5. Mansor M, Ahmad M (2015) Single cell electrical characterization techniques. Int J Mol Sci 16(12):12686–12712
6. Pohl HA (1951) The motion and precipitation of suspensoids in divergent electric fields. J Appl Phys 22(7):869–871
7. Pohl HA (1958) Some effects of nonuniform fields on dielectrics. J Appl Phys 29(8): 1182–1188
8. Pohl HA, Hawk I (1966) Separation of living and dead cells by dielectrophoresis. Science 152(3722):647–649
9. Grimnes S, Martinsen OG (2015) Chapter 7 – electrodes. In: Bioimpedance and bioelectricity basics. Elsevier, San Diego, pp 179–254
10. Pethig R (1985) Dielectric and electrical properties of biological materials. J Bioelectricity 4(2):vii–ix
11. Dukhin AS, Karamushka VI, Ulberg ZR, Gruzina TG (2014) Peculiarities of live cells' interaction with micro- and nanoparticles. In: Makino K, Ohshima H (eds) Colloid and interface science in pharmaceutical research and development. Elsevier, Amsterdam, pp 193–222
12. Salimi E, Bridges G (2019) 1.43 - dielectric properties of cells. In: Moo-Young M (ed) Comprehensive biotechnology, 3rd edn. Pergamon, Oxford, pp 585–598
13. Huang Y, Holzel R, Pethig R, Xiao BW (1992) Differences in the AC electrodynamics of viable and non-viable yeast cells determined through

combined dielectrophoresis and electrorotation studies. Phys Med Biol 37(7):1499
14. Broche LM, Labeed FH, Hughes MP (2005) Extraction of dielectric properties of multiple populations from dielectrophoretic collection spectrum data. Phys Med Biol 50(10)
15. Bunthawin S, Wanichapichart P, Tuantranont A, Coster HGL (2010) Dielectrophoretic spectra of translational velocity and critical frequency for a spheroid in traveling electric field. Biomicrofluidics 4(1)
16. Cetin B, Li D (2011) Dielectrophoresis in microfluidics technology. Electrophoresis 32(18):2410–2427
17. Hoettges KF (2010) Dielectrophoresis as a cell characterisation tool. Methods Mol Biol 583: 183–198
18. Hughes MP (2016) Fifty years of dielectrophoretic cell separation technology. Biomicrofluidics 10(3):032801
19. Cottet J, Fabregue O, Berger C, Buret F, Renaud P, Frénéa-Robin M (2019) MyDEP: a new computational tool for dielectric modeling of particles and cells. Biophys J 116(1):12–18
20. Kaler KV, Jones TB (1990) Dielectrophoretic spectra of single cells determined by feedback-controlled levitation. Biophys J 57(2):173–182
21. Lewis DL, Lechleiter JD, Kim D, Nanavati C, Clapham DE (1990) Intracellular regulation of ion channels in cell membranes. Mayo Clin Proc 65(8):1127–1143
22. Wanichapichart P, Bunthawin S, Kaewpaiboon A, Kanchanapoom K (2002) Determination of cell dielectric properties using dielectrophoretic technique. Sci Asia 28: 113–119
23. Afshar S, Salimi E, Fazelkah A, Braasch K, Mishra N, Butler M et al (2019) Progression of change in membrane capacitance and cytoplasm conductivity of cells during controlled starvation using dual-frequency DEP cytometry. Anal Chim Acta 1059:59–67
24. Trainito C (2015) Study of cell membrane permeabilization induced by pulsed electric field – electrical modeling and characterization on biochip. Universite Paris-Saclay, Paris
25. Hoettges K, Henslee E, Torcal Serrano RM, Jabr R, Abdallat R, Beale A et al (2019) Ten-second electrophysiology: evaluation of the 3DEP platform for high-speed, high-accuracy cell analysis. Sci Rep:9
26. Hughes MP, Fry CH, Labeed FH (2022) Cytoplasmic anion/cation imbalances applied across the membrane capacitance may form a significant component of the resting membrane potential of red blood cells. Sci Rep 12(1):15005
27. Purves D, Augustine GJ, Fitzpatrick D (2001) Neuroscience. Sinauer Associates, Sunderland
28. Kim SY, Silver MR, DeCoursey TE (1996) Ion channels in human THP-1 monocytes. J Mem Biol 152:117–130
29. Henslee EA, Torcal Serrano RM, Labeed FH, Jabr RI, Fry CH, Hughes MP et al (2016) Accurate quantification of apoptosis progression and toxicity using a dielectrophoretic approach. Analyst 23
30. Pethig R (2017) Dielectrophoresis: theory, methodology and biological applications, 1st edn. Wiley
31. Golowasch J, Nadim F (2013) Capacitance, membrane. In: Jaeger D, Jung R (eds) Encyclopedia of computational neuroscience. New York, Springer, pp 1–5
32. Yale AR, Nourse JL, Lee KR, Ahmed SN, Arulmoli J, Jiang AYL et al (2018) Cell surface N-glycans influence electrophysiological properties and fate potential of neural stem cells. Stem Cell Rep 11(4):869–882
33. Hübner Y, Hoettges KF, Kass GE, Ogin SL, Hughes MP (2005) Parallel measurements of drug actions on erythrocytes by dielectrophoresis, using a three-dimensional electrode design. IEE Proc Nanobiotechnol 152(4): 150–154
34. Hoettges KF, Hübner Y, Broche LM, Ogin SL, Kass GE, Hughes MP (2008) Dielectrophoresis-activated multiwell plate for label-free high-throughput drug assessment. Anal Chem 80(6):2063–2068
35. Broche LM, Hoettges K, Ogin S, Kass G, Hughes M (2011) Rapid, automated measurement of dielectrophoretic forces using DEP-activated microwells. Electrophoresis 32(17):2393–2399
36. Mulhall HJ, Labeed FH, Kazmi B, Costea DE, Hughes MP, Lewis MP (2011) Cancer, pre-cancer and normal oral cells distinguished by dielectrophoresis. Anal Bioanal Chem 401(8):2455–2463
37. Liang X, Graham KA, Johannessen AC, Costea DE, Labeed FH (2014) Human oral cancer cells with increasing tumorigenic abilities exhibit higher effective membrane capacitance. Integr Biol 6(5):545–554
38. Graham KA, Mulhall HJ, Labeed FH, Lewis MP, Hoettges KF, Kalavrezos N et al (2015) A dielectrophoretic method of discrimination between normal oral epithelium, and oral and oropharyngeal cancer in a clinical setting. Analyst 14(15):5198–5524
39. Hughes MP, Labeed FH, Hoettges KF, Porter S, Mercadante V, Kalavrezos N et al (2023) Point-of-care analysis for non-invasive

diagnosis of oral cancer (PANDORA): a technology-development proof of concept diagnostic accuracy study of dielectrophoresis in patients with oral squamous cell carcinoma and dysplasia. J Oral Pathol Med 52(4): 305–314

40. Hoque R, Mostafid H, Hughes MP (2020) Rapid, low-cost dielectrophoretic diagnosis of bladder cancer in a clinical setting. IEEE J Transl Eng Health Med 8:1–5
41. Labeed FH, Lu J, Mulhall HJ, Marchenko SA, Hoettges KF, Estrada LC et al (2011) Biophysical characteristics reveal neural stem cell differentiation potential. PLoS One 6(9):e25458
42. Adams TNG, Jiang AYL, Mendoza NS, Ro CC, Lee D-H, Lee AP et al (2020) Label-free enrichment of fate-biased human neural stem and progenitor cells. Biosens Bioelectron 152: 111982
43. Ismail A, Hughes M, Mulhall H, Oreffo R, Labeed F (2015) Characterization of human skeletal stem and bone cell populations using dielectrophoresis. J Tissue Eng Regen Med 9(2):162–168
44. Fry CH, Salvage SC, Manazza A, Dupont E, Labeed FH, Hughes MP et al (2012) Cytoplasm resistivity of mammalian atrial myocardium determined by dielectrophoresis and impedance methods. Biophys J 103(11): 2287–2294
45. Abdallat R, Kruchek E, Matta C, Lewis R, Labeed FH (2021) Dielectrophoresis as a tool to reveal the potential role of ion channels and early electrophysiological changes in osteoarthritis. Micromachines 12(8):949
46. Henslee EA, Crosby P, Kitcatt SJ, Parry JSW, Bernardini A, Abdallat RG et al (2017) Rhythmic potassium transport regulates the circadian clock in human red blood cells. Nat Commun 8(1)
47. Beale AD, Kruchek E, Kitcatt SJ, Henslee EA, Parry JSW, Braun G et al (2019) Casein kinase 1 underlies temperature compensation of circadian rhythms in human red blood cells. J Biol Rhythm 34(2):144–153
48. Labeed FH, Beale AD, Schneider P, Kitcatt SJ, Kruchek EJ, Reece SE (2022) Circadian rhythmicity in murine blood: electrical effects of malaria infection and anemia. Front Bioeng Biotechnol 10:994487
49. Lavi ED, Crivellari F, Gagnon Z (2022) Dielectrophoretic detection of electrical property changes of stored human red blood cells. Electrophoresis 43(12):1297–1308
50. Rohani A, Moore JH, Kashatus JA, Sesaki H, Kashatus DF, Swami NS (2017) Label-free quantification of intracellular mitochondrial dynamics using dielectrophoresis. Anal Chem 89(11):5757–5764
51. Hughes MP, Kruchek EJ, Gibbins JM, Griffiths OV, Abdal B, Labeed F (2022) The platelet electrome: evidence for a role in regulation of function and surface interaction. Bioelectricity 4(3):153–159
52. Matta C, Lewis R, Fellows C, Diszhazi G, Almassy J, Miosge N et al (2021) Transcriptome-based screening of ion channels and transporters in a migratory chondroprogenitor cell line isolated from late-stage osteoarthritic cartilage. J Cell Physiol 236(11):7421–7439
53. Hoettges KF, Dale JW, Hughes MP (2007) Rapid determination of antibiotic resistance in E. coli using dielectrophoresis. Phys Med Biol 52(19):6001–6009
54. Mahabadi S, Labeed FH, Hughes MP (2015) Effects of cell detachment methods on the dielectric properties of adherent and suspension cells. Electrophoresis 36(13): 1493–1498
55. 3DEP Instruction manual 2019. DEPtech. https://deparator.com/
56. Krishnan R, Dehlinger DA, Gemmen GJ, Mifflin RL, Esener SC, Heller MJ (2009) Interaction of nanoparticles at the DEP microelectrode interface under high conductance conditions. Electrochem Commun 11(8):1661–1666
57. Henriksson A, Neubauer P, Birkholz M (2022) Dielectrophoresis: an approach to increase sensitivity, reduce response time and to suppress nonspecific binding in biosensors? Biosensors 12(10)
58. Park J, Komori T, Uda T, Miyajima K, Fujii T, Kim SH (2019) Sequential cell-processing system by integrating hydrodynamic purification and dielectrophoretic trapping for analyses of suspended cancer cells. Micromachines 11(1):47
59. Hughes MP, Kruchek EJ, Beale AD, Kitcatt SJ, Qureshi S, Trott ZP et al (2021) Vm-related extracellular potentials observed in red blood cells. Sci Rep 11(1)
60. NIST/SEMATECH. e-Handbook of statistical methods. http://www.itl.nist.gov/div898/handbook/. March 2024
61. ASTM International. ASTM E178–21: standard practice for dealing with outlying observations. https://doi.org/10.1520/E0178-21
62. Kruchek EJ (2022) Investigating electrophysiological changes in blood cells using dielectrophoresis. University of Surrey

Part III

Techniques for Investigating the Biophysical Characteristics of Biological Membranes

Chapter 14

Biophysical Profiling of Protein Corona on Red Blood Cell-Derived Extracellular Vesicles (REVs): Linear Dichroism and Microfluidic Resistive Pulse Sensing Separate Surface Clearing from Vesicle Disruption

Tasvilla Sonallya, Anikó Gaál, Imola Cs. Szigyártó, Zoltán Varga, and Tamás Beke-Somfai

Abstract

Extracellular vesicles (EVs) have attracted significant scientific attention due to their critical functions in intercellular communication and their possible uses in diverse fields such as immunology, therapeutics, reproductive biology, biotechnology, and medicine. EVs are engulfed in a layer of proteins, also known as protein corona, which is speculated to play roles in several areas, from intercellular communication through immune recognition to cargo delivery. The composition of the protein corona strongly depends on the origin and the biological environment of EVs. Understanding the protein corona opens doors to finding various applications for vesicles by manipulating them. Typical ways of protein corona removal involve applying high salt concentration or the use of surface-active biomolecules, such as peptides, but this can result in EV membrane damage or complete vesicle disruption. Here, we describe a protocol for characterizing the change in protein corona content on red blood cell-derived EVs (REVs) by using linear dichroism spectroscopy (LD) with microfluidic resistive pulse sensing (MRPS). LD can quantify the change in the amount of the surface attached hemoglobins but cannot identify whether these changes are due to vesicle disruption or protein corona removal from intact vesicles. This necessitates the use of MRPS, which counts the number of vesicles before and after adding a surface manipulating compound. Thus it can identify whether the change in LD signal is due to vesicle disruption or due to loss of protein corona. The combined methods can be employed to understand which process takes place in which ratio, allowing the optimization of vesicle engineering toward specific needs.

Key words Extracellular vesicle, Isolation, Linear dichroism spectroscopy (LD), Microfluidic resistive pulse sensing (MRPS)

1 Introduction

Extracellular vesicles (EVs) are nanosized particles enclosed in a lipid bilayer and released by cells through various mechanisms [1–3]. They can be found in a diverse array of bodily fluids, including

Csaba Matta et al. (eds.), *The Surfaceome: Methods and Protocols*, Methods in Molecular Biology, vol. 2908,
https://doi.org/10.1007/978-1-0716-4434-8_14,

serum, plasma, urine, saliva, and breast milk, and they serve important functions in cellular communication by transporting proteins, RNA, and DNA [4–7]. They also have membrane proteins on their surface. Recent research reveals that EVs, like manufactured nanoparticles, can adsorb proteins from a protein-rich environment. These adsorbed proteins may adhere strongly or sparsely onto the EV surface, generating multiple layers of protein termed the "protein corona." The biological environment and the physical and biochemical features of the EVs determine the makeup of this corona.

The protein corona around EVs has garnered significant attention due to its potential implications in biomarker discovery and disease diagnosis and treatment [8]. EV protein corona is critical in mediating biological processes, affecting cargo delivery, and promoting intercellular communication. Furthermore, the heterogeneity of the protein corona has been found to alter the hydrodynamic diameter of EVs and their mobility in the extracellular matrix, highlighting the relevance of knowing its composition and development [9]. Understanding the surface properties also allows for the targeted modification of EVs for increased therapeutic uses, such as enhancing drug transport and boosting efficacy. Additionally, the functional characteristics of the protein corona around EVs have been studied, revealing its significance in promoting angiogenesis [10], skin regeneration [11], and immunomodulation [12]. This illustrates the diverse effect of the protein corona on biological processes [13].

In all, the protein corona around EVs is a complex and dynamic interface with important implications for many biological processes and therapeutic applications. Understanding its structure, formation, and functional activities is critical for exploiting its full potential in diagnostics, treatments, and biomedical research. This chapter outlines a protocol to characterize the surface alterations on EVs, while also following the size distributional changes of the EVs to monitor whether these manipulations could also cause vesicle disruption. In particular, for the initial setup of the protocol, red blood cell-derived EVs (REVs) are employed, as these can be produced in larger quantities suitable for initial optimization procedures [14]. Furthermore, in the case of REVs hemoglobin attaches also to its external surface [15, 16], which contains a heme that has a strong absorption band, the Soret band in the ~406–430 nm range, that offers a direct way to follow REV changes by light spectroscopy. To distinguish the spectral contribution of surface and interior heme groups from each other, we employed polarized light spectroscopy, linear dichroism (LD), which can provide information on the oriented systems, that is in our case, the REVs tested [15, 16]. While the employed solution-phase technique (flow-LD) is described in detail elsewhere [17], a related brief background is presented below in the Methods section for better understanding. Flow-LD can be combined with microfluidic resistive pulse sensing (MRPS), which can describe the particle number and size distribution of EVs [15, 18]. The combination of these two systems is beneficial, as removal of surface proteins can be tracked by LD as the gradual loss of the heme signal can be quantified, however, this loss in intensity could also be due to disrupted vesicles, i.e., due to loss of the oriented system. MRPS in this regard is crucial as it can estimate the vesicle loss, which can then be used to adjust calculations from the LD spectral changes, thus potential artefacts can be avoided. The protocol outlined here relies on the joint use of these techniques offering a reliable and reproducible approach for the preliminary profiling of the surface proteins on the EVs.

2 Materials

2.1 Blood Collection

1. Disinfectants, tourniquets, and needles with a minimum inner diameter of 21G for blood collection.
2. The blood is collected in two 6 mL tri potassium ethylenediaminetetraacetic acid (K3 EDTA) blood collection tubes (Vacuette® Tube, Greiner Bio-One).
3. 12 mL of whole blood.

2.2 Red Blood Cell-Derived EV (REV) Isolation

1. Four 15 mL conical tubes (Eppendorf).
2. Protein LoBind safelock tubes (2 mL, 1.5 mL [Eppendorf]).
3. 1× phosphate buffer saline (PBS) prepared in MilliQ water (set pH to 7.34, filter using 0.22 μm filter).
4. Physiological saline (0.9 wt.%, filter using 0.22 μm filter).
5. 3.5 mL plastic gravity columns filled with Sepharose CL-2B gel (Cytiva).
6. Refrigerated centrifuge, such as Eppendorf 5810R, A-4-62 with adapters for 15 mL centrifuge tubes (800 × g, 2500 × g, and 3000 × g).
7. Refrigerated centrifuge, such as Eppendorf 5415R, with adapter for 2 mL tubes (16,000 × g).

2.3 Microfluidic Resistive Pulse Sensing (MRPS)

1. Running buffer: 1% Tween 20 in PBS.
2. C400 microfluidic cartridge (Spectradyne LLC, USA).
3. Bovine serum albumin (BSA).
4. Polystyrene Beads Styrene 90 (PS90), 1 w/v%, (Polysciences Inc).
5. Ultra Centrifugal filter, 100 kDa NMWCO, 0.5 mL (Amicon).

6. Centrifuge, Eppendorf 5415R, with adapter for 2 mL tubes (1000 × g).
7. nCS1 MRPS instrument (Spectradyne LLC, USA).

2.4 Linear Dichroism Spectroscopy

1. 0.5 mm path length 100 μL cylindrical quartz cuvette.
2. JASCO-1500 spectrometer equipped with a Couette flow cell system.

3 Methods

3.1 Theoretical Background

3.1.1 Linear Dichroism (LD)

To follow the change in surface protein content, we employed solution phase linear dichroism spectroscopy (flow-LD). In flow-LD, the instrument is coupled to a Couette rotating cell, which can generate a shear force in the solution, making soft systems such as liposomes [17] and EVs [16] distorted and thus oriented along the main macroscopic orientation axis, which axis in this case is the circular flow direction. The *LD* signal is defined as the difference between absorption of the light polarized parallel, $A_{||}$, and perpendicular, $A_{\perp}$, relative to the main orientation axis:

$$LD = A - A_{\perp}. \tag{1}$$

Consequently, those chromophore species which are not oriented, i.e., have an anisotropic orientation, do not contribute to the LD spectra as the subtraction in (1) results zero intensity. In our case, this means that in oriented EVs, only the surface-adsorbed heme groups will have a contribution to the LD signal; the interior, unoriented are not seen.

3.1.2 Microfluidic Resistive Pulse Sensing (MRPS)

MRPS is a technique used for the detection and characterization of particles or analytes in a fluidic system. This method is particularly valuable for studying particles at the micro- and nanoscale. MRPS operates as a particle passes through a nanoconstriction leading to lowering the electric current. The presence of the voltage divider that is formed between the fluidic resistor and the nanoconstriction allows for the measurement of the sensing electrode voltage. The varying voltage is proportionate to the volume displaced by a particle passing through a nanoconstriction according to the Coulter principle [19]. The transit time is utilized to calculate the flow rate and, in conjunction with the count rate, is employed to ascertain the number of particles. It provides precise assessments of size measurements of individual nanoparticles providing size distribution with absolute concentration without relying on prior assumptions regarding the sample or particle composition. The particle analysis range lies between 50 nm and 10 μm in diameter, making it ideal for measurements of EVs [19, 20].

3.2 REV Isolation from Whole Blood

1. Obtain 12 mL whole blood from a healthy donor in K3 EDTA collection tubes with informed consent (*see* **Notes 1** and **2**).
2. Transfer the whole blood to a 15 mL centrifuge tube and gently mix.
3. To isolate red blood cells (RBCs) from the blood sample, subject the sample to differential centrifugation at 800 × g for 10 min at 4 °C (*see* **Note 3**).
4. From the separated whole blood, remove the supernatant containing platelet-rich plasma using a transfer pipette and discard. Gently pipette out the buffy coat as much as possible and discard.
5. Add 5 mL of physiological salt solution to the separated RBCs. Centrifuge the sample at 800 × g for 10 min at 4 °C. Discard the supernatant. Repeat this wash step twice to remove the buffy coat completely.
6. Add 6 mL 10× PBS solution (pH 7.4) to the washed RBCs and incubate at 4 °C for 7 days for EV production (*see* **Note 4**).
7. Remove the sample from 4 °C refrigeration and gently mix well. Centrifuge at 2500 × g for 10 min to sediment the cells.
8. Transfer the supernatant using a transfer pipette to another 15 mL falcon tube. Centrifuge at 3000 × g for 30 min at 4 °C. Transfer 6 × 1 mL of the supernatant into Protein LoBind tubes.
9. Centrifuge the supernatant at 16,000 × g for 30 min at 4 °C to pellet the REVs.
10. Discard the supernatant, then resuspend the REV pellet with 100 μL PBS.
11. Filter and purify the REV using size exclusion chromatography with the aid of a 3.5 mL Sepharose CL-2B column (*see* **Note 5**).

3.3 Inferring EV Concentration Via Microfluidic Resistive Pulse Sensing (MRPS)

1. Prepare a working buffer by dissolving 1 mg of BSA in 1 mL of PBS. Transfer 500 μL of the solution to the Amicon® Ultra Centrifugal filter. Centrifuge the sample at 5000 × g for 3 min at room temperature. The filtrate is used as the buffer for the measurements (*see* **Note 6**).
2. Dilute the isolated REV to 1000× folds using the prepared BSA-PBS solution.
3. Make sure to check the calibration of the instrument prior to sample measurement. Calibration beads can be used for that purpose. PS90 is an apt option.
4. 3 μL of the diluted REV is carefully pipetted into the well in the cartridge. Ensure no bubbles are formed. In case of bubbles, remove the sample and reload.

5. Load the cartridge onto the stage of the nCS1 instrument and raise the platform.
6. The experiment can be carried out using nCS1 software (Spectradyne LLC) to control the instrument (*see* **Notes 7** and **8**). Begin acquisition and run the measurement until the event count (N) $\geq$ 2000 is reached. The data can be analyzed using the VIEWER software (Spectradyne LLC). Each file comprises multiple measurements with a 10-s timeframe. Therefore, for efficient file management, opt for "combine data and create run report" before ending the run. Subtract the background (refer to **Note 9**). The size distribution and concentration are generated by the software. However, in the quantification bar, Gaussian fit can be employed either manually by selecting the diameter range (from 100 to 250 nm) or select it graphically. Alternatively, a lognormal fit can also be applied (*see* **Note 10**).

The concentration data gathered from MRPS measurements is critical in determining the dilution strategy for subsequent investigations, especially in the context of linear dichroism. Using the MRPS-derived concentration, we may tune the sample concentration to guarantee the best circumstances for reliable LD measurements. Moreover, MRPS demonstrates versatility by offering insights into variations in size distribution. This potentially could extend its capability to evaluate changes in size distribution that may occur when surface proteins are effectively removed or during processes involving adsorption. MRPS in these scenarios could provide a comprehensive understanding of how the composition and characteristics of particles or molecules evolve, which contributes vital information to experimental research.

3.4 Surface Characterization Using Polarized Light Spectroscopy (Linear Dichroism)

In the methods section, we briefly described the flow-LD spectroscopic approach. LD is often utilized to characterize structural and orientation properties of membrane adsorbed biomolecules [21–23]. The LD spectrum of RBC-derived EVs gives rise to two distinctive positive peaks. The first peak, around ~421 nm, corresponds to the Soret band of hemoglobin associated with the membrane within the protein corona. The second peak, emerging at ~220 nm, results from a combination of n-π* and π-π* amide transitions from helical structures found in REV proteins. The intensity of both of these peaks is sensitive to the amount of proteins attached to the surface of oriented vesicles. Thus, the method provides a means to monitor alterations occurring on the surface of REVs.

1. Pipette 45 μL of REVs to 45 μL of PBS and gently mix well (*see* **Note 11**).
2. Transfer the sample to a cylindrical cuvette.

3. Set the temperature control of the instrument to 25 °C. Purge the instrument with N_2 before starting measurement for a minimum of 5 min (*see* **Note 12**).
4. Place the cuvette into the cuvette holder on the Couette cell and screw the rotor in place tightly.
5. The Couette cell has to be inserted into an LD instrument. The measurement was conducted using a JASCO-1500 spectrometer equipped with an inner rotating cylinder Couette flow cell system. The Spectra Manager program is used to measure and analyze the sample. In measurement parameters, select LD, absorbance, and HT graphs (photomultiplier voltage) to display (*see* **Note 13**). Set wavelength to 205–500 nm for measurement. Ideally, three accumulations are taken for each measurement.
6. Set the parameters as follows: Data interval 0.5 nm, Measure range 600–200 nm, Data pitch 0.5 nm, D.I.T. 4 s, Bandwidth 1.00 nm, Start mode Immediately, Scanning speed: 100 nm/min, Shutter control: Auto, [CD scale 200 mdeg/1.0 dOD, FL scale 200 mdeg/1.0 dOD, NA] (Figs. 1 and 2).
7. Each sample is measured at 0 rpm to measure zero shear gradients followed by 3000 rpm to apply shear to orient the vesicles. The difference is subtracted using the spectrum subtraction tool in the analysis window of the Spectra Manager. This provides an LD signal as shown in Fig. 3.

By employing both MRPS and LD techniques, the LD intensity provides quantitative information on surface clearing of the protein corona, whereas the MRPS size distribution measured before and after treatment provides information on the potential

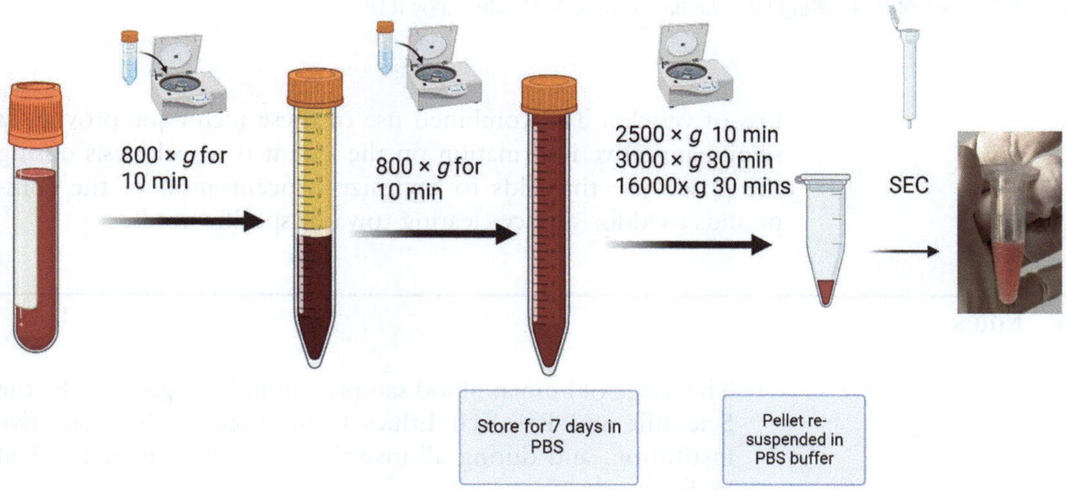

Fig. 1 Illustration of step-by-step protocol for REV isolation using centrifugation process

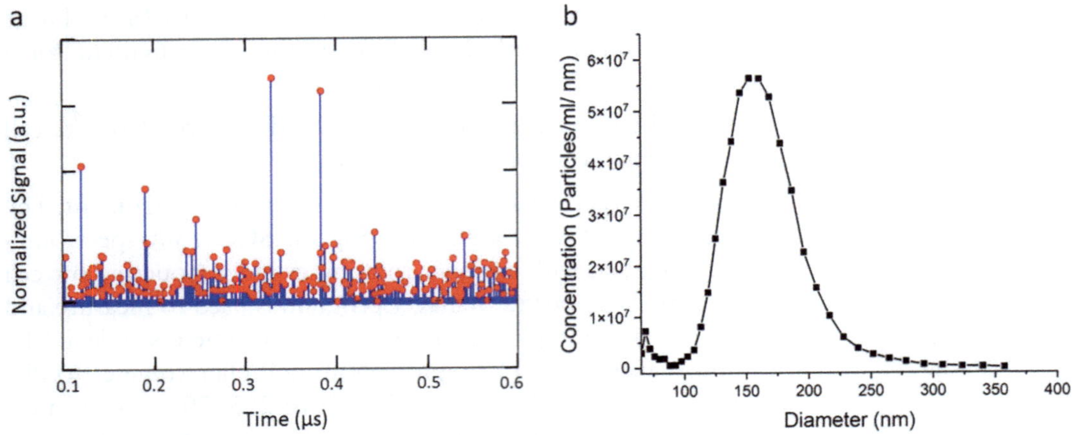

Fig. 2 Analysis of REVs using Microfluidic Resistive Pulse Sensing (MRPS)

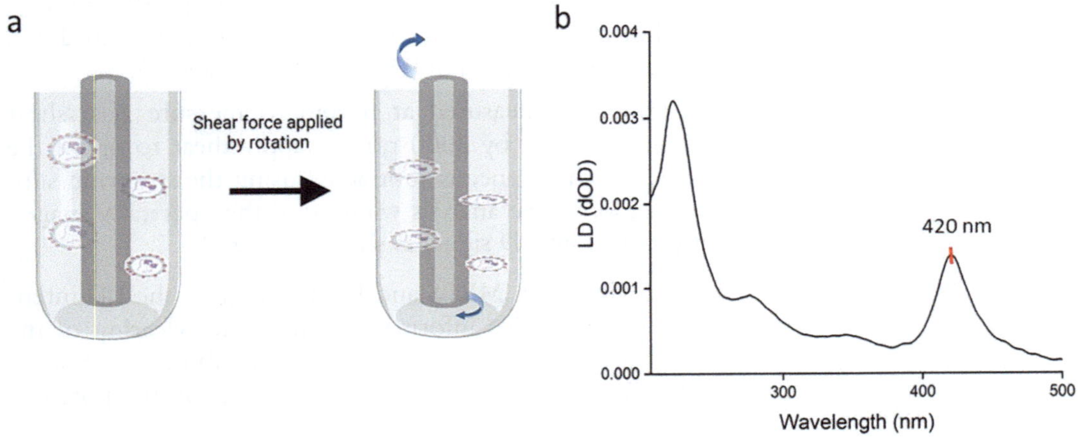

Fig. 3 Analysis of REVs using Flow-Linear Dichroism spectroscopy (LD)

loss of vesicles. The combined use of these technique provides a semiquantitative information on the extent of vesicle lysis during the procedure that aids to optimize concentration of the compounds used for surface clearing towards specific needs.

4 Notes

1. The usage of human blood samples should be approved by the Scientific and Research Ethics Committee of the respective institution, and during all investigations, the guidelines shall be followed.

2. Blood donation is recommended while fasting. Healthy donors who have not taken any medication should be requested to volunteer for blood donation.
3. Using a higher gradient centrifugation such as 2500 × g can also be employed to separate RBCs from whole blood. However, multiple higher gradient centrifugations have been shown to be harsh on the RBCs, leading to hemolysis.
4. It is advised to do an RBC count from suspended RBCs using a hematology analyzer or a flow cytometer before and after incubation to report the cell loss during isolation.
5. Load 100 μL of the resuspended REV onto the almost dry Sepharose column (never let the column run completely dry; however, the PBS must enter the packed gel completely before the loading of the sample) and follow with an additional 900 μl of PBS. This fraction is discarded. Once the added volume entered the packed gel, add 1 mL of the PBS to elute the sample. This fraction contains purified REVs and needs to be collected.
6. To achieve adequate wetting of the microfluidic channels, the BSA solution acts as a detergent to lower the surface tension of the buffer. An alternative is to use 0.03% Tween 20-PBS.
7. The MRPS measurement begins with the priming of the cartridge. This process involves wetting all channels of the cartridge, establishing the direction of fluid flow, and generating an appropriate ionic electric current within the system. Priming is conducted using a 1% Tween 20 (v/v) solution in DPBS [19, 24]. The measurement can only commence once the four wet sensors, located in the cartridge, are verified to be functioning correctly. At this point, the nCS1 program indicator transitions from red to green, signaling the start of the measurements.
8. If a blockage due to a sample that is too concentrated or an air bubble happens, the live measurement window will display a red cross at the time point corresponding to the blockage. First, try selecting "CLEAR CONSTRICTION." If the problem continues, stop the measurement by choosing "STOP" and then "END RUN." Afterward, adjust the dilution of the sample and repeat the measurement in a new cartridge.
9. Upon data combining, peak filters and background subtraction are to be applied, as per the manufacturer, in the nCS1 Data Viewer (Version 2.5.0.249) [20].
10. Measure MRPS initially on the REV sample and subsequently on the sample posttreatment. The same is carried out during the LD measurements.

11. In the case of the concentration range of particle concentration of ~10^{11}, perform a twofold dilution to make sure the HT value does not exceed 700 V (*see* **Note 12**). The measurement is performed at a final sample volume of 90 μL to avoid loss of sample from spilling out of the cuvette on rotation.
12. Using an instrument without purging can cause severe damage to the instrument; hence, this is the most vital step in instrument handling.
13. Protein, buffer, water, and air all absorb light at shorter wavelengths, resulting in fewer photons reaching the detector and increasing the HT voltage. The amount of light reaching the detector becomes insufficient for accurate measurement when the HT voltage exceeds 700 V. Throughout the measurement, make sure the HT value does not exceed 700 V.

Acknowledgments

This work was funded by the Hungarian Momentum Program (LP2016-2), the National Competitiveness and Excellence Program (NVKP_16-1-2016-0007). This work was also funded by the Ministry of Innovation and Technology of Hungary from the National Research, Development and Innovation Fund, TKP2021-EGA-31, KKP22 144180, 2020-1-1-2-PIACI-KFI_2020-00021 and the 2021-1.1.4-GYORSÍTÓSÁV-2022-00072 funding schemes. Support from Eötvös Loránd Research Network, grant no. SA-87/2021 and KEP-5/2021, is also acknowledged. The work of L.G. was financed by the Hungarian National Research, Development, and Innovation Office – NKFIH, FK 134343 project.

References

1. van der Pol E, Böing AN, Gool EL, Nieuwland R (2016) Recent developments in the nomenclature, presence, isolation, detection and clinical impact of extracellular vesicles. J Thromb Haemost 14:48–56
2. Xu R, Greening DW, Zhu H-J, Takahashi N, Simpson RJ (2016) Extracellular vesicle isolation and characterization: toward clinical application. J Clin Invest 126:1152–1162
3. Hartjes T, Mytnyk S, Jenster G, van Steijn V, van Royen M (2019) Extracellular vesicle quantification and characterization: common methods and emerging approaches. Bioengineering 6:7
4. Couch Y, Buzàs EI, Di Vizio D et al (2021) A brief history of nearly EV-erything – the rise and rise of extracellular vesicles. J Extracell Vesicles 10
5. Hu Y, Thaler J, Nieuwland R (2021) Extracellular vesicles in human milk. Pharmaceuticals 14:1050
6. Royo F, Zuñiga-Garcia P, Sanchez-Mosquera P et al (2016) Different EV enrichment methods suitable for clinical settings yield different subpopulations of urinary extracellular vesicles from human samples. J Extracell Vesicles 5
7. Alberro A, Iparraguirre L, Fernandes A, Otaegui D (2021) Extracellular vesicles in blood: sources, effects, and applications. Int J Mol Sci 22:8163
8. Huang D, Chen J, Hu D et al (2021) Advances in biological function and clinical application of

small extracellular vesicle membrane proteins. Front Oncol 11
9. Palviainen M, Saraswat M, Varga Z et al (2020) Extracellular vesicles from human plasma and serum are carriers of extravesicular cargo – implications for biomarker discovery. PLoS One 15:e0236439
10. Todorova D, Simoncini S, Lacroix R, Sabatier F, Dignat-George F (2017) Extracellular vesicles in angiogenesis. Circ Res 120: 1658–1673
11. Ding J-Y, Chen MJ, Wu LF et al (2023) Mesenchymal stem cell-derived extracellular vesicles in skin wound healing: roles, opportunities and challenges. Mil Med Res 10:36
12. Buzas EI (2023) The roles of extracellular vesicles in the immune system. Nat Rev Immunol 23:236–250
13. Wolf M, Poupardin R-W, Ebner-Peking P et al (2022) A functional corona around extracellular vesicles enhances angiogenesis, skin regeneration and immunomodulation. J Extracell Vesicles 11
14. Varga Z, Yuana Y, Grootemaat A-E et al (2014) Towards traceable size determination of extracellular vesicles. J Extracell Vesicles 3
15. Singh P, Szigyártó IC, Ricci M et al (2020) Membrane active peptides remove surface adsorbed protein corona from extracellular vesicles of red blood cells. Front Chem 8
16. Szigyártó IC, Deák R, Mihály J et al (2018) Flow alignment of extracellular vesicles: structure and orientation of membrane-associated bio-macromolecules studied with polarized light. Chembiochem 19:545–551
17. Nordén B, Alison Rodger TD (2010) Linear dichroism and circular dichroism: a textbook on polarized-light spectroscopy. Royal Society of Chemistry
18. Singh P, Szigyártó IC, Ricci M et al (2023) Removal and identification of external protein corona members from RBC-derived extracellular vesicles by surface manipulating antimicrobial peptides. J Extracell Biol 2
19. Cimorelli M, Nieuwland R, Varga Z, van der Pol E (2021) Standardized procedure to measure the size distribution of extracellular vesicles together with other particles in biofluids with microfluidic resistive pulse sensing. PLoS One 16:e0249603
20. Spectradyne LLC (2017) Spectradyne nCS1 – operation manual
21. Hicks MR, Damianoglou A, Rodger A, Dafforn TR (2008) Folding and membrane insertion of the pore-forming peptide gramicidin occur as a concerted process. J Mol Biol 383: 358–366
22. Ardhammar M, Mikati N, Nordén B (1998) Chromophore orientation in liposome membranes probed with flow dichroism. J Am Chem Soc 120:9957–9958
23. Kogan M, Beke-Somfai T, Nordén B (2011) Flow-alignment of bicellar lipid mixtures: orientations of probe molecules and membrane-associated biomacromolecules in lipid membranes studied with polarized light. Chem Commun 47:7356
24. Varga Z, Madai M, Kemenesi G, Beke-Somfai-T, Jakab F (2022) Single-particle detection of native SARS-CoV-2 virions by microfluidic resistive pulse sensing. Colloids Surf B Biointerfaces 218:112716

Chapter 15

ATR-Infrared Spectroscopy and Acoustic Sensing in Characterization of Blood and Pancreatic Ductal Adenocarcinoma-Derived Extracellular Vesicles

Imola Cs. Szigyártó, Priyanka Singh, Tasvilla Sonallya, Loránd Románszki, Judith Mihály, Zoltán Wiener, Tímea Bebesi, Gábor Mészáros, Zsófia Keresztes, Michael Thompson, Zoltán Varga, and Tamás Beke-Somfai

Abstract

Extracellular vesicles (EVs) are phospholipid bilayer-enclosed nanoparticles secreted by most cells into the biofluids. They have an important role in intercellular communication by carrying bioactive components (e.g., lipids, proteins, nucleic acids, and other metabolites) and reflecting the biochemical and metabolic processes of their parent cell. Due to their widespread molecular cargo, they have potential diagnostic and therapeutic roles, so they can be considered as new generation biomarkers. With the drastic increase in the number of publications and the comparability of the obtained results, the development of standardized protocols has become necessary. Here we aim to demonstrate the applicability of two, non-conventional techniques, such as ATR-infrared spectroscopy (ATR-FTIR) and electromagnetic piezoelectric acoustic sensor (EMPAS) in the characterization of compositional changes of blood- and pancreatic tumor-derived vesicles. In addition, we provide guidelines in sample collection, isolation, and separation of EVs. IR spectroscopy, as a fast and label-free technique, gives biochemical insight into the sample composition, while EMPAS serves information on vesicle quantity and in membrane integrity changes during storage.

Key words Extracellular vesicles, Red blood cell, Platelet, Pancreatic cell line, Isolation, Size-exclusion chromatography, Infrared spectroscopy, Electromagnetic piezoelectric acoustic sensor

1 Introduction

Extracellular vesicles are biological nanoparticles with a size distribution ranging from 30 nm to 5 μm, released by most types of cells and are found in blood, saliva, breast milk, amniotic fluid, semen, urine, cerebrospinal fluid, lymphatic fluid, nasal secretions, tears, and sweat [1]. Based on their origin, size, and morphology, they can be classified into three main groups: exosomes, microvesicles, and apoptotic bodies [2]. It is known that both healthy cells and

Csaba Matta et al. (eds.), *The Surfaceome: Methods and Protocols*, Methods in Molecular Biology, vol. 2908, https://doi.org/10.1007/978-1-0716-4434-8_15,

tumor cells release vesicles; however, their quantity and composition may differ [3]. Besides proteins and lipids, EVs can also contain RNA characteristic of the particular cells producing them, which can contribute to transferring genetic information from the donor to the recipient cells [4, 5]. The online databases, Vesiclepedia and ExoCarta, contain lipid, protein, and RNA data entries that have been identified in various sub-types of EVs [6, 7]. EVs have emerged as important mediators of intercellular communication and can provide information for early diagnosis and treatment of various diseases [8–10]. In addition, they exhibit a beneficial role in regenerative medicine and potential applications in tissue engineering [11] and they show potential roles in immune regulation by delivering inflammatory factors and receptors [12]. Due to the multiple potentials in EVs, the number of publications is steadily increasing, which necessitates the development of standardized protocols. The International Society of Extracellular Vesicles (ISEV) has published guidelines to assist researchers in characterizing EVs [2, 34]. The MISEV2018 and the newly updated MISEV2023 contain advice in EV nomenclature, sample collection, separation, storage, characterization, and functional studies. Scientists are encouraged to report the EV data of their studies in the EV-TRACK knowledgebase [13]. The widely used techniques for EV characterization such as western blotting [14], nanoparticle tracking analysis [15], electron microscopy [16], dynamic light scattering [17], flow cytometry [18], protein assays [19], and others [20] give fundamental insight to these species. However, the biophysical characterization of EVs at the molecular level is lagging despite their widespread investigation in biology and medical fields. Consequently, additional techniques that could rapidly provide further structural information on these systems are highly desirable [21–23]. In this respect, the purpose of the present protocol is to demonstrate the applicability of two, non-conventional techniques, such as attenuated total reflection Fourier-transform infrared spectroscopy (ATR-FTIR) and electromagnetic piezoelectric acoustic sensor (EMPAS), to characterize the compositional changes of different EVs. ATR-FTIR is a fast and reliable method to provide information on protein-to-lipid ratio and on sample purities without any complicated sample preparation [24–26]. The adsorption of extracellular vesicles onto various surfaces can be monitored using a sensor-based technique, EMPAS [27]. Brief descriptions on the principle of these techniques are provided in the Methods section. In this protocol chapter, the isolation, separation, and purification of EVs from three different sources are presented together with the applicability of the two aforementioned techniques. EVs were isolated from red blood cells (REVs), platelets (PEVs), and human pancreatic tumor cells

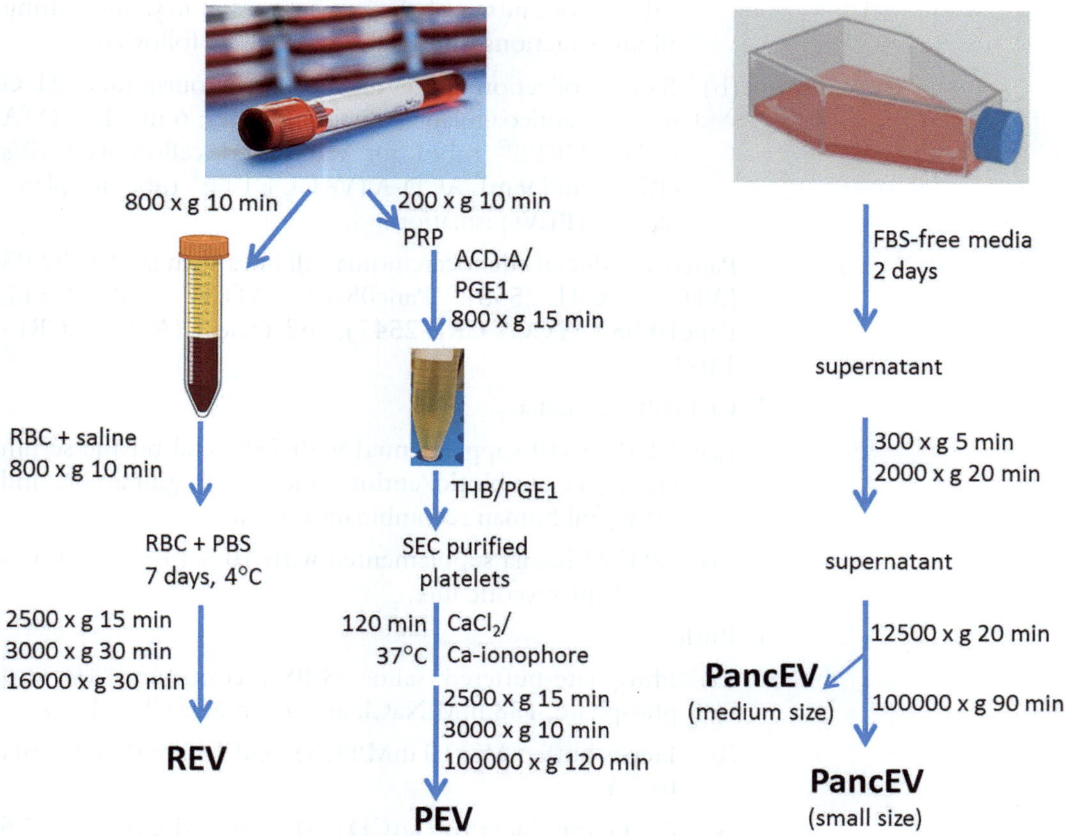

Fig. 1 General workflow for isolation of red blood cells (REVs), platelets (PEVs), and human pancreatic tumor cells (PancEVs)-derived vesicles

(PancEVs) (*see* Fig. 1). Red blood cells produce vesicles in large quantities [28, 29] making them an ideal reference material for biophysical characterization methods. In contrast, the quantity of EVs isolated from pancreatic tumors is orders of magnitude smaller, making their detection and characterization challenging. By fine-tuning the previously applied isolation protocols, it is possible to eliminate the non-EV-derived components, e.g., protein aggregates, that sediment along with the vesicles.

2 Materials

Prepare all solutions in Milli-Q water and sterile filter the solutions on 0.2 μm filter. All reagents and solvents utilized are of analytical grade purity for all experiments.

1. Blood Collection
   (a) Human blood from healthy donors (usage of human blood should be approved by the Scientific and Research

Ethics Committee of the respective country, and during all investigations, the guidelines shall be followed).

(b) Blood collection using disinfectants, tourniquet, 21-G needles, anticoagulant-containing tubes: 6 mL K3EDTA (VACUETTE® tube) for red blood cell-derived EVs (REVs) and 9 mL ACD-A (VACUETTE® tube) for platelet EVs (PEVs) isolation.

2. Pancreatic ductal adenocarcinoma cell lines such as Panc02.03 (ATCC® CRL-2553), Panc08.13 (ATCC® CRL-2551), Panc10.05 (ATCC® CRL-2547), and Panc-1 (ATCC® CRL-1469).
3. Cell culture media
   (a) RPMI 1640 supplemented with 15% fetal bovine serum (FBS), 1x antibiotic/antimycotic mix, L-glutamine, and 10 U/ml human recombinant insulin.
   (b) DMEM media supplemented with 10 % FBS and antibiotic/anti-mycotic mix.
4. Buffers:
   (a) Phosphate-buffered saline (PBS), containing 10 mM phosphate, 138 mM NaCl, and 2.7 mM KCl, pH 7.4.
   (b) Hepes buffer: Mix 10 mM Hepes and 140 mM NaCl, pH to 7.4.
   (c) Acid citrate dextrose (ACD): Mix 39 mM citric acid, 75 mM sodium citrate, and 135 mM dextrose, pH to 7.4.
   (d) Tyrode/Hepes buffer (THB): Mix 137 mM NaCl, 0.3 mM NaH_2PO_4, 3.5 mM Hepes, 5.5 mM D(+)-glucose, pH 7.35; filter through 0.1 μm filter; and store at 4 °C.
   (e) Ca^{2+}-Tyrode buffer: Mix 100 μL of Tyrode/Hepes buffer with 50 μL of 4 M $CaCl_2$.
5. Prostaglandin E1, 100 μg/mL of PGE1 in ethanol and calcium ionophore (A23187), 10 μM in ethanol.
6. Sterile, disposable materials: Falcon-type tubes: 15 and 50 mL; LoBind tubes: 0.5 and 1.5 mL; 0.1 and 0.22 μm pore size syringe filters; petri dishes; and 25 cm^2 or 75 cm^2 cell culture flask.
7. 3.5 mL gravity column filled with Sepharose CL-2B.
8. 1 cm path-length disposable cuvettes.
9. Teflon tubes (1.58 mm od, 0.8 mm id).
10. Centrifuge with adapters for 15 and 2 mL tubes.
11. Thermo-shaker for microtubes.

12. Dynamic light scattering:
    (a) AvidNano W130i dynamic light scattering device (DLS, Avid Nano Ltd., High Wycombe, UK) equipped with a diode laser (660 nm) and a photodiode detector.
    (b) Low volume, 1 cm path-length cuvette.
13. Fourier transform infrared spectroscopy:
    (a) Varian 2000 FTIR Scimitar Series spectrometer (Varian Inc., USA) equipped with a liquid nitrogen cooled mercury-cadmium-telluride (MCT) detector and a "Golden Gate" single reflection diamond ATR accessory (Specac Ltd, UK).
14. Home-built electromagnetic piezoelectric acoustic sensor [27]:
    (a) Plexiglas flow-through cell (~78 μL internal volume).
    (b) 5 mm diameter hand-wound coil of a 105 μm diameter polyurethane-coated copper wire (Goodfellow).
    (c) Frequency generator (Hewlett Packard 8648B).
    (d) Trimmer (muRata, Seminole, FL, USA) place in parallel with the coil terminals.
    (e) Lock-in amplifier (SR510, Stanford Research Systems).
    (f) Diode detector (house made) for the voltage drop developed across the coil terminals at acoustic resonance, with the output being fed into the lock-in amplifier.
    (g) Digital oscilloscope (Tektronix TDS 210).
    (h) Syringe pump (Harvard Apparatus Pump 11) equipped with a 60 mL plastic syringe (Henke-Sass, Wolf GmbH, Germany) providing a flow rate of 25 μL/min.
    (i) AT-cut quartz crystal wafer resonators to minimize the temperature effect on the resonant frequency.
    (j) Crystals with 13 mm diameter and 83 mm thickness (Laptech Precision Inc., Bowmanville, ON, Canada) operate at the 49th overtone of their $f_0 = 20$ MHz fundamental frequency ($f_{49} \approx 984$ MHz).
    (k) A code running under LabView 6.0 is used to control the frequency generator and the lock-in amplifier, as well as for data acquisition.

3 Methods

3.1 REVs Isolation from Whole Blood (see Note 1)

1. Collect 15 mL of blood from a healthy adult donor in K3EDTA tubes.
2. Mix the blood gently, make aliquots in 15 mL Falcon tubes, and centrifuge the tubes at 800 × g for 10 min at 4 °C.

3. Remove supernatant and the buffy coat from the red blood cells (RBCs).
4. Add physiological salt solution (0.9% NaCl) to RBCs to double the volume and centrifuge at 800 × g for 10 min, 4 °C to completely remove platelets and buffy coat. Discard the supernatant and repeat the washing step twice (*see* **Note 2**).
5. Dilute buffy coat-free RBCs to triple volume with 0.22 μm filtered phosphate buffered saline (PBS, pH 7.4) and store for 7 days at 4 °C (*see* **Note 3**).
6. After the end of the incubation period, mix gently and centrifuge the sample at 2500 × g for 15 min. Transfer the supernatant into a new tube and centrifuge again at 3000 × g for 30 min at 4 °C.
7. Aliquot the supernatant into 2 mL Eppendorf tubes and centrifuge at 16,000 × g for 30 min at 4 °C.
8. Discard the supernatant and resuspend the pellet in 100 μL PBS.
9. Use size-exclusion chromatography to further purify REVs (*see* **Note 4**).
10. Load 100 μL of REV sample onto the top of the 3.5 mL gravity column.
11. Add 900 μL of PBS while discarding the flow-through.
12. Elute the sample with 1 mL PBS, collect the fraction, and store at 4 °C before further use (*see* **Note 5**).

3.2 PEVs Isolation from Whole Blood (see Note 6)

1. Collect 15 mL of blood from a healthy adult donor in ACD-A tubes.
2. Mix the blood gently, aliquot into 15 mL Falcon tubes, and centrifuge the tubes at 200 × g for 10 min at room temperature (RT).
3. Transfer the platelet rich plasma (PRP) to a new 15 mL Falcon tube and add ACD-A in a 9:1 volume ratio (adjust based on the PRP volume) and 10 μL of PGE1 solution (100 μg/mL) to prevent platelet activation.
4. Gently mix by inverting the Falcon tube.
5. Centrifuge the mixture at 800 × g for 15 min at RT.
6. Remove and discard the supernatant and resuspend gently the white layer of settled platelet in 1 mL of Ca^{2+}-free Tyrode/Hepes buffer supplemented with 10 μL of PGE1 solution (*see* **Note 7**).
7. Load maximum 1 mL of platelet suspension onto the 3.5 mL SEC column (*see* **Note 8**).

8. Collect platelets in fractions 1 & 2 (total 1 mL), measure the platelet concentration using flow cytometry and dilute with Ca^{2+}-THB buffer to 2.5 x 10 particles / mL.
9. Activate the platelets for EV release by adding 2 mM CaCl and 10 μM Ca-ionophore and incubate for 2 h at 37 °C and 320 rpm.
10. Centrifuge the sample at 2500 × g for 15 min at RT. Collect and transfer the supernatant and centrifuge at 3000 x g for 10 min.
11. Transfer the supernatant into a new tubes and ultracentrifuge at 100,000 × g for 2h at 4 °C.
12. Remove and discard the supernatant and resuspend the pellet in 100 μL PBS.
13. Store the samples at 4 °C until further use (*see* **Note 5**).

3.3 PancEVs Isolation from PDAC Cell Line

1. Culture PDAC cells (Panc02.03, Panc08.13, and Panc10.05) in RPMI-1640 media and Panc-1 in DMEM media.
2. Split cells into the desired culturing plate and culture them until the culture reaching confluency.
3. Wash cells three times with PBS.
4. Add FBS-free medium and culture them 2 days.
5. Centrifuge cell supernatant at 300 × g for 5 min at RT, remove the supernatant and centrifuge at 2000 × g for 20 min at 4 °C (*see* **Note 9**).
6. Centrifuge the supernatant at 12,500 × g for 20 min at 16 °C for medium-sized EVs (*see* **Note 10**).
7. Ultracentrifuge the separated supernatant at 100,000 × g for 90 min at 4 °C for small-sized vesicles and discard the supernatant.
8. Resuspend the pellet with PBS and ultracentrifuge again as a washing step; remove and discard the supernatant.
9. Resuspend the pellet in 100 μL of PBS and store at 4 °C until further use (*see* **Note 11**).

3.4 Dynamic Light Scattering (DLS)

Dynamic light scattering is used for the characterization of particle sizes in solutions. Place 80 μL of the sample in a low-volume disposable cuvette with 1 cm path length and measure the average particle size and size distribution of the isolated EVs (REVs, PEVs, PancEVs). The time dependent autocorrelation functions are measured for 10 s, repeated 10 times. Measurement data analysis is performed with iSize 3.0 software.

3.5 Attenuated Total-Reflection Fourier-Transform Infrared Spectroscopy (ATR-FTIR)

Infrared spectroscopy (IR) is based on the interaction of electromagnetic radiation (infrared light, the least invasive radiation) with matter. The vibrational transitions of functional groups and bonds within the molecules can be detected and identified. The applicability of IR spectroscopy (in particular ATR-FTIR) on EV characterization was, to our best knowledge, first demonstrated by our research group [24]. This fast, efficient, and label-free method can provide structural information about the biomolecular composition of samples. Characteristic bands of amide I and amide II vibrations (~ 1654 and 1542 cm^{-1}), the symmetric and antisymmetric methylene stretching vibrations (~ 2850 and 2923 cm^{-1}), the carbonyl-stretching vibrations of the phospholipids, cholesterol, and triglycerides esters (~ 1734 cm^{-1}), and the phosphate (PO_2^-) vibrations from the fingerprint region (~1300–1000 cm^{-1}) can be observed in one spectrum and are characteristic of proteins, lipids, carbohydrates, and phosphates, respectively. The amide I region can be used to determine the secondary structure of proteins, whereas the ratio of amide I and amide II bands provide insights into the orientation and tertiary structure of proteins. Since proteins and lipids show distinctive absorption bands, the spectrum can be used to calculate the spectroscopic protein-to-lipid ratio, by dividing the integrated area of the amide I band by the integrated area of C–H stretching vibrations (from 3042 to 2800 cm^{-1}) [30].

3.5.1 IR Spectrum Collection

1. Set the nitrogen purge stream of the IR spectrometer up to 6 L/min.
2. Switch-on the ResolutionPro measuring program and check the interferogram (Collect/Rapid Scan/Setup).
3. Fill the detector with liquid nitrogen until the interferogram value is 5.4 and wait approx. 10 min.
4. Set the measurement parameters: electronics speed: 10 kHz, filter: 5, UDR: 2, resolution: 2, sensitivity: 1.5, Optics IR source: SCR1, beam: internal, Detector: bb-MCT, Aperture: open.
5. Clean the ATR-crystal with ethanol and then collect the background (Collect/RapidScan/Background) with a new file name (128 scans are recommended).
6. Add 3 μL of the sample on the top of the diamond and collect the spectra (64 scans are recommended), this will give the "wet-spectrum."
7. Let it drying the sample by slowly evaporation of solution at RT and collect the spectrum.
8. Between measurements, clean the ATR crystal using ethanol.
9. Repeat **steps 6** and 7 with buffer solution.

10. After each data acquisition, perform the ATR correction.
11. Use the Origin 2018 software package (OriginLab Corporation, Northampton, MA, USA) for all spectra analysis.

3.5.2 Protein-to-Lipid Ratio Calculation

1. Import data in Origin program and plot the spectrum.
2. Smooth the spectrum using Savitsky-Golay method (polynomial grade: 5, smoothing points: 10).
3. Make a baseline correction.
4. Repeat **steps 2** and **3** for buffer sample.
5. Subtract the buffer dry film spectrum from the samples' spectrum.
6. Estimate the protein content by fitting the amide I band (A (Amide I)) using a mixed Gaussian/Lorentzian function (*see* **Note 12**).
7. Assess the lipid content by total integrated intensity of CH_2/CH_3 stretching vibration from 3042 to 2800 cm^{-1} (A(CH_2/CH_3)).
8. Calculate the spectroscopic protein-to-lipid ratio as follows: P/L spectr = A(Amide I)/A(CH_2/CH_3).

3.6 Electromagnetic Piezoelectric Acoustic Sensor (EMPAS)

The EMPAS is a highly sensitive method due to the applicable ultrahigh frequency overtones [27]. The device operates on the principle of piezoelectricity, similar to quartz crystal microbalances (QCM). It measures the temporal change (Δf) in the resonance frequency (f) of a vibrating quartz crystal plate over time, and this change depends on the quantity of material adsorbed onto the quartz. The advantage of EMPAS lies in the fact that the crystal's vibration occurs not through direct electrical contact (as QCM), but rather through an induced electromagnetic field around a planar coil near the crystal. From a chemical point of view, the unique operating mode of the instrument allows the investigation of adsorption processes on non metallic surfaces (e.g., the unmodified quartz surface without any labeling of the samples). It permits the use of thinner crystals with higher fundamental frequencies (f_0 up to 20 MHz) as compared to those typically used in QCM and measurements at unusually high overtones (up to 1 GHz). As the sensitivity of mass measurement based on vibrating quartz depends squarely to the fundamental frequency and linearly to the number of the applied overtone for a vacuum, solid, thin, and well-adhered layer to the crystal, the theoretical sensitivity of EMPAS can be even three orders of magnitude higher than that of a conventional QCM. It should be noted that the increased sensitivity does not necessarily means the increase of resolution in mass measurement to the same extent, since it depends on the other factors, such as the signal-to-noise ratio of the overall setup under the actual experimental conditions.

3.6.1 EMPAS Data Acquisition

1. Place a 20 MHz AT-cut quartz sensor crystal in a microfluidic cell and seal the cell.
2. Ensure temperature control (thermostat), vibration-free conditions, and electromagnetic shielding.
3. Introduce filtered and degassed PBS solution using a syringe pump into the system (*see* **Note 13**).
4. Determine the resonance frequency of the crystal at the selected overtone (*see* **Note 14**).
5. Start the measurement by tracking the crystal frequency over time. The resulting curve reaches a stable baseline after 1–2 h, which will be the reference frequency.
6. Introduce 2 mL of EVs solutions into the microfluidic cell using a syringe pump.
7. The Δf curves plotted as a function of time (t) approach to a constant value, which can be correlated to the mass and thickness of the monolayer adsorbed on the crystal surface (*see* **Note 15**).
8. Use a new quartz crystal for each experiment, for cleaning protocol (*see* **Note 16**).
9. Storage conditions of EVs will influence the r_{max} value (*see* **Note 17**).

4 Notes

1. For a more detailed description, see *T. Sonallya* et al. protocol chaps. 14 and 15.
2. More than three washing steps resulted in cloudy, opalescent supernatant, due to the bursting of RBCs, making the removal of the upper layer (buffy coat) difficult.
3. The seven days storage time in PBS buffer resulted in the formation of mainly monodisperse vesicles, in contrast to the shorter storage time.
4. Before adding the sample, equilibrate the column with at least 15 mL of PBS and wait until the buffer disappears from the top of gel, but prevent drying.
5. Sample characterization is performed freshly, as soon as possible after isolation, until then, the sample is stored at 4 °C. However, for the later measurements, the samples are stored at −80 °C.
6. *See* reference *M. Aatonen* et al., Exosomes and Microvesicles: Methods and Protocols, 2017 [31].
7. Do not pipette up and down and avoid suspending the red blood cells at the very bottom of the tube.

8. Discard the void volume, collect fractions and judge based on opacity or verify with DLS for platelet detection.
9. The pellet contains the largest EVs, such as apoptotic bodies, and the supernatant can be further centrifuged.
10. If large and medium-sized EVs are required, dissolve the pellets after 2000 × g centrifugation for large-sized EVs and after 12,500 × g for medium-sized EVs in PBS, and centrifuge them again as a washing step. Discard the supernatant and dissolve the pellets in PBS.
11. The exact amount of the PBS depends on the requested EV concentration for further application and it must be experimentally tested. However, as a starting point, we recommend 100 μL.
12. For EVs samples two component fit is sufficient for the amide band: one centered ~ 1653 cm^{-1}, the other ~ 1600–1620 cm^{-1}. The latest might belong to aggregated proteins, amino acid residues, nonvesicular contaminants, etc. For protein-to-lipid calculation, the integrated area of the band component centered at ~ 1650 cm^{-1} was applied (A(Amide I)) (for more details, *see* reference Mihaly et al., BBA—Biomembranes, 2017 [24]). Representative IR spectra of isolated and purified vesicles are presented in Fig. 2.
13. Air bubbles can be completely removed from the cell by introducing a small amount of ethanol, which usually solves this problem. The recommended flow rate is 25 μL/min in case of one syringe pump, whereas set the flow rate to smaller and larger values for multiple ones.

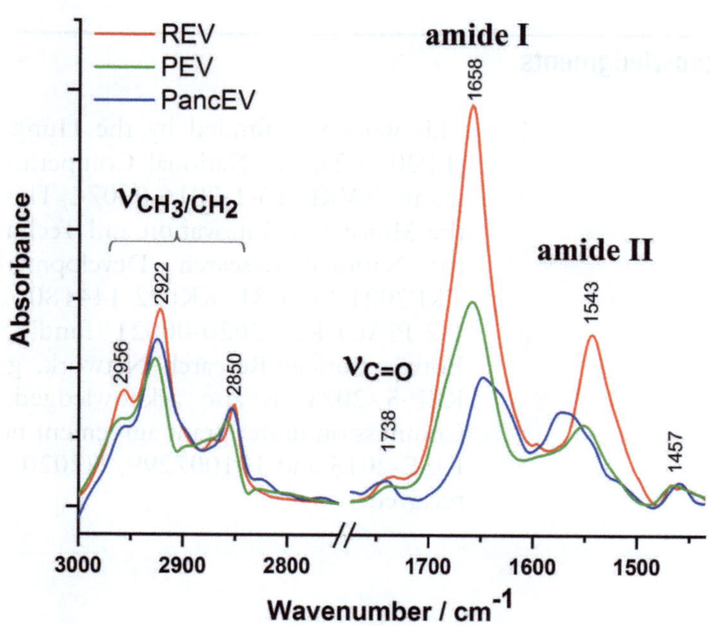

Fig. 2 ATR-FTIR spectra of EVs isolated from different sources

14. The resonance frequency is typically around the product of the fundamental frequency and the harmonic number (e.g., ~ 984 MHz for the 49th harmonic).
15. Repeating the measurement using different EVs concentrations will result in different adsorption signals. It is common that the sample volume does not permit a long measurement, therefore during this period (1–2 min) the curve does not reach the equilibrium value. In such cases, the maximum rate of frequency decrease ($r_{max} = -df/dt$) has to be calculated. The r_{max} values obtained from different sample concentrations will give us a calibration curve which is independent from the flow rate.
16. For a detailed cleaning description, *see* reference L. Románszky et al., Sensors and Actuators B: Chemical, 2018 [32].
17. There is no consensus in the literature on how to store EVs solutions derived from different sources to preserve their structure and functionality over time. Conflicting results suggest different stability at 4, RT, or −80 °C. For REVs, we observed a significant increase in measured Δf and r_{max} even after a short freeze-thaw cycle. This phenomenon can be attributed to the detachment of proteins adsorbed onto the external surface of EVs, as well as to the partial rupture of the membrane, which allows the release of internal contents. Particles sizes influence their mobility, and consequently, their adsorption to the crystal surface, which also strongly depends on sample concentration, flow rate, and temperature [33]. Keeping these parameters at constant value, reproducibility of the results can be observed (*see* Fig. 3).

Acknowledgments

This work was funded by the Hungarian Momentum Program (LP2016-2), the National Competitiveness and Excellence Program (NVKP_16-1-2016-0007). This work was also funded by the Ministry of Innovation and Technology of Hungary through the National Research, Development and Innovation Fund, TKP2021-EGA-31, KKP22 144180, K131594, and the 2020-1-1-2-PIACI-KFI_2020-00021 funding schemes. Support from Eötvös Loránd Research Network, grant no. SA-87/2021 and KEP-5/2021, is also acknowledged. Funding from European Commission under grant agreement no. 690898/H2020-MSCA-RISE-2015 and 101007299/H2020-MSCA-RISE-2020 was also received.

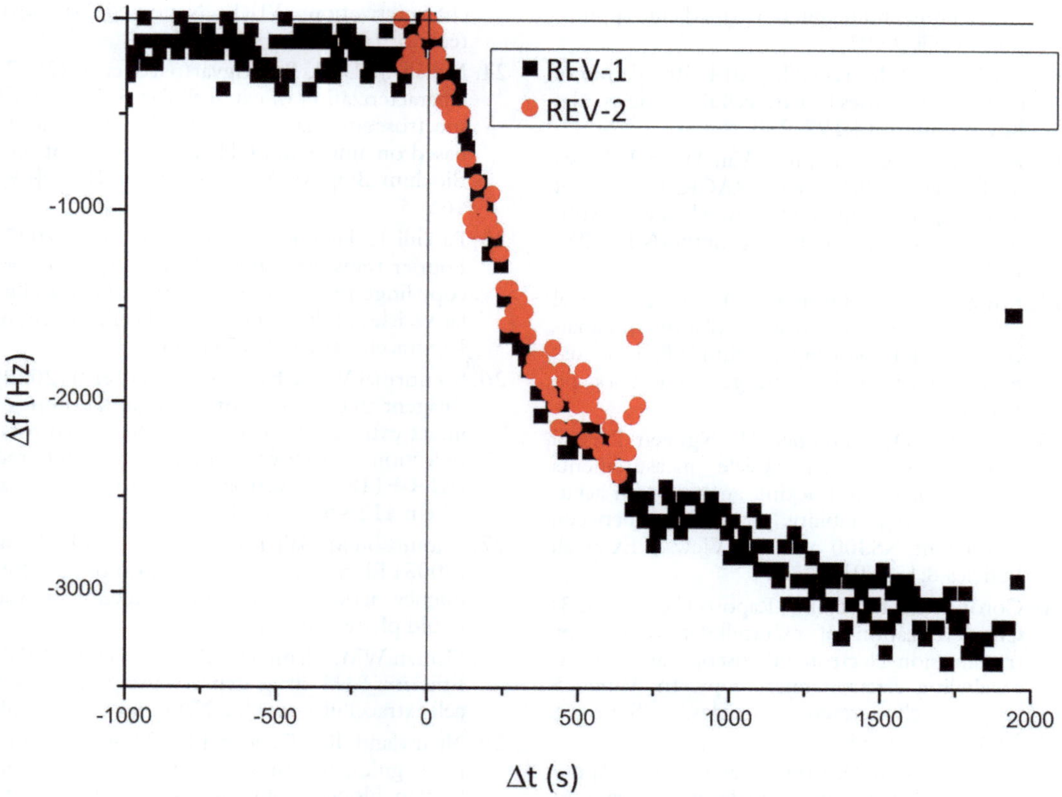

Fig. 3 Repeatability of the adsorption of REVs measured with EMPAS method. Time course of frequency shift at the 49th overtone of two hydrophobic quartz crystals with two REV samples (100× dilution of the stock solution) after a short exposure (1 h) to −30 °C followed by thermal equilibration at RT (1 h). The slopes of the lines represent the maximum rate of frequency shift $(df/dt)_{max}$, for REV-1 (black line): 7.3839 Hz/s^{-2} and for REV-2 (red line): 7.37548 Hz/s^{-2}, respectively

References

1. Cheng L, Hill AF (2022) Therapeutically harnessing extracellular vesicles. Nat Rev Drug Discov 21:379–399
2. Théry C, Witwer KW, Aikawa E et al (2018) Minimal information for studies of extracellular vesicles 2018 (MISEV2018): a position statement of the International Society for Extracellular Vesicles and update of the MISEV2014 guidelines. J Extracell Vesicles 7:1535750
3. Chang W-H, Cerione RA, Antonyak MA (2021) Extracellular vesicles and their roles in cancer progression. In: Robles-Flores M (ed) Cancer cell signaling. Springer, New York, pp 143–170
4. Skotland T, Sagini K, Sandvig K et al (2020) An emerging focus on lipids in extracellular vesicles. Adv Drug Deliv Rev 159:308–321
5. Harmati M, Bukva M, Böröczky T et al (2021) The role of the metabolite cargo of extracellular vesicles in tumor progression. Cancer Metastasis Rev 40:1203–1221
6. Pathan M, Fonseka P, Chitti SV et al (2019) Vesiclepedia 2019: a compendium of RNA, proteins, lipids and metabolites in extracellular vesicles. Nucleic Acids Res 47:D516–D519
7. Keerthikumar S, Chisanga D, Ariyaratne D et al (2016) ExoCarta: a web-based compendium of exosomal cargo. J Mol Biol 428:688–692
8. Szabo G, Momen-Heravi F (2017) Extracellular vesicles in liver disease and potential as biomarkers and therapeutic targets. Nat Rev Gastroenterol Hepatol 14:455–466
9. Herrmann IK, Wood MJA, Fuhrmann G (2021) Extracellular vesicles as a next-generation drug delivery platform. Nat Nanotechnol 16:748–759
10. Du S, Guan Y, Xie A et al (2023) Extracellular vesicles: a rising star for therapeutics and drug delivery. J Nanobiotechnol 21:231
11. Tsiapalis D, O'Driscoll L (2020) Mesenchymal stem cell derived extracellular vesicles for tissue

engineering and regenerative medicine applications. Cells 9:991

12. Robbins PD, Morelli AE (2014) Regulation of immune responses by extracellular vesicles. Nat Rev Immunol 14:195–208
13. EV-TRACK Consortium, Van Deun J, Mestdagh P et al (2017) EV-TRACK: transparent reporting and centralizing knowledge in extracellular vesicle research. Nat Methods 14:228–232
14. Kowal EJK, Ter-Ovanesyan D, Regev A et al (2017) Extracellular vesicle isolation and analysis by western blotting. In: Kuo WP, Jia S (eds) Extracellular vesicles. Springer, New York, pp 143–152
15. Bachurski D, Schuldner M, Nguyen P et al (2019) Extracellular vesicle measurements with nanoparticle tracking analysis – an accuracy and repeatability comparison between NanoSight NS300 and ZetaView. J Extracell Vesicles 8:1596016
16. Corona ML, Hurbain I, Raposo G et al (2023) Characterization of extracellular vesicles by transmission electron microscopy and immunolabeling electron microscopy. In: Vainio S (ed) Cell-secreted vesicles. Springer, New York, pp 33–43
17. Kogej K, Božič D, Kobal B et al (2021) Application of dynamic and static light scattering for size and shape characterization of small extracellular nanoparticles in plasma and ascites of ovarian cancer patients. Int J Mol Sci 22:12946
18. Welsh JA, Van Der Pol E, Arkesteijn GJA et al (2020) MIFlowCyt-EV: a framework for standardized reporting of extracellular vesicle flow cytometry experiments. J Extracell Vesicles 9: 1713526
19. Cloet T, Momenbeitollahi N, Li H (2021) Recent advances on protein-based quantification of extracellular vesicles. Anal Biochem 622:114168
20. Gardiner C, Vizio DD, Sahoo S et al (2016) Techniques used for the isolation and characterization of extracellular vesicles: results of a worldwide survey. J Extracell Vesicles 5:32945
21. Vaz R, Serrano VM, Castaño-Guerrero Y et al (2022) Breaking the classics: next-generation biosensors for the isolation, profiling and detection of extracellular vesicles. Biosens Bioelectron X 10:100115
22. Birhanu AG (2023) Mass spectrometry-based proteomics as an emerging tool in clinical laboratories. Clin Proteomics 20:32
23. Suthar J, Taub M, Carney RP et al (2023) Recent developments in biosensing methods for extracellular vesicle protein characterization. WIREs Nanomed Nanobiotechnol 15:e1839
24. Mihály J, Deák R, Szigyártó IC et al (2017) Characterization of extracellular vesicles by IR spectroscopy: fast and simple classification based on amide and CH stretching vibrations. Biochim Biophys Acta Biomembr 1859:459–466
25. Paolini L, Federici S, Consoli G et al (2020) Fourier-transform infrared (FT-IR) spectroscopy fingerprints subpopulations of extracellular vesicles of different sizes and cellular origin. J Extracell Vesicles 9:1741174
26. Szentirmai V, Wacha A, Németh C et al (2020) Reagent-free total protein quantification of intact extracellular vesicles by attenuated total reflection fourier transform infrared (ATR-FTIR) spectroscopy. Anal Bioanal Chem 412:4619–4628
27. Thompson M, Ballantyne SM, Cheran L-E et al (2003) Electromagnetic excitation of high frequency acoustic waves and detection in the liquid phase. Analyst 128:1048
28. Usman WM, Pham TC, Kwok YY et al (2018) Efficient RNA drug delivery using red blood cell extracellular vesicles. Nat Commun 9:2359
29. Nieuwland R, Siljander PR (2024) A beginner's guide to study extracellular vesicles in human blood plasma and serum. J Extracell Vesicles 13:e12400
30. Bebesi T, Kitka D, Gaál A et al (2022) Storage conditions determine the characteristics of red blood cell derived extracellular vesicles. Sci Rep 12:977
31. Aatonen M, Valkonen S, Böing A et al (2017) Isolation of platelet-derived extracellular vesicles. In: Hill AF (ed) Exosomes and microvesicles. Springer, New York, pp 177–188
32. Románszki L, Tatarko M, Jiao M et al (2018) Casein probe–based fast plasmin determination in the picomolar range by an ultra-high frequency acoustic wave biosensor. Sens Actuators B Chem 275:206–214
33. Románszki L, Varga Z, Mihály J et al (2020) Electromagnetic piezoelectric acoustic sensor detection of extracellular vesicles through interaction with detached vesicle proteins. Biosensors 10:173
34. Welsh JA, Goberdhan DCI, O'Driscoll L, Buzas EI et al (2024) Minimal information for studies of extracellular vesicles (MISEV2023): From basic to advanced approaches. J Extracell Vesicles. 2024 Feb;13 (2):e12404. https://doi.org/10.1002/jev2.12404

Chapter 16

Floating Films of Biological Importance at the Liquid–Gas Interface: Langmuir Monolayer Approach

Anita Wnętrzak, Anna Chachaj-Brekiesz, and Patrycja Dynarowicz-Latka

Abstract

Langmuir films are monolayers of water-insoluble molecules formed on aqueous subphases due to the forces of self-assembly. The increased interest in Langmuir monolayers is primarily related to their use as models of biological membranes. The possibility of studying the incorporation and behavior of bioactive compounds (i.e., drugs) in the mem0brane structure seems to be particularly important. The Langmuir technique enables the investigation of such aspects as orientation of molecules at the interface and interactions with lipids present in membranes. Here, we present the research methodology on an example of a chain-oxidized cholesterol derivative using Langmuir monolayers: from single-component to mixed films. The protocol also describes the application of related complementary in situ techniques, such as Brewster angle microscopy (BAM; film texture imaging), electric surface potential change (ΔV measurements), and polarization modulation-infrared reflection-absorption spectroscopy (PM-IRRAS; IR spectra).

Key words Langmuir monolayers, Model membranes, Surface behavior, Membrane interactions, Oxysterols

1 Introduction

Monolayers at the air–water interface provide a useful experimental tool for studying interactions occurring in biological membranes, such as lipid–lipid, lipid–water, and lipid–drug interactions. This method is based on compression isotherms, obtained by measuring the surface pressure (π), as a function of the molecular area (A) [1]. The surface pressure is defined as the difference between the surface tension of pure subphase γ_0 and the surface tension of the monolayer-covered subphase γ. During a routine experiment, a specific amount of substance(s) dissolved in a volatile solvent is dropped onto the subphase surface using a microsyringe. After solvent evaporation, an insoluble monolayer is formed on the subphase due to self-assembly forces and can be compressed using movable barrier(s). Upon compression, a gradual decrease in

Csaba Matta et al. (eds.), *The Surfaceome: Methods and Protocols*, Methods in Molecular Biology, vol. 2908,
https://doi.org/10.1007/978-1-0716-4434-8_16,

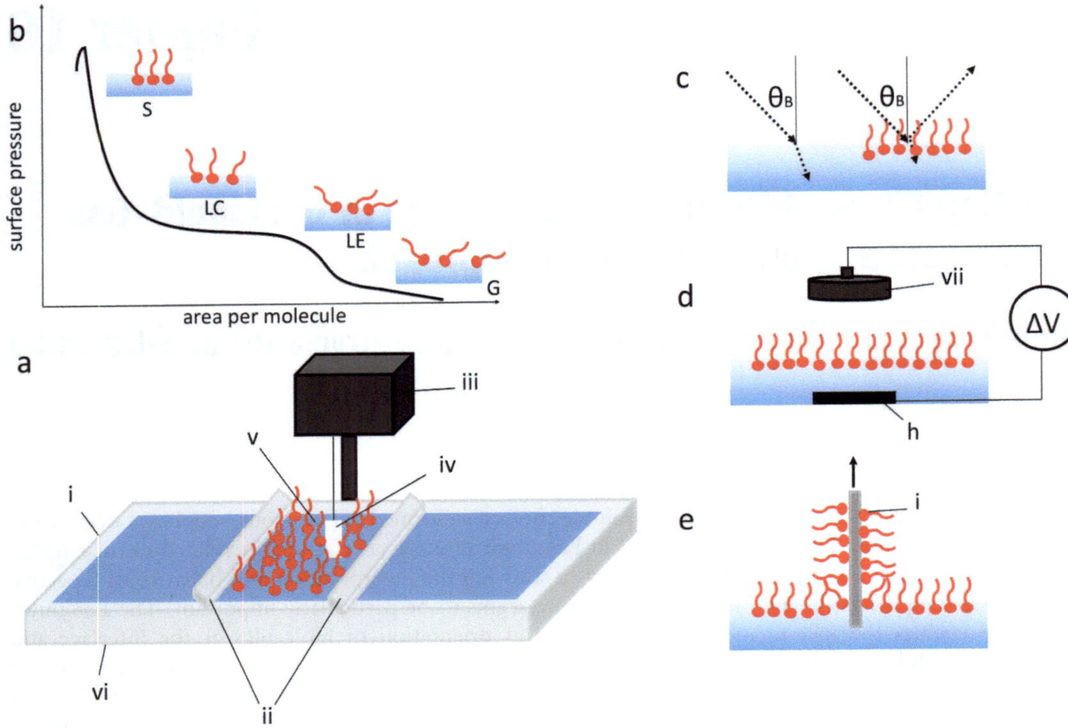

Fig. 1 (**a**) Ideal schema of Langmuir trough; (i) Langmuir trough, (ii) barriers, (iii) surface pressure sensor, (iv) Wilhelmy plate, (v) monolayer, (vi) subphase. (**b**) The course of the model π-A isotherm showing various phase states of the Langmuir monolayer together with a schematic illustration of molecular order at the phase boundary: G—gas state, LE—liquid expanded state, LC—liquid condensed state, S—solid state. (**c**) The principle of operation of the Brewster angle microscope (θ_B—Brewster angle). (**d**) The principle of surface potential measurement; (vii) vibrating electrode, (viii) reference electrode. (**e**) Langmuir–Blodgett deposition scheme, (ix) solid support

surface area per molecule is observed and the pressure increases (*see* Fig. 1a, b).

The course of the isotherm is determined by the type of molecules forming the film and additionally depends on experimental conditions, such as temperature, compression rate, or subphase composition. The thermodynamic analysis of the isotherm allows us to draw conclusions about the physical state of the monolayer at a particular stage of compression, which results from molecular ordering. In turn, in the case of multicomponent systems, the strength and nature of interactions between the components, their mutual miscibility (or phase separation), and stability can be determined [2].

In recent years, an increase in interest in Langmuir films has been observed, mainly due to their application as a model of biological membranes. By using this methodology, the incorporation of many drugs into the membrane can be studied [2–7]. The most interesting achievements concern the selectivity of the action of drugs that act at membrane level, such as new generation anticancer alkylphosphocholines [8, 9], antifungal polyene antibiotics, such as amphotericin B [10–12], the induction of phase separation by cyclosporine A [13–15], and some cholesterol derivatives (e.g., oxysterols) [16] as well as the intramembrane transport of oxysterols [17–19]. Apart from the biological aspect, Langmuir films deposited on solid supports are used in chemical analytics (in the production of chemical sensors), optics (in the production of antireflective coatings), and in molecular electronics, where the defect-free structure of layers with precise molecular thickness is particularly important [20–25]. The increased interest in Langmuir films has also contributed to the development of many modern techniques that expanded the possibilities of analyzing surface layers and their full characterization. Thus, in addition to the classical thermodynamic analysis of interactions [26], it is now possible to study the Langmuir films by microscopic (in situ with Brewster angle [27] and fluorescence microscopy [28], as well as atomic force microscopy (AFM) [29] after film transition onto solid support), spectroscopic (PM-IRRAS and SFG [29]), and scattering (grazing incidence X-ray diffraction, GIXD [30]) techniques. Despite its long history and the simplicity of research, the Langmuir monolayer technique remains unknown to many researchers. Moreover, obtaining repeatable results with this method requires strict adherence to the methodology and ensuring high purity of materials, equipment, and the laboratory environment. Therefore, this protocol describes the research methodology, paying attention to the basic factors that can affect the accuracy of the experiments performed. 25-hydroxycholesterol (25-OH), an oxidized derivative of cholesterol (chol), was chosen as an exemplary research material. Due to the presence of an additional hydroxyl group (OH) in the chain, this molecule exhibits unique properties at the water–air interface [18]. The SM/25-OH mixture was presented as an example of a two-component system [31]. Furthermore, to investigate 25-OH activity in membrane environment, the myelin sheath model was used as an exemplary multicomponent system [32]. Here, a systematic description of experiments for single- and multicomponent systems complemented by imaging techniques (BAM), surface potential measurements (ΔV), and PM-IRRAS is provided.

2 Materials

2.1 Chemical Compounds

All chemicals used in the experiments should be of high purity (≥99%). Take care of their proper storage.

Film forming lipids: DPPE (1,2–dipalmitoylglycero–3–phosphoethanolamine), DPPS (1,2–dipal mitoylglycero–3–phospho–L-serine sodium salt), egg SM (containing 86% of

N–(hexadecanoyl)sphing–4–enine–1–phosphocholine), chol (cholesterol), and 25−OH (25–hydroxycholesterol) of >99% purity. Compounds should be stored in tightly closed boxes according to the manufacturer's instructions (lipids usually in freezer at −20 °C).

Solvents: ethanol of ≥99% purity and chloroform of spectroscopic purity stabilized with ethanol (*see* **Note 1**).

Subphase: Deionized ultrapure water with a minimum resistivity of 18.2 MΩ·cm and surface tension of 72.8 mN/m at 20 °C (*see* **Note 2**).

2.2 Apparatus

Classical Langmuir experiment: A Teflon Langmuir KSV 2000 trough (KSV) with a surface area of 700 cm^2, equipped with two movable Teflon barriers with adjustable speed is used (*see* **Note 3**). The Langmuir trough is placed on an antivibration table in a dust protection casing. Surface pressure is measured with an accuracy of 0.01 mN/m using a Wilhelmy plate made from ashless chromatography paper (Whatman Chr1) as a surface pressure sensor (*see* **Note 4**). The temperature is maintained at 20.0 ± 0.1 °C by a circulating water system underneath the trough connected to a thermostat.

Brewster angle microscopy: The texture of floating monolayers is visualized using the ultraBAM instrument (Accurion GmbH) installed over the Langmuir trough. The minimum reflection is established with a *p*-polarized laser beam (λ = 658 nM) incident on the pure aqueous surface at the Brewster angle (53.15°). The light reflected from the monolayer is collected through a 10× objective and lens system to a CCD camera. BAM images show monolayers fragments of 720 μm × 400 μm. Schematic diagram of the principle of Brewster angle microscopy is presented in Fig. 1c.

Surface potential: Surface potential measurements are performed with the Kelvin probe (model KP2, NFT) mounted on a Langmuir trough. The crucial elements of the measuring apparatus are the vibrating electrode based on a piezoelectric transducer and a platinum reference electrode. Schematic diagram of the principle of Brewster angle microscopy is presented in Fig. 1d.

PM-IRRAS: The PM-IRRAS spectra of Langmuir monolayers, compressed to the selected surface pressure values, are recorded with the KSV PM-IRRAS instrument installed over Langmuir trough (KSV NIMA, KN 1002, total area = 98 cm^2).

Langmuir–Blodgett trough: The transfer of selected Langmuir monolayers onto solid substrate is carried out using NIMA 612D trough (total area = 600 cm^2). Schematic diagram of the Langmuir–Blodgett method is shown in Fig. 1e.

3 Methods

3.1 Basic Experiment

1. To prepare stock solutions of single compounds, a few milligrams of the substance (typically 1–4 mg) should be weighed on an analytical balance (with minimum accuracy of ±0.01 mg) in a volumetric flask (of 5 or 10 mL capacity and class A) (*see* **Note 5**).
2. Dissolve the weighed amount of substance (25-OH, POPC, SM, Chol) in chloroform or chloroform/ethanol mixture (DPPS and DPPE) (*see* **Note 6**). Solutions should have concentrations within the range of 0.1–0.4 mg/mL. After preparing the solution, protect the flasks with Parafilm or other sealing foil. Specific concentrations of solutions used in the experiment presented here are provided in Tables 1 and 2.
3. Prepare mixed solutions by mixing the appropriate volumes of the respective stock solutions (*see* **Note 7**) in 2 mL vials. Prepare mixed solutions with oxysterol mole fractions of X = 0.25, X = 0.5, and X = 0.75 (*see* Table 3).
4. To prepare a multicomponent mixture constituting a model of the myelin sheath, mix the lipids in mole proportion of Chol:PE:SM:PS:PC 50:20:12:9:9 by measuring the appropriate volumes of respective stock solutions and mixing in 2 mL vial (*see* Table 4). The composition of model membranes should be carefully selected based on data from the literature.

Table 1
Data required to prepare solutions: molar weights (MW) and concentrations (C) for selected lipids

Compound	25-OH	POPC	DPPE	DPPS	Chol	SM (egg)
M (g/moL)	402.653	760.076	691.959	757.960	386.66	703.03
C (mg/mL)	0.244	0.228	0.200	0.252	0.250	0.322

Table 2
Data required to prepare solutions and the volume of solutions to be applied to the subphase in the case of a trough with a total area of 700 cm^2 for single-component systems (on the example of 25-OH and SM); C—concentration, MW—molecular weight, V—volume

	C (mg/mL)	MW (g/moL)	V (μL)
25-OH	0.244	402.653	365
SM	0.322	703.03	255

Table 3
Data required to prepare solutions and the volume of solutions to be applied to the subphase in the case of a trough with a total area of 700 cm^2 for two-component 25-OH/SM system; C—concentration, MW—molecular weight, V—volume

Film composition		Volume to prepare the mixture		Data to be entered into the program			Actual volume of the applied solution
Mole fractions		V (μL)					
SM	25-OH	SM	25-OH	C (mg/mL)	V (μL)	MW (g/moL)	V (μL)
0.75	0.25	306.0	694.0	1	100	742.85	240
0.5	0.5	569.5	430.5			547.15	350
0.25	0.75	798.8	201.2			570.47	360

Table 4
Data required to prepare solutions and the volume of solutions to be applied to the subphase in the case of a trough with a total area of 700 cm^2 for multicomponent system mimicking the myelin sheath in the presence of and without 25-OH oxysterol; C—concentration, MW—molecular weight, V—volume; 1 is a mixture of Chol/DPPE/SM/DPPS/POPC in proportion of 50:20:12:9:9 (model of myelin sheath), 2 is a mixture of Chol/DPPE/SM/DPPS/POPC/25-OH in proportion of 25:20:12:9:25, and 3 is a mixture of DPPE/SM/DPPS/POPC/25-OH in proportion of 20:12:9:9:25:50

Film composition	Volume to prepare the mixture V (μL)						Data to be entered into the program			Actual volume of the applied solution
System	Chol	DPPE	SM (egg)	DPPS	POPC	25-OH	C (mg/mL)	V (μL)	MW (g/moL)	V (μL)
1	336.5	301.1	114.0	117.8	130.6	0.0	1	100	850.83	220
2	163.8	293.1	111.0	114.7	127.1	190.4			840.50	280
3	0.00	66.24	25.08	25.91	28.72	86.04			835.72	290

5. Applying the same methodology, prepare mixtures where cholesterol is partially (50%) or fully replaced with oxysterol in myelin sheath system (*see* Table 4 and **Note 8**).
6. Just after preparing the solutions and mixtures, close the flasks and vials tightly and secure them with Parafilm or other sealing foil. Store solutions in the refrigerator.
7. Before each measurement, thoroughly clean the Langmuir trough with ashless paper wipes or cotton pads soaked in

chloroform and, after evaporation, repeat the cleaning procedure with wipes or cotton pads soaked with ethanol. Remember about cleaning the edges and barriers (*see* **Note 9**).

8. Fill the trough with ultrapure water. The subphase should form a convex meniscus.
9. Remove the water by suction using a pipette tip linked with aspiration pump. The tip should be oriented perpendicular to the surface.
10. Fill the trough with the subphase. The subphase level cannot be too low or too high. The subphase should extend 1–2 mm beyond the edge of the trough (*see* **Note 10**).
11. Switch on trough controller and run the Langmuir trough program.
12. Immerse the paper surface pressure sensor in subphase. Wait a few minutes until the paper absorbs the water well.
13. Reset the tensiometer value to zero.
14. Check the cleanliness of the trough with the tensiometer while compressing the surface of the clean subphase. The trough is clean when the pressure during compression does not exceed 0.1 mN/m after compression to the closed barrier position. In case of higher surface pressure, the cleaning procedure must be repeated (**steps 7–9**).
15. Complete all required data in the Langmuir trough program, such as the name of the sample, type of Wilhelmy plate and its dimension, type of subphase and its temperature, mass of the tested compound and its molecular weight, concentration of the solution, volume of the solution applied during the experiment, and compression rate (*see* **Note 11**).
16. Apply the appropriate amount of the solution (one-component or a mixture, *see* Tables 1, 2, 3, and 4) onto the surface of the subphase using a glass microsyringe (*see* **Note 12**). The solution should be applied slowly, drop-by-drop, over the entire surface of the trough. Monitor tensiometer readings during dropping. The pressure should not increase rapidly.
17. Wait ~10 min for the solvent to evaporate (*see* **Note 13**).
18. Start measurement; in a classic experiment, the barriers move symmetrically at a given speed. During the measurement, a π-A isotherm is recorded that shows the dependence of surface pressure on the area per molecule (*see* Fig. 2a, g, and i).
19. After the measurement is completed, save the data.
20. Remove the monolayer by suction using a pipette tip linked with aspiration pump. For this purpose, hold the tip inlet perpendicular to the subphase surface, removing the monolayer from between the barriers. Then, remove the entire

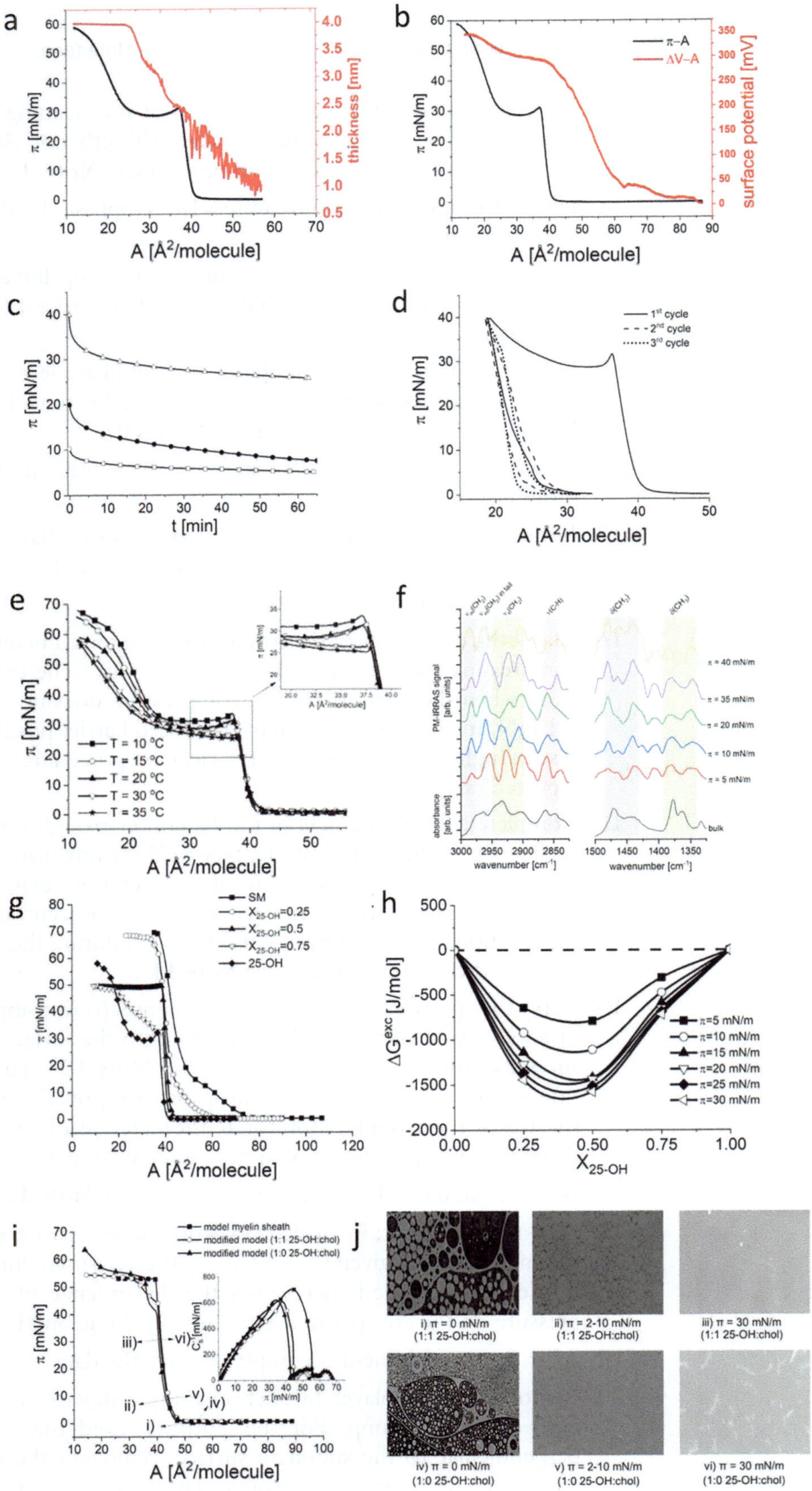

Fig. 2 (**a**) π-A isotherm together with monolayer thickness values for 25-OH; (**b**) π-A isotherm together with electric surface potential changes (ΔV) for 25-OH; (**c**) Stability of Langmuir monolayers from 25-OH at 20 °C; (**d**) Isotherms of compression–expansion cycles registered for 25-OH at 20 °C for compression to 40 mN/m;

subphase and clean the trough according to the procedure described above (**step 7**).

21. Due to the significant dependence of the results on numerous and variable measurement conditions, the characterization of monolayers in terms of the influence of experimental conditions on the course of π-A isotherms is an important issue. On the basis of the procedures described above, a number of preliminary experiments can also be performed (*see* **Note 14** and Fig. 2c, d, and e).

3.2 Brewster Angle Microscopy

1. Clean the trough according to the instructions given above and run the Langmuir program (Subheading 3.1, **steps 7–14**).
2. Switch on BAM electronic control.
3. Clean black glass plate (supplied by the producer) with alcohol (ethanol or methanol) using a dust-free wipe.
4. Place a plate at the bottom of the trough. Position the plate underneath the BAM objective.
5. Fill the trough with subphase (be careful not to contaminate the objective). The subphase should form a convex meniscus.
6. Switch on electronic controller and the active vibration isolation controller.
7. Follow Subheading 3.1, **steps 15–17** from the procedure described above.
8. Check whether the displayed values of polarizer (1°) and analyzer angels (10°) are correct. The Brewster angle for the water–air interface is 53.15°.
9. Switch on the laser beam.
10. Move the z-lift down to the moment when the surface of water is displayed and a reflected spot in the middle of the image display is noticeable.
11. Gently change the polarizer and analyzer angle values and make a fine adjustment of the z-position to sharpen the image.
12. Start the isotherm measurement.

Fig. 2 (continued) (**e**) Temperature dependence of π-A isotherms for 25-OH; (**f**) PM-IRRAS spectra for the 25-OH monolayer registered at 20 °C compared to the ATR-FTIR spectrum of 25-OH in bulk: hydrocarbon stretching vibrations region (i) and hydrocarbon scissoring vibrations region (ii); (**g**) π-A isotherms measured for mixed two-component system of 25-OH/SM; (**h**) Changes in free energy of mixing (ΔG^{exc}) as a function of 25-OH content for 25-OH/SM mixed system; (**i**) π-A isotherms for Chol:PE:SM:PS:PC (50:20:12:9:9) mixed system mimicking myelin sheath, as well as modified mixed system with addition of 25-OH together with compressibility modulus dependencies for the studied systems; (**j**) BAM images at different surface pressure values for mixed systems showed in Figure 2i. Results adapted from [18, 31, 32] with permission from ASC (**a–f**), Royal Society (**g–h**), and Elsevier (**i–j**)—Creative Commons CC BY 4.0

13. While measuring the isotherm, monitor the changes of the film texture and take appropriate images at the given pressures (*see* Fig. 2j). The software mainly allows for both taking photos and recording compresion videos (*see* **Note 15**).
14. Save the data.
15. Turn off the laser.
16. Move the z-lift up to its starting position.
17. Remove the monolayer by suction using a pipette tip linked with aspiration pump. For this purpose, hold the tip perpendicular to the subphase surface, removing the monolayer from between the barriers. Then, remove the entire subphase and clean the trough according to the procedure described above.
18. Remove the glass plate and clean it with methanol and distilled water.

3.3 Electric Surface Potential Change Measurement

1. Clean the trough according to the instructions given above (Subheading 3.1, **step** 7).
2. Clean the reference electrode with chloroform and ethanol using a dust-free wipe and wash several times with distilled water.
3. Place the reference electrode at the bottom of the Langmuir trough and connect it with Kelwin probe.
4. Fill the trough with the subphase. The subphase should form a convex meniscus.
5. Immerse the paper surface pressure sensor in the subphase. Wait a few minutes for the paper to absorb water properly.
6. Reset the tensiometer value to zero.
7. Check the cleanliness of the trough with the tensiometer while compressing the surface of the clean subphase. The trough is clean when the pressure during compression does not exceed 0.1 mN/m. In case of higher surface pressure, the cleaning procedure must be repeated (Subheading 3.1, **step** 7).
8. Place a vibrating electrode at a distance of approximately 2 mm above the subphase surface.
9. Zero the electric surface potential change value.
10. Follow Subheading 3.1, **steps 15–18** from the basic procedure described above (*see* **Note 16**).
11. Be careful not to get the vibrating electrode wet during measurement.
12. After the measurement is completed, save the data.
13. Remove the monolayer by suction using a pipette tip linked with aspiration pump. For this purpose, hold the tip perpendicular to the subphase surface, removing the monolayer from

between the barriers. Then, remove the entire subphase and clean the trough according to the procedure described above.

14. Clean the reference electrode with chloroform and ethanol.

3.4 PM-IRRAS Spectra Measurement

1. Clean the trough according to the instructions given above (Subheading 3.1, **steps 7–14**).
2. Fill the trough with subphase. The subphase should form a convex meniscus.
3. Immerse the paper surface pressure sensor in subphase. Wait a few minutes for the paper to absorb water well.
4. Reset the tensiometer value to zero.
5. Check the cleanliness of the trough with the tensiometer while compressing the surface of the clean subphase. The trough is clean when the pressure during compression does not exceed 0.1 mN/m. In case of higher surface pressure, the cleaning procedure must be repeated.
6. Switch on PEM electronic control.
7. Using the back knob at the top of goniometer set the value of incidence angle (*see* **Note 17**).
8. Optimize the interferogram height by adjusting manually the goniometer height with the front knob at the top of goniometer and/or setting values of "Input gain," "Sigma gain," "Predelta gain," and "Delta gain" in the software. The value of interferogram height should be equal to approximately 8 V. Take spectrum for pure subphase with measuring time of 10 s to make sure that for the obtained preliminary spectra (1) min delta/sigma values are over −30,000, (2) max delta/sigma values are under 30,000, and (3) interferogram maximum height is equal to ~8 V (*see* **Note 18**).
9. Measure background spectrum with maximum spectral resolution (8 cm^{-1}) and at least 6000 scans (measurement time 600 s). Save the obtained background spectrum.
10. Run the Langmuir trough program. Complete all required data, such as the name of the sample, type of Wilhelmy plate and its dimensions, type of subphase and its temperature, mass of the tested compound and concentration of the solution, volume of the solution applied during the experiment, and compression rate.
11. Apply the appropriate amount of the solution (one-component or a mixture) onto the surface of the subphase using a glass microsyringe. The solution should be applied slowly, drop-by-drop, over the entire surface of the trough. Monitor tensiometer readings during application. The pressure should not increase rapidly.
12. Wait ~10 min for the solvent to evaporate.

13. Compress the film to the given value of surface pressure and wait a few minutes for the film to stabilize at the selected surface pressure.
14. Measure the spectrum applying the same settings as for the background measurement (maximum spectral resolution (8 cm^{-1}) and at least 6000 scans (measurement time 600 s). Save the obtained spectrum (*see* Fig. 2f).
15. Remove the monolayer by suction using a pipette tip linked with aspiration pump. For this purpose, hold the tip perpendicular to the subphase surface, removing the monolayer from between the barriers. Then, remove the entire subphase and clean the trough according to the procedure described above.
16. To assure reproducibility of the results, perform measurements at least twice.

3.5 Transfer of Monolayer onto Mica (see Note 19)

1. Prepare the solutions and the trough according to Subheading 3.1, **steps 1–14** of the basic procedure (*see* **Note 20**).
2. Using adhesive tape, clean the surface of mica by removing a few top layers of the mineral (*see* **Note 21**). To do this, put a piece of mica on the adhesive side of the tape, stick it, and then gently lift the mica sheet by prying it with tweezers. On the adhesive tape, only a few mica layers should remain. Repeat two times on each side (*see* **Note 22**).
3. Use a tweezer with a very fine tip to attach the prepared mica to the dipper and immerse it in the clean subphase.
4. Check the cleanliness of the trough once again by compressing the barriers. If necessary, remove contaminations from the subphase surface by aspiration.
5. Apply the appropriate amount of the solution (one-component or a mixture) onto the surface of the subphase using a glass microsyringe and compress the film to the desired surface pressure (*see* **Note 23**).
6. Wait a few minutes for the film to stabilize at the selected surface pressure.
7. Start raising the sample with a dipper with the appropriate speed (*see* **Note 24**).
8. After the sample is completely emerged from subphase, carefully place it in the appropriate container using tweezers.
9. Secure the sample for further measurements (*see* **Note 25**).
10. Clean the trough according to the basic procedure (*see* **Note 26**).

3.6 Results: Analysis of π-A Isotherms

3.6.1 Single-Component Films

Changes in surface pressure occurring with the decrease in the surface area per molecule are related to changes in the fluidity of the monolayer, which, in turn, can be attributed to the different interfacial elasticity of the film. To determine the elastic properties of a monolayer, a more commonly used quantity is the compressibility modulus: $C_s^{-1} = -A\left(\frac{d\pi}{dA}\right)$ [33]. It is convenient to calculate the derivative using commonly available spreadsheets (i.e., Origin or Excel). C_s^{-1} values below 25 mN/m suggest that the film is in a low-density liquid phase; the ranges of 25–50 mN/m and 100–250 mN/m are characteristic for the liquid expanded and liquid condensed states (LE and LC, respectively), while for C_s^{-1} above 500 mN/m, the film is in the solid state [34]. With a series of π-A isotherm measurements at different temperatures, phase transition in monolayer can be analyzed [18, 26]. The entropy $\Delta S_t = \left(\left(\frac{\partial \pi_t}{\partial T}\right) - \left(\frac{\partial \gamma_0}{\partial T}\right)\right)\Delta A$ and enthalpy $\Delta H_t = T\Delta S_t$ of the phase transition can be calculated, where $((\partial \pi_t)/\partial T)$ is the slope of the dependence of the $\pi(T)$ (values obtained from π-A isotherms measured at different temperatures), and $\Delta A = A_e – A_b$, where A_e is the area per molecule at the end of plateau and A_b is the area per molecule at its beginning. The value of $((\partial \gamma_0)/\partial T)$, where γ_0 is the surface tension of pure water, is equal to -0.153 mN·m^{-1}·K^{-1} for temperatures between 10 and 35 °C [18]. When the calculated ΔH_t and ΔS_t are negative, it suggests that the LE–LC transition is exothermic and that the formed LC phase is more ordered compared with the LE state. Positive values ΔH_t and ΔS_t suggest that the phase transition is endothermic and spontaneous.

3.6.2 Multicomponent Films

Mixed monolayers exhibit nonideal behavior, caused by significant intermolecular interactions, when certain properties or simple functions *f* describing them do not depend linearly on the composition of the film *X*, and positive or negative deviations from ideality are observed in the course of *f*(*X*). Analysis based on additivity rules uses two equations:

$$A^{id}_{1\ldots N} = \sum_{i=1}^{N} A_i X_i$$

$$\pi^{id}_{c\,1\ldots N} = \sum_{i=1}^{N} \pi_{c\,i} X_i$$

where A_i is the mean molecular areas observed for each component and X_i is the mole ratio of component. The linear course of the above functions indicates perfect miscibility of the components resulting from the lack of interactions between molecules or their complete immiscibility, while deviations from ideality indicate the existence of interactions in these systems.

The interactions and stability of the system can also be determined using changes in the excess Gibbs free energy of mixing [26]:

$$\Delta G^{\text{exc}} = N_A \int_{\pi*}^{\pi} A^{\text{exc}} d\pi,$$

where A^{exc} is the excess area per molecule ($A^{\text{exc}} = A_{1\ldots N} - A^{\text{id}}_{1\ldots N}$, where $A_{1\ldots N}$ is the mean molecular area in the mixed monolayer and $A^{\text{id}}_{1\ldots N}$ is the area in an ideal mixture described by the additivity rule) at a particular surface pressure, N_A is the Avogadro constant, and π^* is a surface pressure below which the components can be assumed to be ideally mixed. In practice, a following formula is used:

$$\Delta G^{\text{exc}} = 6.02(P_{12} - (P_1X_1 + P_2X_2),$$

where P_{12}, P_1, and P_2 are the areas under the A-π curves for given pressures for the mixed film and reference isotherms, respectively (for example, 0–10 mN/m, 0–20 mN/m, or 0–30 mN/m) and X_1 and X_2 are the mole ratio of reference component. The value of ΔG^{exc} provides information on whether the particular interaction is energetically favorable ($\Delta G^{\text{exc}} < 0$) or not ($\Delta G^{\text{exc}} > 0$), while for $\Delta G^{\text{exc}} = 0$ ideal mixing (or immiscibility) occurs [26]. The value of ΔG^{exc} allows to conclude about the strength of mutual interactions and possible affinity of the components to each other. Strong interactions are considered to be those for which the function has absolute values greater than 500 J/mol. For a more detailed analysis, please see [26].

3.7 Analysis of Δ V-A Curve

The ΔV-A curves measured for single component films allow us to obtain information on electrical properties of surface film and track changes in molecular orientation and organization (formation of domains or multilayers) during compression of molecules at the water surface. From the values of ΔV and molecular area, assuming an appropriate model (1, 2, or 3-plane capacitor), the apparent dipole moment can be calculated, and the orientation angle of the molecules can be estimated (for more details, see [35]). Moreover, the ΔV meets the additivity rule and allows us to investigate miscibility and interactions in multicomponent films [36, 37].

3.8 PM-IRRAS Spectra Processing

The result of PM-IRRAS experiments are PM-IRRAS spectra which are dependencies of PM-IRRAS signal intensity from the wavenumber. In the first step, subtract background spectrum from the spectrum measured for monolayer at the subphase. Any software dedicated for IR spectra processing is suitable (i.e., OPUS or OMNIC), as well as commonly available spreadsheets (i.e., Origin or Excel). If it is necessary, perform other manipulations on obtained monolayer spectra such as smoothing or signal normalization.

In the next step, identify molecular vibrations which yield bands observed in spectrum. It can be done based on the literature data on molecular vibrations in the investigated molecules or by analysis of theoretical vibrational spectra calculated using density functional theory method. Analyze variations of the spectral positions and mutual intensities of bands to asses information on distribution, orientation, conformation, as well as functional groups responsible for molecular interactions [29].

3.9 BAM Images Analysis

Morphological analysis of Langmuir monolayers by BAM is possible due to differences in the surface density in the monolayers and the existence of the phenomenon of anisotropy of boundary phases. BAM exploits the nonreflective state that occurs for the angle of incidence specified by the first Snell's law. The presence of the monolayer at the air–water interface modifies the local refractive index, which causes that a small amount of light is reflected and registered by the detector. As a result, we obtain an image of the texture as areas of different brightness which is determined by individual molecules (or domains) and packing density in the sampling area. The analysis of BAM results is comparative and is based on the monitoring of the textures with the changes in pressure. Moreover, the changes in film texture containing active substance in respect to the reference devoid of this substance can result from intermolecular interactions.

4 Notes

1. Solvents used to prepare solutions should be of high purity (≥99%). The solvent should be a neutral volatile liquid substance, immiscible with water. Solvents with a boiling point of 40–80 °C are most suitable for experiments under normal conditions. Solvents should also not react with the substance dissolved as well as with the subphase. The solvent used should also be able to spread evenly on the aqueous subphase. The parameter that determines this is the so-called initial spreading coefficient, S_i, which should be greater than zero [38]. Currently, the most commonly used is chloroform ($S_i = 13$) of chromatographic grade stabilized with ethanol. However, it should be noted that it is a liquid with a higher density than water, so if the solution is applied carelessly, single drops of the solution containing substance of interest may settle at the bottom of the Langmuir balance [39] and false the measurement results. Other applied solvents include hexane, cyclohexane, heptane, or dimethylsulfoxide. Solvent mixtures can also be used. Sometimes it may be necessary to add small amount

(up to 5–10%) of polar solvent (methanol or ethanol) to apolar solvent to assure good solvent spreading (polar solvent increases the value of S_i) and solubility of the studied substance. Increased quantity of polar solvent can result in desorption of the studied substance from the surface to the bulk phase. Please note that the kind of a solvent used may alter the shape of the isotherm, especially for macromolecules [40].

2. The subphase onto which the film is spread also requires particular attention. Ultrapure water is used in routine experiments. Water should be stored in clean, tightly closed glass containers. Any contamination affects the measurement results and may lead to their falsification. Dedicated buffers (e.g., PBS) or aqueous NaCl or HCl solutions are also often used as subphase. However, it should be remembered that the presence of ions and a change in pH may affect the results of experiments, among others, by changing the solubility of the studied substance. In the case of water-soluble substances, aqueous NaCl solutions can be used to decrease their water solubility. The temperature of the subphase is also an important factor and should be kept constant during the experiment. Its changes may drastically influence the surface behavior of the studied compounds.

3. Most troughs in use today are made of single-piece solid Teflon (which is material based on polytetrafluoroethylene, PTFE) and have a rectangular shape (round or square shapes can also be found). PTFE is hydrophobic and oleophobic, which guarantees nonwetting of the trough and prevents spilling over of the subphase liquid. Sometimes, metal or glass troughs covered with a thin layer of PTFE are used; however, they are not as durable as solid PTFE. The area of the trough may vary depending on the model, but the general rule is that the larger the area, the better the stability of the formed films. Unfortunately, larger areas require the use of more subphase and chemicals, which is a disadvantage for expensive compounds. The Langmuir trough can be equipped with single or double (symmetrical) barriers, the speed of which must be easily controlled during measurement. The most common barriers are made of Teflon or Delrin (polyoxomethylene, hydrophilic material) [41]. The choice of barrier material may influence on the characteristics of the isotherm—although no changes were observed for stearic acid, the isotherms differed for DPPC [41]. The barriers should fit snugly to the trough to prevent leakage both under them and around the edges of the trough. Some trough models also have a dipping well, which enables the transfer of films onto a solid surface using the Langmuir–Blodgett technique (vertical transfer). It is also necessary to ensure that the Langmuir trough is inside a cover (box) that

protects against external contaminations and air mass movements. In fact, the best solution is to conduct Langmuir experiments in an environment free from dust and other contaminants, so-called *clean room.*

4. Most troughs measure surface pressure using the Wilhelmy method. In this method, a plate made of fully wetted material is suspended in the aqueous phase. When the plate is immersed in liquid, a concave meniscus is created. The net downward force on the plate is the sum of the gravitational force and surface tension forces, partially balanced by the buoyant force. The size, shape, and preparation of the Wilhelmy plate depend on the experiments to be performed with it. Plates are made of quartz, glass, mica, platinum, or ashless filter paper. In fact, the results of the experiment do not depend on the material used, provided that the material is well wetted with liquid. During measurements, some of the monolayer compounds may adsorb on the plate, so it must be well cleaned (e.g., by heating as in the case of platinum plate). The advantage of using filter paper is that there is no need to clean it; after experiments, the paper sensor can be replaced with a new one. Paper sensors with dimensions of 1.0 cm × 2.5 cm are used most frequently. It is also important to remember to periodically calibrate the pressure sensor according to the manufacturer's instructions.

5. Maintaining cleanliness of laboratory equipment is extremely important, since contaminations present on the glass equipment affects the measurement results. Therefore, volumetric flasks must be thoroughly cleaned. Typically, we clean the flasks with chromic acid cleaning solution. To do this, we fill flasks with the chromic acid cleaning solution and leave for several minutes, then we rinse them thoroughly with distilled water and leave them to air dry. The use of a stream of inert gas, e.g., nitrogen, accelerates the drying process. Wash the plastic or glass bottle stopper with detergent, rinse thoroughly and leave to dry. It is recommended to wipe them finally with ethanol with dust-free wipes. It should also be remembered that chromic acid is corrosive and carcinogenic, so it must be used with appropriate protective clothing, laboratory glasses, and gloves. Waste must be disposed of in accordance with the safety procedures.

6. In the case of poor solubility of the compounds in chloroform, a mixture of solvents can be used, e.g., chloroform–ethanol (for example, in a volume ratio of 9:1 or 4:1). The compound should be first dissolved in ethanol and in the next step chloroform should be added. In the case of very poorly soluble substances, the solution can be sonicated. However, you must make sure that the sonication process does not influence the

structure of compound (i.e., DNA molecules undergo fragmentation during sonification).

7. In order to prepare solutions, precise calculations must be performed to obtain mixtures with a specific composition (*see* Tables 3 and 4). At each time, we take into account the mole composition of the mixture, the molecular weights of the substances, the concentration of the stock solutions, and the area of the trough. You can use a microsyringe or automatic pipette for measuring respective volumes of stock solutions. After measuring of each solution, rinse the microsyringe with chloroform or apply new sterile tip on the pipette. It is important to close the flasks and vials immediately after using the solution to limit solvent evaporation (this may affect the concentrations of the solutions). After preparing the solution, protect the flasks with parafilm and store in refrigerator. In the case of mixed monolayers, the preparation procedure can be carried out in two ways [42]. The first one is to prepare solutions of each of the components of the monolayer separately and apply them separately on the surface in the amounts necessary to obtain the desired composition of the monolayer. The second method is to prepare a solution containing a ready-made mixture of the tested substances, in appropriate molar ratios, and apply the prepared mixed solution on the surface of the subphase. Due to better repeatability of isotherms, the second of the presented techniques is more often used, in which mixed monolayers are created from ready-made mixtures of their components [43].
8. There are two approaches to examine the effects of active compounds on model biological membranes. In the approach described in the protocol, we replace one of the components of the model membrane (Chol) with an active substance (25-OH). In the second approach, the composition of the model membrane is kept constant, and the tested substance is added in appropriate mole fractions.
9. The best is to use ashless paper wipes or 100% cotton pads. One of the problems associated with cleaning the trough using cotton pads is the presence of surfactants on their surface, which may result in the registration of the so-called *ghost isotherms*. You should ensure that the cotton pads do not leave threads on the surface. After experiments, we clean Teflon parts with chloroform, ethanol, and then chloroform again, and leave it to dry, while Delrin parts only with ethanol.
10. If there is low subphase level on the trough, the studied material may leak under the barriers. On the contrary, if the subphase level is too high, the subphase together with the research material may spill beyond the edges of the trough.

11. The compression rate should be selected and fixed for each experiment individually. In most experiments for membrane lipids, a speed of 20 cm^2/min is used. Too slow compression may cause molecules to desorb into the subphase and result in smaller surface areas per molecule in the monolayer. Too fast compression does not allow to attain the equilibrium conditions and may affect the stability of the formed films.
12. The capacity of the microsyringe is individually selected for each measurement. For large volumes of solutions, it is best to use syringes with a capacity of 250 μL or 500 μL. The microsyringes should have fixed needles, preferably with a bevel tip. Instead of a microsyringe, you can also use a micropipette for dropping. There is also an approach where, instead of preparing mixtures and applying ready-made mixed solutions to the subphase, each of the compounds is applied onto subphase separately. However, experimental data shows that preparing mixtures gives more repeatable results.
13. The evaporation time should be selected for a specific solvent. In the case of chloroform, it is usually 10 min. Before the main measurements, it is a good idea to perform a series of preliminary measurements to check how waiting time before the start of compression affects the course of the measured π-A curve.
14. The parameters checked most frequently are the compression speed, solvent evaporation time, subphase temperature, number of molecules dropped onto the subphase, type of barrier material, and position of the paper in relation to the barriers. To check the influence of the amount of the applied substance, in **step 16** of the described protocol, change the volume of the dropped solution with each subsequent measurement. In the case of the 25-OH solution (C = 0.244 mg/mL), the valueswere: 300 μL, 365 μL, and 400 μL. Differences in the course of π-A isotherms resulting from the presence of different amount of molecules in the monolayer may result from aggregation processes and the formation of condensed domains on the surface. Unfortunately, significant changes in this aspect exclude the possibility of investigating the molecule in mixed systems. In turn, the compression rates (for a constant set volume of 365 μL) were 5 cm^2/min, 20 cm^2/min, and 50 cm^2/min. To check the impact of the paper position, place it vertically and parallel to the barriers before starting the measurement. To check the influence of the subphase temperature on the course of π-A isotherms, the subphase should be heated or cooled using a thermostat. After the desired temperature is obtained, the substance is dropped onto the subphase, and the measurement is carried out according to the protocol. For 25-OH, experiments were carried out at the following temperatures: 10 °C, 15 °C, 20 °C, 30 °C, and

35 °C. Each isotherm should be repeated at least three times to ensure the experimental curves' reproducibility to ±2 Å^2/molecule.

An important step is also to check the stability of the monolayer. Many substances are characterized by poor stability and undergo desorption to the subphase or evaporation. Stability can be tested in two ways. Static stability measurement involves measuring changes in surface pressure as a function of time in a constant surface area. To check the static stability, in **step 23 1**, in the program interface, set the pressure to which the monolayer should be compressed (for 25-OH, these pressures were 10 mN/m, 20 mN/m, and 40 mN/m, respectively), fix the position of the barriers and monitor pressure changes as a function of time for 1 h. Pressure drop means low stability. Another method is to measure dynamic stability, in which successive cycles of compression and decompression of the monolayer are recorded. The experiment can be performed for the pressures set in the program. The shifting of isotherms toward a smaller surface area per molecule indicates aggregation or desorption of monolayer material into the subphase.

15. In some BAM models, the software allows you to measure the film thickness (*see* Fig. 2a). To do this, select the region of interest (ROI) on the water subphase image, set the analyzer and polarizer angles to 0° and start calibration (automatic process according to the producer's instructions). The apparatus measures gray level (GL) value as a function of the angle of incidence and converts it to reflectivity and film thickness. To perform the measurement, it is necessary to know the reflectance ratio of the subphase and the monolayer.
16. In case of neutral molecules analysis, it is extremely important to wait after each applied drop of solution to ensure that the electric surface potential change values return to the initial value (0 mV). The rate of deposition should be slow and the value of electric surface potential change after application of every solution drop should not exceed ±25 mV. The measurement should be started with approximately twice the surface area per molecule than for the π-A isotherm measurement.
17. The value of incidence angle should be selected for each compound individually, however, the values from the range 76–80° are used in standard [44–47].
18. The maximum height of interferogram is important as it informs us about the intensity of measured signal. This value is higher for subphase covered with monolayer and lower for pure subphase (background). Interferogram height value should not exceed 8 V for pure subphase (background) as

there is risk that the signal for subphase covered with monolayer will become oversaturated.

19. Mica is a hydrophilic material which enables to deposit lipid molecules with hydrophilic groups adsorbed to the surface and hydrophobic chains oriented to the air. However, with this technique you can use two types of surfaces: hydrophobic or hydrophilic. Deposition onto hydrophobic surface always takes place on the down path, while for hydrophilic surface deposition takes place on the up path. The most frequently used hydrophilic substrates include mica, glass slides, and polished silicon wafers, while the hydrophobic ones are highly ordered pyrolytic graphite (HOPG) and silanized SiO_2. Mica is easy to clean, which is why it is the most frequently chosen substrate. In turn, silicon or glass should be subjected to a cleaning procedure that includes washing the slides with detergents, then rinsing in an ultrasonic bath in deionized water and then with isopropanol. Instead of using ultrasonic bath, an argon plasma cleaning can be applied.
20. A trough equipped with a special dipping well and a dipper should be used to carry out the measurements.
21. In the case of mica sheets, it should be carefully cut into smaller pieces with an area of approximately 1 cm by 1 cm and 0.15–0.21 mM thick. You can also use ready-made disks. Always use the highest quality grade V1 mica. The quality of substrate surface determines the deposition process.
22. The substrate should be flat and smooth.
23. Before starting the transfer, the stability of the monolayer at the desired surface pressure should be checked in a separate experiment. Spread the monolayer on the trough subphase and compress it to the selected surface pressure. Stop the barriers. Monitor the change in surface pressure with time. Usually, an initial drop in surface pressure is observed but with time the value of surface pressure stabilizes. For film stabilization, you should allow ~10–20 min. If the pressure drops continuously without stabilizing, the monolayer is unsuitable for Langmuir–Blodgett transfer. When samples are transferred to a solid substrate, ensure that the Langmuir film is tightly packed. Therefore, films compressed to a surface pressure of 30 mN/m are transferred most often. Control the surface pressure while raising the sample. It should remain constant.
24. The rise speed should be between 1 and 3 mm/min. It should be set depending on the stability of the Langmuir film. For monolayers characterized by high stability, it should be lower (1 mm/min). The Langmuir–Blodgett technique also makes it possible to transfer more than one layer to a solid substrate. Such experiment involves sequential emergence and immersion of the sample in the subphase. The surface of substrate after

coating with Langmuir film changes from hydrophilic to hydrophobic with every layer. A parameter that allows you to assess the coverage of the substrate with the film is transfer ratio (TR), defined as the ratio between the decrease in monolayer area during a deposition stroke and the area of the substrate immersed in water. The best samples have TR close to 1.

25. It is convenient to store samples in transparent plastic Petri dish sets. To prevent the sample from moving and damage, the substrate can be glued to the bottom of the dish using Pritt self-adhesive paste. Samples prepared in this way can be analyzed using atomic force microscopy.

26. Remember to clean the dipping well.

References

1. Gaines GL (1966) Insoluble monolayers at liquid-gas interfaces. Interscience, New York
2. Martins BA, Deffune E, Oliveira ON, de Moraes ML (2022) Penicillin-binding proteins (PBPs) determine antibiotic action in Langmuir monolayers as nanoarchitectonics mimetic membranes of methicillin-resistant Staphylococcus aureus. Colloids Surf B Biointerfaces 214. https://doi.org/10.1016/j.colsurfb.2022.112447
3. Arczewska M, Gagoś M (2011) Molecular organization of antibiotic amphotericin B in dipalmitoylphosphatidylcholine monolayers induced by K+ and Na+ ions: the langmuir technique study. Biochim Biophys Acta Biomembr 1808:2706–2713. https://doi.org/10.1016/J.BBAMEM.2011.07.027
4. Pedrosa M, Maldonado-Valderrama J, Gálvez-Ruiz MJ (2022) Interactions between curcumin and cell membrane models by Langmuir monolayers. Colloids Surf B Biointerfaces 217: 112636. https://doi.org/10.1016/J.COLSURFB.2022.112636
5. Dynarowicz-Latka P, Hac-Wydro K (2014) Edelfosine in membrane environment – the Langmuir monolayer studies. Anti Cancer Agents Med Chem 14:499–508. https://doi.org/10.2174/18715206146661403092307 22
6. Feder-Kubis J, Wnętrzak A, Chachaj-Brekiesz A (2020) Terpene-based ionic liquids from natural renewable sources as selective agents in antifungal therapy. ACS Biomater Sci Eng 6: 3832–3842. https://doi.org/10.1021/acsbiomaterials.0c00447
7. Wnętrzak A, Szymczuk D, Chachaj-Brekiesz A et al (2024) Lithocholic acid-based oligomers as drug delivery candidates targeting model of lipid raft. Biochim Biophys Acta Biomembr 1866. https://doi.org/10.1016/J.BBAMEM.2024.184294
8. Wnętrzak A, Latka K, Dynarowicz Latka P (2013) Interactions of alkylphosphocholines with model membranes – the langmuir monolayer study. J Membrane Biol 246:453–466. https://doi.org/10.1007/s00232-013-9557-4
9. Wnętrzak A, Lipiec E, Łątka K et al (2014) Affinity of alkylphosphocholines to biological membrane of prostate cancer: studies in natural and model systems. J Membr Biol 247:581–589. https://doi.org/10.1007/s00232-014-9674-8
10. Seoane R, Miñones J, Conde O et al (1999) Interaction between amphotericin B and sterols in monolayers. Mixed films of ergosterol-amphotericin B. Langmuir 15:3570–3573. https://doi.org/10.1021/la980541q
11. Seoane R, Miñones J, Conde O et al (1999) Interactions between amphotericin B and sterols in monolayers. Mixed films of amphotericin B-cholesterol. Langmuir 15:5567–5573. https://doi.org/10.1021/LA980571T
12. Dynarowicz-Łątka P, Seoane R, Miñones J et al (2003) Study of penetration of amphotericin B into cholesterol or ergosterol containing dipalmitoyl phosphatidylcholine Langmuir monolayers. Colloids Surf B Biointerfaces 27:249–263. https://doi.org/10.1016/S0927-7765(02)00099-1
13. Dynarowicz-Łątka P, Wnętrzak A, Makyła-Juzak K (2015) Cyclosporin A in membrane lipids environment: implications for antimalarial activity of the drug – the langmuir monolayer studies. J Membr Biol 248:1021–1032. https://doi.org/10.1007/S00232-015-9814-9

14. Wnętrzak A, Makyła-Juzak K, Chachaj-Brekiesz A et al (2018) Cyclosporin A distribution in cholesterol-sphingomyelin artificial membranes modeled as Langmuir monolayers. Colloids Surf B Biointerfaces 166:286–294. https://doi.org/10.1016/j.colsurfb.2018.03.031
15. Szafran K, Jurak M, Wiącek AE (2022) Effect of chitosan on the interactions between phospholipid DOPC, cyclosporine A and lauryl gallate in the Langmuir monolayers. Colloids Surf A Physicochem Eng Asp 652:129843. https://doi.org/10.1016/J.COLSURFA.2022.129843
16. Wnętrzak A, Chachaj Brekiesz A, Stępniak A et al (2022) Different effects of oxysterols on a model lipid raft – langmuir monolayer study complemented with theoretical calculations. Chem Phys Lipids 244. https://doi.org/10.1016/j.chemphyslip.2022.105182
17. Chachaj-Brekiesz A, Wnętrzak A, Kobierski J et al (2023) Site of the hydroxyl group determines the surface behavior of bipolar chain-oxidized cholesterol derivatives – langmuir monolayer studies supplemented with theoretical calculations. J Phys Chem B 127:2011–2021. https://doi.org/10.1021/acs.jpcb.2c08629
18. Wnętrzak A, Chachaj-Brekiesz A, Kobierski J et al (2020) Unusual behavior of the bipolar molecule 25-hydroxycholesterol at the air/-water interface – langmuir monolayer approach complemented with theoretical calculations. J Phys Chem B 124:1104–1114. https://doi.org/10.1021/acs.jpcb.9b10938
19. Wnętrzak A, Chachaj-Brekiesz A, Kuś K et al (2021) 25-hydroxycholesterol interacts differently with lipids of the inner and outer membrane leaflet – the langmuir monolayer study complemented with theoretical calculations. J Steroid Biochem Mol Biol 211. https://doi.org/10.1016/J.JSBMB.2021.105909
20. Agina EV, Mannanov AA, Sizov AS et al (2017) Luminescent organic semiconducting langmuir monolayers. ACS Appl Mater Interfaces 9:18078–18086. https://doi.org/10.1021/ACSAMI.7B01919
21. Giancane G, Valli L (2012) State of art in porphyrin Langmuir-Blodgett films as chemical sensors. Adv Colloid Interf Sci 171–172:17–35. https://doi.org/10.1016/j.cis.2012.01.001
22. Valli L (2005) Phthalocyanine-based Langmuir-Blodgett films as chemical sensors. Adv Colloid Interf Sci 116:13–44. https://doi.org/10.1016/J.CIS.2005.04.008
23. Gorbachev I, Smirnov A, Ivanov G et al (2022) Langmuir-Blodgett films of arachidic and stearic acids as sensitive coatings for chloroform HF SAW sensors. Sensors 23:100. https://doi.org/10.3390/S23010100
24. Dumur F, Reculusa S, Mruczkiewicz M et al (2016) Multilayer Langmuir-Blodgett films as diffractive external 3D photonic crystal in blue OLEDs. Opt Express 24:27184. https://doi.org/10.1364/OE.24.027184
25. Hussain SA (2009) Langmuir-Blodgett films a unique tool for molecular electronics
26. Dynarowicz-Latka P, Wnętrzak A, Chachaj-Brekiesz A (2024) Advantages of the classical thermodynamic analysis of single – and multi-component Langmuir monolayers from molecules of biomedical importance – theory and applications. J R Soc Interface 21. https://doi.org/10.1098/RSIF.2023.0559
27. Mobius D (1998) Morphology and structural characterization of organized monolayers by Brewster angle microscopy. Curr Opin Colloid Interface Sci 3:137–142. https://doi.org/10.1016/S1359-0294(98)80005-8
28. Stottrup BL, Nguyen AH, Tüzel E (2010) Taking another look with fluorescence microscopy: image processing techniques in Langmuir monolayers for the twenty-first century. Biochim Biophys Acta Biomembr 1798:1289–1300. https://doi.org/10.1016/J.BBAMEM.2010.01.003
29. Sofińska K, Lupa D, Chachaj-Brekiesz A et al (2022) Revealing local molecular distribution, orientation, phase separation, and formation of domains in artificial lipid layers: towards comprehensive characterization of biological membranes. Adv Colloid Interf Sci 301
30. Dynarowicz-Łątka P, Wnętrzak A, Broniatowski M, Flasiński M (2013) Miscibility and phase separation in mixed erucylphosphocholine–DPPC monolayers. Colloids Surf B Biointerfaces 107:43–52. https://doi.org/10.1016/J.COLSURFB.2013.01.044
31. Kobierski J, Wnętrzak A, Chachaj-Brekiesz A et al (2021) How the replacement of cholesterol by 25-hydroxycholesterol affects the interactions with sphingolipids: the Langmuir monolayer study complemented with theoretical calculations. J R Soc Interface 18: rsif.2021.0050. https://doi.org/10.1098/rsif.2021.0050
32. Chachaj-Brekiesz A, Wnętrzak A, Włodarska S et al (2020) Molecular insight into neurodegeneration – langmuir monolayer study on the influence of oxysterols on model myelin sheath. J Steroid Biochem Mol Biol 202. https://doi.org/10.1016/j.jsbmb.2020.105727

33. Davies JT, Rideal EK (1961) Interfacial phenomena. Academic Press
34. Chachaj-Brekiesz A, Kobierski J, Wnętrzak A et al (2024) Insight into the molecular mechanism of surface interactions of phosphatidylcholines –langmuir monolayer study complemented with molecular dynamics simulations. J Phys Chem B 128:1473–1482. https://doi.org/10.1021/ACS.JPCB.3C06810
35. Chachaj-Brekiesz A, Kobierski J, Griñón Echaniz R et al (2022) Comprehensive approach to the interpretation of the electrical properties of film-forming molecules. J Phys Chem B 126: 7037–7046. https://doi.org/10.1021/ACS.JPCB.2C04526
36. Nakahara H, Minamisono M, Shibata O (2019) Lateral interaction of cholesterol with a semifluorinated amphiphile at the air–water interface. Colloids Surf B Biointerfaces 181: 1035–1040. https://doi.org/10.1016/J.COLSURFB.2019.06.072
37. Nakahara H, Nakamura S, Nakamura K et al (2005) Cerebroside Langmuir monolayers originated from the echinoderms: II. Binary systems of cerebrosides and steroids. Colloids Surf B Biointerfaces 42:175–185. https://doi.org/10.1016/J.COLSURFB.2005.01.013
38. Harkins WD (1952) Physical chemistry of surface films. Reinhold, New York
39. Brzozowska I, Figaszewski ZA (2002) The equilibrium of phosphatidylcholine–cholesterol in monolayers at the air/water interface. Colloids Surf B Biointerfaces 23:51–58. https://doi.org/10.1016/S0927-7765(01)00209-0
40. Vila Romeu N, Miñones J, Iribarnegaray E, et al (1997) Influence of the solvent on the spreading of poly[(D,L-lactic acid)-co-(glycolic acid)] monolayers. Colloid Polym Sci 275: 580–586. https://doi.org/10.1007/S003960050121
41. Hardy NJ, Richardson TH, Grunfeld F (2006) Minimising monolayer collapse on Langmuir troughs. Colloids Surf A Physicochem Eng Asp 284–285:202–206. https://doi.org/10.1016/J.COLSURFA.2006.02.001
42. Miñones J, Miñones J, Conde O et al (2000) Influence of a spreading method on the properties of amphotericin B-dipalmitoyl phosphatidic acid mixed films at the air/water interface. Langmuir 16:5743–5748. https://doi.org/10.1021/LA9915116
43. Gonçalves Da Silva AM, Viseu MI, Campos CS, Rechena T (1998) Effect of the spreading procedure on the formation of cationic–anionic mixed monolayers. Thin Solid Films 320:236–240. https://doi.org/10.1016/S0040-6090(97)00948-6
44. Fidalgo Rodriguez JL, Caseli L, Minones Conde J, Dynarowicz-Latka P (2019) New look for an old molecule – solid/solid phase transition in cholesterol monolayers. Chem Phys Lipids 225:104819. https://doi.org/10.1016/J.CHEMPHYSLIP.2019.104819
45. Chachaj-Brekiesz A, Wnętrzak A, Kobierski J et al (2023) Site of the hydroxyl group determines the surface behavior of bipolar chain-oxidized cholesterol derivatives – langmuir monolayer studies supplemented with theoretical calculations. J Phys Chem B 127:2011–2021. https://doi.org/10.1021/ACS.JPCB.2C08629
46. Rosa ME, Alves Conserva GA, Passero LFD et al (2022) Unsaturated lipids modulating the interaction of the antileishmanial isolinderanolide E with models of cellular membranes. Bioorg Chem 124:105814. https://doi.org/10.1016/J.BIOORG.2022.105814
47. Aoki PHB, Morato LFC, Pavinatto FJ et al (2016) Molecular-level modifications induced by photo-oxidation of lipid monolayers interacting with erythrosin. Langmuir 32:3766–3773. https://doi.org/10.1021/ACS.LANGMUIR.6B00693

Chapter 17

Protein Analysis by FT-IR/ATR Spectroscopy

Gabriela Graziani, Giovanni D'Atri, Elena Gabusi, and Gina Lisignoli

Abstract

Fourier transform infrared spectroscopy (FT-IR) is a powerful tool for the study of inorganic and organic materials. In the past years, many efforts have been put forward for the use of FT-IR in living cells and tissues, and as a diagnostic tool to study pathology in vivo, ex vivo, and in vitro, due to its peculiar advantages, which include high sensitivity, small volume of samples, fast acquisition time, and broad range of compounds that can be analyzed. In orthopedic field, tissues, cells, and exosomes have been studied for applications in cartilage and osteoarthritis. For this application, FT-IR allows a fingerprint of the analyzed material to be obtained, providing information on the biological molecules that are contained (i.e., proteins, lipids, nucleic acids), in terms of quantity and organization, furnishing important information for studies on tissue pathophysiology, disease progression, and therapeutic evaluation.

Key words FT-IR, Protein analysis, Spectroscopy, Chondrocytes, Orthopedics

1 Introduction

Fourier transform infrared spectroscopy (FT-IR) is a powerful technique to investigate both the organic and inorganic molecular structure of materials [1–3]. In FT-IR, the acquisition is based on identifying the functional groups that compose a sample, by exploiting their different vibration modes depending on the IR frequency (0.78 and 1000 mM). When IR radiation is absorbed, the molecule to which the photon is transferred is excited to a higher energy state, resulting in the vibration of the molecular bonds, detectable at different wavelength. The molecular vibrations are of two main types: stretching and bending, where stretching (normally results in the bands of higher intensity) is the change in the bond length (i.e., the interatomic distance along the bond axis), while bending entails the change of bond angles (while the length remains constant). Depending on the rotation axis, bending in turn can be divided into rocking, scissoring, wagging, or twisting [4].

Csaba Matta et al. (eds.), *The Surfaceome: Methods and Protocols*, Methods in Molecular Biology, vol. 2908,
https://doi.org/10.1007/978-1-0716-4434-8_17,

For FT-IR, different detectors, sources of lights, and components are used, which influence the acquisition range. In the past years, attenuated total reflectance (ATR) FT-IR has emerged as particularly promising, thanks to high resolution and easy and fast acquisition of the samples, that can be put directly in contact with the detector, without the need for preparation (for instance, mixing with potassium bromide-KBr and pelleting). In ATR, the detector is a crystal having a high index of refraction higher than the sample (and normally >2). Light is directed through the crystal to the sample interface, and the incidence angle is selected so that it can be completely reflected at the crystal–sample interface and can penetrate the sample, at a depth that depends on the angle of incidence and the ratio between the refraction indices of the crystal and the sample [5, 6].

Largely used for the study of materials, FT-IR is now being increasingly used for the analysis of biological materials [7, 8], both in vitro, in vivo, and ex vivo, including cells [9–14], tissues [9, 15–20], and extracellular vesicles [21–23].

The increasing use of FT-IR is boosted by some key advantages, which include versatility and large range of applications, the possibility to analyze a very small volume of samples, the high sensitivity to the chemical composition and the architecture of molecules, short measuring time, and the relatively low costs (top class spectrometers for 40,000 € or $) [1, 24].

In biological materials, different biomolecules can be detected, such as lipids, proteins, carbohydrates, and nucleic acids, which have a unique chemical structure and thus distinctive IR spectra [25, 26]. In the case of proteins, FT-IR spectroscopy appears particularly promising, because of its broad applicability, that spans from small soluble proteins to large membrane proteins, merged with high molecular information content and sensitivity high enough to detect environmental changes. Consequently, it is an important tool to study protein structure, molecular mechanisms of protein reactions, and protein folding [1, 27].

In the musculoskeletal system, FT-IR spectroscopy has been proposed for the study of different tissues [28] and pathologies, including cancer [21, 29], osteomyelitis [30], and osteoarthritis [24, 31–36]. In particular, FT-IR has been used in vivo and ex vivo to detect modifications in articular cartilage, subchondral bone, and synovial fluids as markers of osteoarthritis, by investigating dissolution and compositional changes, including degradation of macromolecules, modifications in the shape and size of chondrocytes, as well as alterations in mineralization and/or in collagen and proteoglycans content or structure, between healthy and pathological conditions [37].

While the study of cartilage by FT-IR has been widely investigated, only a few studies apply FT-IR to the study of chondrocytes,

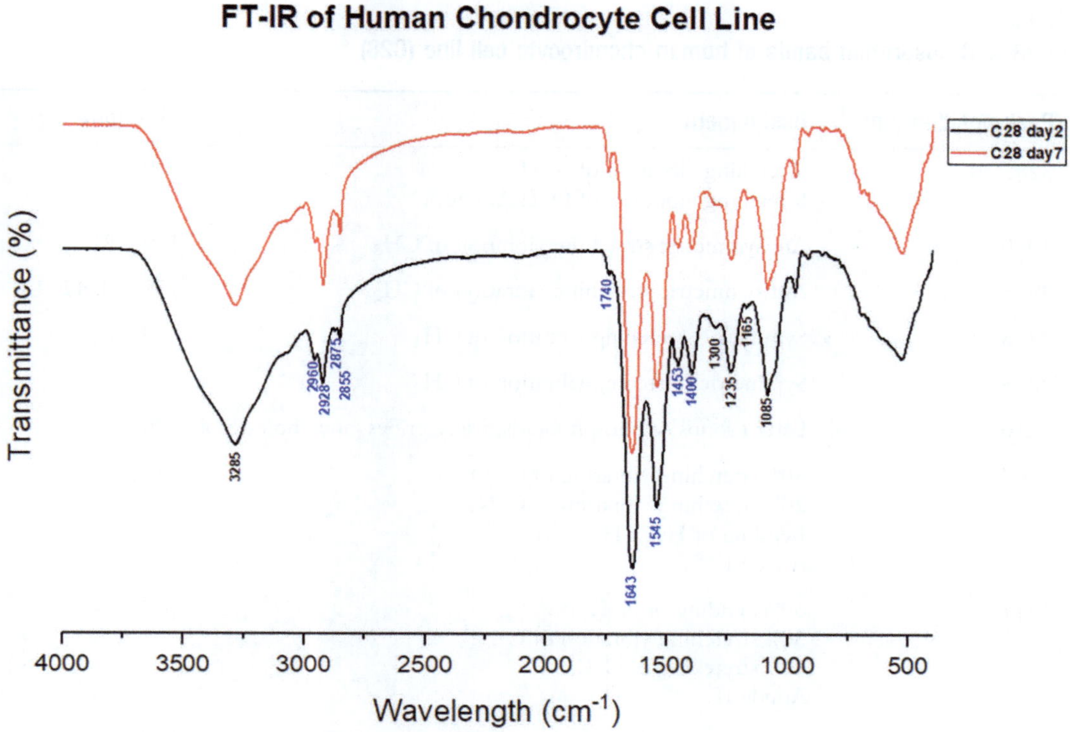

Fig. 1 Representative spectra of C28 after subtracting PBS curve. Bands in red are those relevant to lipid and protein structure

showing that lipids, GAGs, collagen, and aggrecan bands are present in different areas of the spectrum [26, 38].

We evaluated the spectrum of human chondrocyte cell line (C28) at two time points of cell culture (day 2 and day 7), as shown in Fig. 1. FT-IR analysis evidenced different bands in the spectrum belonging to different functional groups and biomolecules.

In Table 1, we summarized the peak position of functional groups and biomolecules.

The exact position of the bands correlates with protein type, but also with their orientation and form (i.e., helical, sheet or random) [44]. Hence, switches in peak positions of the bands are indicative of structural molecule modification and integrity. Indeed, for instance, it has been shown that, in cartilage, collagen integrity can be monitored by the ratio between the peak areas at 1338 cm^{-1} and that of Amide II band and correlates with Mankin grade [37].

In addition, the band peak intensity is also important, as it is correlated with the amount of each compound, and their modifications provide evidence sign of structural changes. Focusing on protein:lipid ratio, we can characterize protein–membrane interactions, the orientation of membrane-bound proteins, the secondary

Table 1
FT-IR/ATR absorption bands of human chondrocyte cell line (C28)

Peak position (cm^{-1})	Assignment	Literature
3285	Stretching vibration of N–H; Stretching vibration of O–H Amide A	[39]
2960	Antisymmetric stretching vibration of CH_3	[39–42]
2928	Antisymmetric stretching vibration of CH_2	[39, 40, 42, 43]
2875	Symmetric stretching vibration of CH_2	[39–42]
2855	Symmetric stretching vibration of CH_2	[39–42]
1740	Ester groups of phospholipids, triglycerides, and cholesterol	[39]
1643	80% stretching vibration of C=O; 20% stretching vibration of C–N; Bending of H–O–H; Amide I	[39]
1545	60% bending of N–H; 30% stretching vibration of C–N; 10% stretching of C–C; Amide II	[39]
1453	Antisymmetric deformation of CH_3; Antisymmetric deformation of CH_2	[39]
1235	Antisymmetric stretching vibration of PO_2^-	[39]
1083	Symmetric stretching vibration of PO_2^-	[39]

structure, and the dynamic properties. Previous reports have evidenced that the variations in the protein:lipid ratio may indicate changes in the fluidity of membranes and their ability to perform crucial functions, such as membrane transport and cell signaling [45, 46]. In Fig. 2, the peak zones of protein (1700–1350 cm^{-1}) and lipid (3000–2700 cm^{-1}) in chondrocytes C28 (at day 2 and 7) are shown.

However, bands of different molecules can overlap, thus preventing a correct interpretation of this information. For this reason, second derivatives can be used to improve spectral resolution of the peaks and avoid overlapping [47]. In particular, the second derivative in FTIR analysis could help to increase the sensitivity, reduce the baseline effects, facilitate the identification of spectral features by separating neighboring peaks, identifying more details in the data obtained [48, 49]. This also permits to differentiate between potentially overlapping compounds, such as collagen and proteoglycans (PG) [47]. In Fig. 3, we show how it is possible to use the second derivative calculated by the Savitzky–Golay algorithm, implemented in Matlab R2023b, with a second-order polynomial and a 15-point window, in order to facilitate the identification of

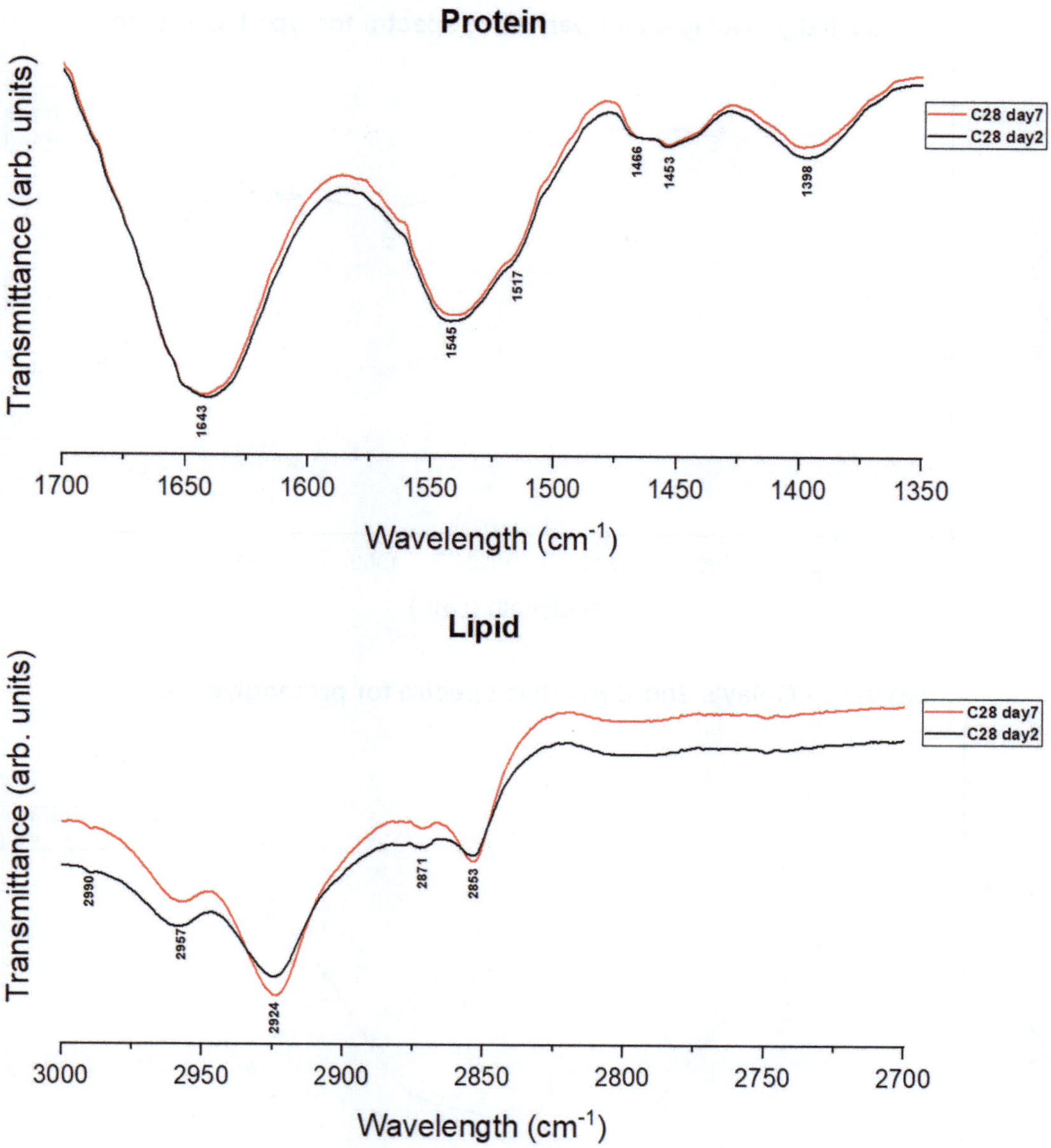

Fig. 2 Band region corresponding to protein and lipid zones, respectively

the peaks around 1338 cm^{-1} and 1072 cm^{-1}, corresponding to the band of type II collagen and PG, respectively, as suggested in [35, 36]. By comparing Figs. 1 and 3, it is clear how bands concealed by overlapping can be clearly detected by the used of second derivatives.

Finally, FT-IR spectroscopy can be applied to exosomes [44], including osteoarthritic plasma extracellular vesicles [50]. As in the case of cells, several parameters can be assessed, permitting a structural characterization of proteins and protein–membrane interactions in biological matrices.

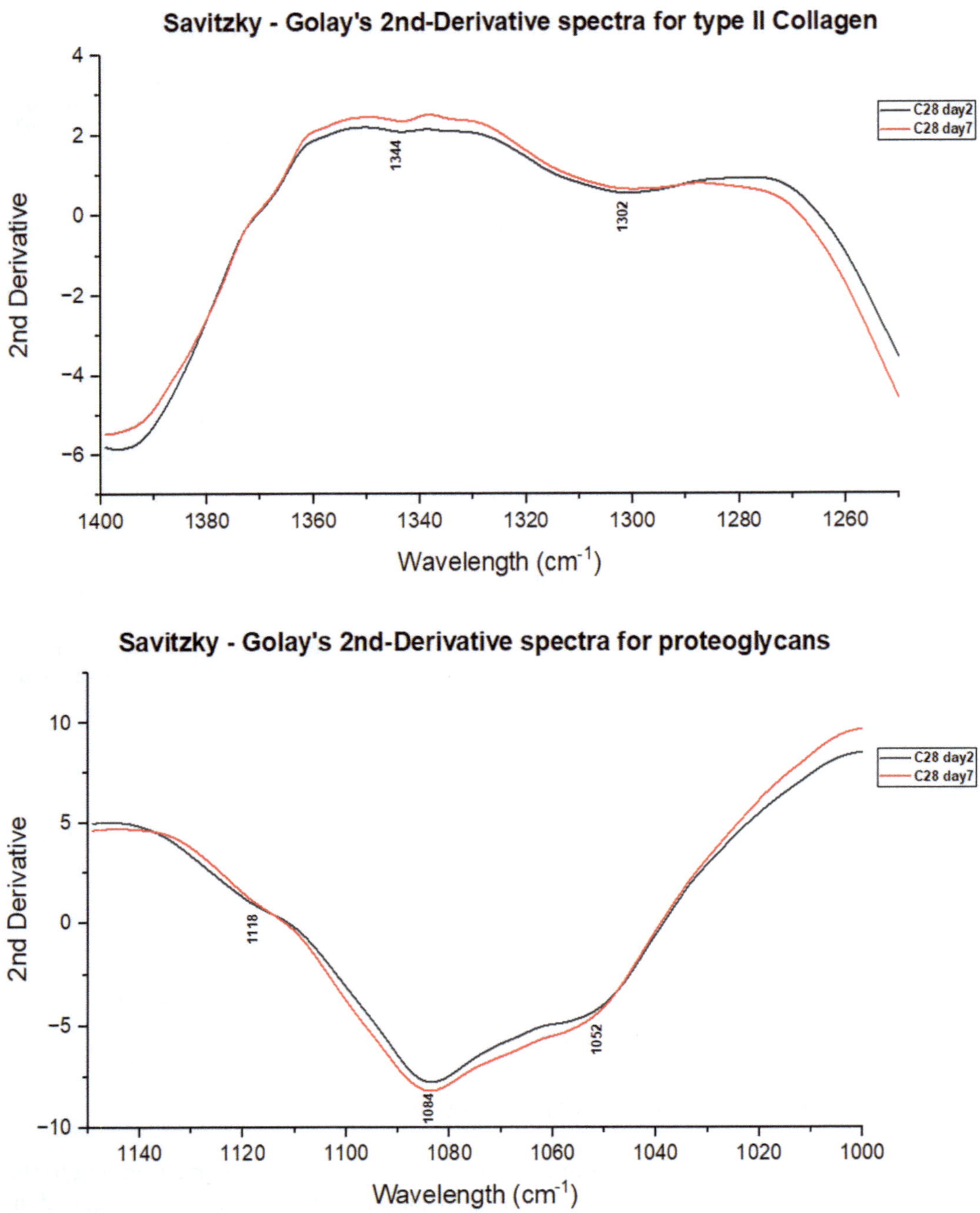

Fig. 3 Savitzky–Golay's second-derivative spectra for type II collagen and proteoglycans in the 1400–1250 cm^{-1} and 1150–1000 cm^{-1} spectral windows

Being applicable to cells, tissues, and exosomes, FT-IR appears a powerful and versatile tool for the study of tissues, permitting to use small volumes and reduce processing time and costs. When applied to the study of proteins, it permits to evaluate their presence, amount, conformation, and ratio to lipids. Moreover, it appears important to study the structure of proteoglycans and collagen in chondrocytes. However, in spite of its wide range of applicability, the use of FT-IR for proteins study is at its early stages and several parameters are yet to be investigated to fully implement its use for the study of the surfaceome.

2 Methods

2.1 Sample Preparation and Data Acquisition

Data are acquired in the whole spectrum, that is 400–4000 cm^{-1} for diamond ATR crystal and 600–4000 cm^{-1} for germanium crystal. In our case, the acquisition was performed by a PerkinElmer Spectrum 2 instrument, equipped with a diamond crystal ATR and by a PerkinElmer Spotlight i2 microscope, coupled with the spectrum 2 instrument. The same acquisition procedure can be applied to extracellular vesicles.

2.2 Cells

For the acquisition, human chondrocyte cell line (C28/I2) in micromass culture was prepared.

1. Thaw and expand C28/I2 cells in T150 flasks in DMEM High Glucose containing 10% FBS.
2. Place the flasks at 37 °C and 5% CO_2 in an incubator.
3. Pellet 5 × 10^5 C28/I2 cells in conical tubes to form micromasses ($n = 3$) and culture from day 2 to day 7 in DMEM High Glucose medium supplemented with 10% FBS and 50 μg/mL ascorbic acid.

As an alternative, cells can be seeded in monolayer on coverslips and cultured for the relevant time. Cell pellets are then dried and detached from the glass substrate by gently peeling with a spatula before acquisition.

1. Cells must be washed in PBS to avoid possible inference of the bands of crystallized medium in the 400–1200 cm^{-1} area (*see* **Note 1**).
2. Blank is acquired using the same acquisition parameters used for the samples.
3. Spectrum of PBS is acquired using the same acquisition parameters used for the samples.
4. Cells (micromasses or detached pellets) are deposited on the ATR detector and left to dry in contact with the crystal until complete evaporation of the medium (*see* **Note 2**, **3**, and **4**).
5. Completeness of drying can be monitored by acquisition of serial spectra, checking the progressive disappearance/sharpening of the two bands of water, and the emergence of the bands characteristic of the sample (*see* **Note 5**).

6. Spectra are acquired using the following acquisition parameters: acquisition range (for diamond crystal): 4000–400 cm^{-1}, resolution 4 cm^{-1}, acquisition: 128 scans, step size: 0.5 cm^{-1}.

2.3 Exosome Acquisition

Extracellular vesicle (EV) spectra are acquired starting from 5 μL drops containing 9 ± 2 μg total protein.

1. EVs are resuspended in PBS to avoid possible inference of the bands of crystallized medium in the 400–1200 cm^{-1} area.
2. Blank is acquired.
3. Bands of PBS are acquired at each time point, following the same conditions used for EVs samples.
4. EV-containing droplets are deposited onto the ATR crystal and left to dry.
5. Complete drying is verified.
6. Spectra are acquired using the same acquisition parameters used for the cells: acquisition range (for diamond crystal): 4000–400 cm^{-1}, resolution 4 cm^{-1}, acquisition: 128 scans, step size: 0.5 cm^{-1}.

2.4 FT-IR Microscopy

Cells (5000 cell/cm^2) are seeded on Thermanox coverslips and cultured for the relevant time.

1. Cells are washed in PBS to avoid possible inference of the bands of crystallized medium in the 400–1200 cm^{-1} area.
2. Dried cells on coverslips are mounted on the microscope. After finding the focus and acquiring the optical image of the sample, blank is acquired using the same acquisition parameters and spot size used for the samples.
3. Spectrum of dried PBS is acquired using the same acquisition parameters used for the samples.
4. Spectra are acquired, using the following acquisition parameters: acquisition range (for germanium crystal): acquisition range 600–4000 cm^{-1}, resolution 4 cm^{-1}, 16 scans and step size 1 cm^{-1}, spot size 100 × 100 μm^2.
5. Maps can be acquired to investigate uniformity within one sample (*see* **Note 6**).

2.5 Data Processing

Irrespective of the type of sample (cells vs exosomes) and acquisition (FT-IR/ATR or FT-IR microscopy), data processing is carried out as follows:

2.5.1 Normalization and Subtraction of the PBS Bands, Bands Labeling

1. All bands are scaled with respect to the external band of PBS at 1084 cm^{-1}, so as to normalize the curve intensity, which depends on the exact amount of cells and on their contact with the crystal.

2. The PBS curve is then subtracted from each specimen.
3. The position of all bands is recorded (*see* **Note 7**).
4. Curves of different samples can be compared visually or subtracted to visualize differences in the bands position or intensity.
5. A first indication of the total amount of proteins and lipids contained in the samples can be achieved, as the area underlying the bands correlates with the amount of each component.

2.5.2 Protein:Lipid Ratio

To calculate the protein:lipid ratio, it is necessary to follow these steps:

1. Curves must be scaled with respect to the Amide I band at 1650 cm^{-1}.
2. The integral under the curves in the protein and lipid spectrum (in the ranges between 1700–1350 cm^{-1} and 3000–2700 cm^{-1}, respectively) must be calculated by selecting the protein and lipid wavelength ranges.
3. The ratio between the area subtended by the protein and lipid curves must be calculated as the ratio of the values obtained in **step 2** using a Matlab code to obtain the area subtended by the protein and lipid curves, respectively.

2.5.3 Secondary Structure

The second derivative can be calculated with the Savitzky–Golay algorithm via Matlab code using the following steps:

1. The wavelength range of type II collagen (around the value of 1338 cm^{-1}) and of PG (around the value of 1072 cm^{-1}) is selected.
2. The polynomial degree and filter size is defined to apply the algorithm.
3. The second derivative is calculated using the Savitzky–Golay algorithm.
4. Conclusions.

FT-IR is a powerful and versatile tool for the study of tissues, as well as cells and exosomes, permitting to use small volumes. When applied to the study of proteins, it permits to evaluate their presence, amount, conformation, and ratio to lipids. Moreover, it appears important to study the structure of proteoglycans and collagen in chondrocytes. However, the use of FT-IR for proteins study is at its early stages and several parameters are yet to be investigated to fully implement its use for the study of the surfaceome.

3 Notes

1. The PBS spectrum shows characteristic sharp bands at 1076, 976, 857 cm^{-1} (ν P-O) and broad bands at 1660 and ~ 3000 cm^{-1}, which overlap with some important bands of functional groups of proteins. Hence, it is important to always acquire a PBS curve and subtract it from the samples curves, prior to further analyses.
2. Cells in monolayer are detached from the glass substrate by gently peeling with a spatula and transferred in conical tubes.
3. Both PBS-washed cell detached monolayer and cell pellets are transferred on ATR detector. To do so, monolayers are gently detached with a spatula, while pellets are pipetted. Excess liquid surrounding the samples is accurately removed to reduce drying time.
4. One of the main limitations of the analysis is that FT-IR can be only performed on dried samples. Therefore, in aqueous solutions, the strong absorbance of water in the mid-infrared spectral region (near 1645 cm^{-1}) overlaps the important Amide I band of proteins and some side chain bands [1]. When the sample is wet, only two broad bands are visible at ~1600 cm^{-1} and 3000–3500 cm^{-1}.
5. To monitor cells monolayers and pellets drying, the progressive disappearance of the bands at 3000–3500 cm^{-1} band and appearance of those relevant to the sample is visualized by using "preview" or by acquiring spectra at a low number of accumulation. The latter can be done by stopping the acquisition, so as to avoid modifying parameters of background. The sample can be considered dry when: (1) the 3000–3500 cm^{-1} is flattened and sharpened, (2) clear bands emerge in the areas relevant for the sample (in this case, the proteins and lipid region), (3) the sample bands do not show significant modifications in two subsequent acquisitions, acquired at least at 2 min distance. Drying requires approximately 3–5 min for cells monolayers and 15 min for cell pellets.
6. Acquisition of at least five nonoverlapping 100 × 100 μm^2 regions is recommended. Spots below 25 × 25 μm^2 shall be avoided as the intensity of the signal progressively decreases for increasing reduction of the acquisition spots.
7. The position of the bands can be directly acquired through FT-IR incorporated software.

Acknowledgments

Prof. Enrico Sassoni (University of Bologna, Bologna, Italy) is gratefully acknowledged for support to FT-IR acquisitions. This research was supported by Italian Ministry of Health- 5xmille anno 2019 redditi 2018 "Medicina rigenerativa e riparativa personalizzata per le patologie dei tessuti muscolo-scheletrici e la chirurgia ricostruttiva ortopedica."

References

1. Barth A (2007) Infrared spectroscopy of proteins. Biochim Biophys Acta 1767(9): 1073–1101. https://doi.org/10.1016/j.bbabio.2007.06.004
2. Pradhan S, Rajamani S, Agrawal G, Dash M, Samal SK (2017) NMR, FT-IR and Raman characterization of biomaterials. In: Tanzi MC, Farè S (eds) Characterization of polymeric biomaterials. Woodhead, pp 147–173. https://doi.org/10.1016/B978-0-08-100737-2.00007-8
3. Marcott C, Padalkar M, Pleshko N (2017) Infrared and raman microscopy and imaging of biomaterials at the micro and nano scale, pp 498–518. https://doi.org/10.1016/B978-0-12-803581-8.10183-3
4. Ojeda JJ, Dittrich M (2012) Fourier transform infrared spectroscopy for molecular analysis of microbial cells. Methods Mol Biol 881:187–211. https://doi.org/10.1007/978-1-61779-827-6_8
5. Kaur H, Rana B, Tomar D, Kaur S, Jena KC (2021) Fundamentals of ATR-FTIR spectroscopy and its role for probing in-situ molecular-level interactions. In: Modern techniques of spectroscopy. Progress in optical science and photonics. Springer, Singapore, pp 3–37. https://doi.org/10.1007/978-981-33-6084-6_1
6. Ramer G, Lendl B (2013) Attenuated total reflection fourier transform infrared spectroscopy. In: Encyclopedia of analytical chemistry. Wiley, Chichester. https://doi.org/10.1002/9780470027318.a9287
7. Zohdi V, Whelan DR, Wood BR, Pearson JT, Bambery KR, Black MJ (2015) Importance of tissue preparation methods in FTIR micro-spectroscopical analysis of biological tissues: 'traps for new users'. PLoS One 10(2): e0116491. https://doi.org/10.1371/journal.pone.0116491
8. Poonprasartporn A, Chan KLA (2021) Live-cell ATR-FTIR spectroscopy as a novel bioanalytical tool for cell glucose metabolism research. Biochim Biophys Acta Mol Cell Res 1868(7):119024. https://doi.org/10.1016/j.bbamcr.2021.119024
9. Naumann D (2013) Infrared spectroscopy of cells, tissues, and biofluids. In: Roberts GCK (ed) Encyclopedia of biophysics. Springer, Berlin/Heidelberg. https://doi.org/10.1007/978-3-642-16712-6_120
10. Sabbatini S, Conti C, Orilisi G, Giorgini E (2017) Infrared spectroscopy as a new tool for studying single living cells: is there a niche? Biomed Spectrosc Imaging 6:85–99. https://doi.org/10.3233/BSI-170171
11. Faramarzi B, Moggio M, Diano N, Portaccio M, Lepore M (2023) A brief review of FT-IR spectroscopy studies of sphingolipids in human cells. Biophysica 3:158–180. https://doi.org/10.3390/biophysica3010011
12. Kumar S, Salammal ST, Goormaghtigh E (2014) A FTIR imaging characterization of fibroblasts stimulated by various breast cancer cell lines. PLoS One 9:e111137. https://doi.org/10.1371/journal.pone.0111137
13. Vazquez-Zapien GJ, Mata-Miranda MM, Sanchez-Monroy V, Delgado-Macuil RJ, Perez-Ishiwara DG, Rojas-Lopez M (2016) FTIR spectroscopic and molecular analysis during differentiation of pluripotent stem cells to pancreatic cells. Stem Cells Int 2016:6709714. https://doi.org/10.1155/2016/6709714
14. Perut F, Graziani G, Columbaro M, Caudarella R, Baldini N, Granchi D (2020) Citrate supplementation restores the impaired mineralisation resulting from the acidic micro-environment: an in vitro study. Nutrients 12(12):3779. https://doi.org/10.3390/nu12123779
15. Boskey A, Pleshko CN (2007) FT-IR imaging of native and tissue-engineered bone and cartilage. Biomaterials 28(15):2465–2478. https://doi.org/10.1016/j.biomaterials.2006.11.043

16. Koziol P, Liberda D, Kwiatek WM, Wrobel TP (2020) Macromolecular orientation in biological tissues using a four-polarization method in FT-IR imaging. Anal Chem 92(19):13313–13318. https://doi.org/10.1021/acs.analchem.0c02591
17. Chaber R, Łach K, Depciuch J, Szmuc K, Michalak E, Raciborska A, Koziorowska A, Cebulski J (2017) Fourier transform infrared (FTIR) spectroscopy of paraffin and deparafinnized bone tissue samples as a diagnostic tool for Ewing sarcoma of bones. Infrared Phys Technol 85:364–371. https://doi.org/10.1016/j.infrared.2017.07.017
18. Kyriakidou M, Anastassopoulou J, Tsakiris A, Koui M, Theophanides T (2017) FT-IR spectroscopy study in early diagnosis of skin cancer. In Vivo 31(6):1131–1137. https://doi.org/10.21873/invivo.11179
19. Li L, Bi X, Sun H, Liu S, Yu M, Zhang Y, Weng S, Yang L, Bao Y, Wu J, Xu Y, Shen K (2018) Characterization of ovarian cancer cells and tissues by Fourier transform infrared spectroscopy. J Ovarian Res 11(1):64. https://doi.org/10.1186/s13048-018-0434-8
20. Graziani G, Govoni M, Vivarelli L, Boi M, De Carolis M, Bianchi M, Bignozzi MC, Carnevale G, Marmi F, Maltarello MC, Dallari D (2020) A comprehensive microstructural and compositional characterization of allogenic and xenogenic bone: application to bone grafts and nanostructured biomimetic coatings. Coatings 10(6):522. https://doi.org/10.3390/coatings10060522
21. Perut F, Graziani G, Roncuzzi L, Zini N, Avnet S, Baldini N (2022) FT-IR spectral signature of sensitive and multidrug-resistant osteosarcoma cell-derived extracellular nanovesicles. Cells 11(5):778. https://doi.org/10.3390/cells11050778
22. Mihály J, Deák R, SzigyártóI C, Bóta A, Beke-Somfai T, Varga Z (2017) Characterization of extracellular vesicles by IR spectroscopy: fast and simple classification based on amide and C H stretching vibrations. Biochim Biophys Acta Biomembr 1859:459–466. https://doi.org/10.1016/j.bbamem.2016.12.005
23. Paolini L, Federici S, Consoli G, Arceri D, Radeghieri A, Alessandri I, Bergese P (2020) Fourier-transform infrared (FT-IR) spectroscopy fingerprints subpopulations of extracellular vesicles of different sizes and cellular origin. J Extracell Vesicles 9:1741174. https://doi.org/10.1080/20013078.2020.1741174
24. Miller LM, Bourassa MW, Smith RJ (2013) FTIR spectroscopic imaging of protein aggregation in living cells. Biochim Biophys Acta 1828(10):2339–2346. https://doi.org/10.1016/j.bbamem.2013.01.014
25. Mata-Miranda MM, Martinez-Cuazitl A, Guerrero-Robles CI, Noriega-Gonzalez JE, Garcia-Hernandez JS, Vazquez-Zapien GJ (2019) Biochemical similarity between cultured chondrocytes and in situ chondrocytes by chemometric analysis from FTIR microspectroscopy. Biotechnol Rep (Amst) 9(24):e00391. https://doi.org/10.1016/j.btre.2019.e00391
26. Linus A, Ebrahimi M, Turunen MJ, Saarakkala S, Joukainen A, Kröger H, Koistinen A, Finnilä MAJ, Afara IO, Mononen ME, Tanska P, Korhonen RK (2021) High-resolution infrared microspectroscopic characterization of cartilage cell microenvironment. Acta Biomater 134:252–260. https://doi.org/10.1016/j.actbio.2021.08.001
27. Tatulian SA (2013) Structural characterization of membrane proteins and peptides by FTIR and ATR-FTIR spectroscopy. Methods Mol Biol 974:177–218. https://doi.org/10.1007/978-1-62703-275-9_9
28. Mavrogenis AF, Megaloikonomos PD, Panagopoulos GN, Papagelopoulos PJ, Theophanides T, Anastassopoulou J (2015) Side effects of radiation in bone and cartilage: an FT-IR analysis. J Long-Term Eff Med Implants 25(4):289–295. https://doi.org/10.1615/jlongtermeffmedimplants.2015011782
29. Kar S, Katti DR, Katti KS (2019) Fourier transform infrared spectroscopy based spectral biomarkers of metastasized breast cancer progression. Spectrochim Acta A Mol Biomol Spectrosc 208:85–96. https://doi.org/10.1016/j.saa.2018.09.052
30. Chaber R, Arthur CJ, Depciuch J, Łach K, Raciborska A, Michalak E, Cebulski J (2018) Distinguishing Ewing sarcoma and osteomyelitis using FTIR spectroscopy. Sci Rep 8(1):15081. https://doi.org/10.1038/s41598-018-33470-3
31. Yu C, Zhao B, Li Y, Zang H, Li L (2021) Vibrational spectroscopy in assessment of early osteoarthritis-a narrative review. Int J Mol Sci 22(10):5235. https://doi.org/10.3390/ijms22105235
32. Das Gupta S, Shrestha P, Ali N, Turkiewicz A, Hughes V, Jonsson E, Tjörnstand J, Neuman P, Önnerfjord P, Rieppo L, Englund M, Saarakkala S (2022) Infrared spectroscopy can detect end-stage osteoarthritis from human synovial fluid: a pilot study. Osteoarthr Cartil 30(1):S108. https://doi.org/10.1016/j.joca.2022.02.137
33. Rieppo L, Rieppo J, Jurvelin JS, Saarakkala S (2012) Fourier transform infrared

spectroscopic imaging and multivariate regression for prediction of proteoglycan content of articular cartilage. PLoS One 7(2):e32344. https://doi.org/10.1371/journal.pone.0032344

34. Afara IO, Prasadam I, Arabshahi Z, Xiao Y, Oloyede A (2017) Monitoring osteoarthritis progression using near infrared (NIR) spectroscopy. Sci Rep 7(1):11463. https://doi.org/10.1038/s41598-017-11844-3
35. Yin J, Xia Y, Lu M (2012) Concentration profiles of collagen and proteoglycan in articular cartilage by Fourier transform infrared imaging and principal component regression. Spectrochim Acta A Mol Biomol Spectrosc 88:90–96. https://doi.org/10.1016/j.saa.2011.12.002
36. Rieppo L, Saarakkala S, Närhi T, Helminen HJ, Jurvelin JS, Rieppo J (2012) Application of second derivative spectroscopy for increasing molecular specificity of Fourier transform infrared spectroscopic imaging of articular cartilage. Osteoarthr Cartil 20(5):451–459. https://doi.org/10.1016/j.joca.2012.01.010
37. Bi X, Yang X, Bostrom MP, Camacho NP (2006) Fourier transform infrared imaging spectroscopy investigations in the pathogenesis and repair of cartilage. Biochim Biophys Acta 1758(7):934–941. https://doi.org/10.1016/j.bbamem.2006.05.014
38. Yagami K, Suh JY, Enomoto-Iwamoto M, Koyama E, Abrams WR, Shapiro IM, Pacifici M, Iwamoto M (1999) Matrix GLA protein is a developmental regulator of chondrocyte mineralization and, when constitutively expressed, blocks endochondral and intramembranous ossification in the limb. J Cell Biol 147(5):1097–1108. https://doi.org/10.1083/jcb.147.5.1097
39. Gieroba B, Przekora A, Kalisz G, Kazimierczak P, Song CL, Wojcik M, Ginalska G, Kazarian SG, Sroka-Bartnicka A (2021) Collagen maturity and mineralization in mesenchymal stem cells cultured on the hydroxyapatite-based bone scaffold analyzed by ATR-FTIR spectroscopic imaging. Mater Sci Eng C Mater Biol Appl 119:111634. https://doi.org/10.1016/j.msec.2020.111634
40. Portaccio M, Errico S, Chioccarelli T, Cobellis G, Lepore M (2020) Fourier-transform infrared microspectroscopy (FT-IR) study on caput and cauda mouse spermatozoa. PRO 42(1):19. https://doi.org/10.3390/ecsa-6-06537
41. Toyran N, Lasch P, Naumann D, Turan B, Severcan F (2006) Early alterations in myocardia and vessels of the diabetic rat heart: an FTIR microspectroscopic study. Biochem J 397(3):427–436. https://doi.org/10.1042/BJ20060171
42. Toyran N, Turan B, Severcan F (2007) Selenium alters the lipid content and protein profile of rat heart: an FTIR microspectroscopic study. Arch Biochem Biophys 458(2):184–193. https://doi.org/10.1016/j.abb.2006.12.012
43. Cakmak G, Togan I, Severcan F (2006) 17Beta-estradiol induced compositional, structural and functional changes in rainbow trout liver, revealed by FT-IR spectroscopy: a comparative study with nonylphenol. Aquat Toxicol 77(1):53–63. https://doi.org/10.1016/j.aquatox.2005.10.015
44. Rondelli V, Helmy S, Passignani G, Parisse P, Di Silvestre D (2022) Integrated strategies for a holistic view of extracellular vesicles. ACS Omega 7(23):19058–19069. https://doi.org/10.1021/acsomega.2c01003
45. Los DA, Murata N (2004) Membrane fluidity and its roles in the perception of environmental signals. Biochim Biophys Acta Biomembr 1666(1):142–157. https://doi.org/10.1016/j.bbamem.2004.08.002
46. Szalontai B, Nishiyama Y, Gombos Z, Murata N (2000) Membrane dynamics as seen by Fourier transform infrared spectroscopy in a cyanobacterium, synechocystis PCC 6803: the effects of lipid unsaturation and the protein-to-lipid ratio. Biochim Biophys Acta Biomembr 1509(1):409–419. https://doi.org/10.1016/s0005-2736(00)00323-0
47. Wang X, Zhai M, Zhao Y, Yin J (2018) A review of articular cartilage and osteoarthritis studies by Fourier transform infrared spectroscopic imaging. Ann Joint 3:210. https://doi.org/10.21037/aoj.2018.01.04
48. Zou Y, Ma G (2014) A new criterion to evaluate water vapor interference in protein secondary structural analysis by FTIR spectroscopy. Int J Mol Sci 15(6):10018–10033. https://doi.org/10.3390/ijms150610018
49. Dong A, Huang P, Caughey WS (1990) Protein secondary structures in water from second-derivative amide I infrared spectra. Biochemistry 29(13):3303–3308. https://doi.org/10.1021/bi00465a022
50. Clarke EJ, Lima C, Anderson JR, Castanheira C, Beckett A, James V, Hyett J, Goodacre R, Peffers MJ (2022) Optical photothermal infrared spectroscopy can differentiate equine osteoarthritic plasma extracellular vesicles from healthy controls. Anal Methods 14: 3661–3670. https://doi.org/10.1101/2022.03.11.483922

Index

Csaba Matta et al. (eds.), *The Surfaceome: Methods and Protocols*, Methods in Molecular Biology, vol. 2908, https://doi.org/10.1007/978-1-0716-4434-8,

MIX
Papier aus verantwortungsvollen Quellen
Paper from responsible sources
FSC® C105338

If you have any concerns about our products,
you can contact us on
ProductSafety@springernature.com

In case Publisher is established outside the EU,
the EU authorized representative is:
Springer Nature Customer Service Center GmbH
Europaplatz 3, 69115 Heidelberg, Germany

Printed by Libri Plureos GmbH
in Hamburg, Germany